Capstone Engineering Design Workbook

Volume 1

Define, Conceive, Prove

Bahram Nassersharif, Ph.D.
Distinguished University Professor
Department of Mechanical, Industrial, and Systems Engineering
University of Rhode Island

Table of Contents

List of Figures

List of Tables

List of Assignments

Preface

The journey through engineering design is one marked by creativity, innovation, and rigorous problem-solving. Volume 1 of the "Engineering Capstone Design Workbook" is dedicated to guiding students through the foundational stages of this journey: defining the design problem, creating solution concepts, selecting the top concept, and proving that it meets both the problem definition and the customer requirements. This volume is an essential companion for any engineering student embarking on their capstone project, providing structured methodologies and practical insights to navigate these critical phases of the design process.

In the initial chapters, we delve into the very essence of engineering design. The introduction offers an overview of the engineering design process, distinguishing it from the scientific method and emphasizing the importance of understanding the problem before seeking solutions. We explore the steps involved in defining the problem, gathering information, and developing design specifications, ensuring that students build a solid foundation upon which to base their creative efforts.

The workbook emphasizes the importance of iterative development and effective communication, both vital for successful engineering design. Students are introduced to various techniques for generating and evaluating design concepts, including brainstorming, morphological analysis, and the TRIZ method. By providing these tools, we aim to foster innovative thinking and enable students to develop robust solutions that are not only technically sound but also aligned with customer needs.

A significant portion of this volume is dedicated to prototyping and modeling, allowing students to translate their concepts into tangible forms. This hands-on approach is crucial for validating design choices and ensuring that the final product meets all specified requirements. We also cover the critical aspects of testing and iteration, highlighting the necessity of refining and optimizing designs based on empirical data and feedback.

Volume 1 further explores the collaborative nature of capstone design projects. We discuss the roles of students, faculty, and industry mentors, underscoring the value of teamwork and professional interactions. Effective communication, both written and oral, is given due attention, preparing students to articulate their ideas clearly and persuasively to diverse audiences.

Ethical and professional responsibilities are integral to engineering practice, and this volume addresses these topics comprehensively. Students are encouraged to consider ethical implications in their design decisions and adhere to professional standards throughout their projects.

By the end of Volume 1, students will have a clear understanding of how to define a design problem, generate and evaluate solutions, and demonstrate that their chosen concept meets all requirements. This volume lays the groundwork for the subsequent stages of detailed design and implementation, covered in Volume 2, ensuring a seamless transition as students progress in their capstone projects.

We hope that this workbook serves as a valuable resource, guiding you through the initial stages of your capstone design project with confidence and clarity. Embrace the opportunities and challenges that lie ahead, and may your journey through engineering design be both rewarding and enlightening.

Bahram Nassersharif, Ph.D.

July 2024

1 Introduction

Welcome to the "Engineering Capstone Design Workbook," a comprehensive guide crafted for students, mentors, sponsors, and professors involved in capstone design projects across various engineering disciplines. This workbook aims to provide a structured approach to engineering design, emphasizing problem-solving, innovation, and the practical application of theoretical knowledge. Capstone design projects serve as a culminating academic experience where students synthesize and apply their engineering education to real-world problems, bridging the gap between academia and industry.

This workbook is not just a textbook or guide; it is also a logbook for you to document and record your design work in an organized manner. As you work through your capstone project, use this workbook to log your progress, thoughts, and findings. Proper documentation is crucial for tracking your design process and ensuring that all aspects of your project are thoroughly recorded.

1.1 Understanding Design in Engineering

The term "design" is ubiquitous across different fields, each interpreting it through the lens of their specific needs and methodologies. According to the dictionary, as a verb, design means to devise, contrive, have a purpose, create for a specific function or end, make a drawing, or draw plans. As a noun, design is described as a mental project or scheme, a preliminary sketch or outline of the arrangement of elements in a product, and the creative art of executing aesthetic or functional designs. However, these definitions need to encapsulate the practice and essence of engineering design fully.

In engineering, design—often referred to as the engineering method—is a formal, rigorous, and systematic process aimed at optimizing solutions to problems. These problems may involve creating new solutions or improving existing ones, whether they are processes, devices, or concepts. The nature of these problems is inherently complex and ill-defined, requiring a comprehensive approach to understand and address them effectively. Figure 1-1 illustrates the capstone engineering preliminary design process diagram. Figure 1-2 This shows the final phase of the engineering capstone design, which encapsulates this iterative and multifaceted approach, typically conducted during the final semester of study.

This workbook serves as both a guide and a record of the engineering design process. It includes sections where you can log your design decisions, sketches, research, and iterations. This dual-purpose approach ensures that you are not only learning about the engineering design process but also actively engaging with it and creating a comprehensive record of your work.

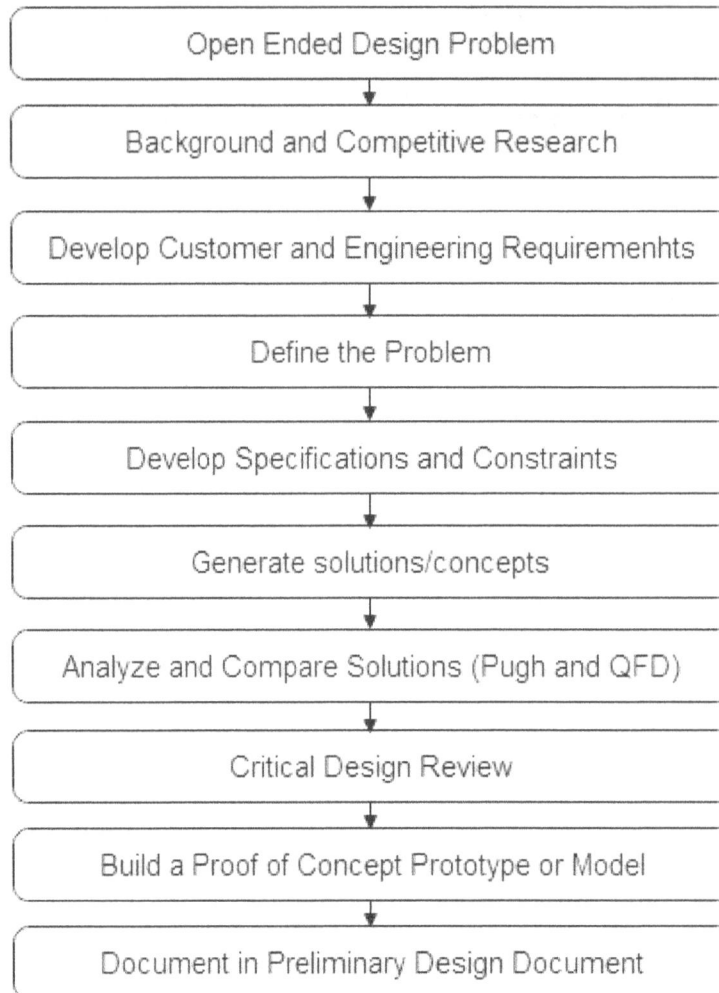

```
┌─────────────────────────────────────────────────────┐
│              Open Ended Design Problem                │
└─────────────────────────────────────────────────────┘
                          │
┌─────────────────────────────────────────────────────┐
│          Background and Competitive Research          │
└─────────────────────────────────────────────────────┘
                          │
┌─────────────────────────────────────────────────────┐
│      Develop Customer and Engineering Requiremenhts   │
└─────────────────────────────────────────────────────┘
                          │
┌─────────────────────────────────────────────────────┐
│                  Define the Problem                   │
└─────────────────────────────────────────────────────┘
                          │
┌─────────────────────────────────────────────────────┐
│          Develop Specifications and Constraints       │
└─────────────────────────────────────────────────────┘
                          │
┌─────────────────────────────────────────────────────┐
│                Generate solutions/concepts            │
└─────────────────────────────────────────────────────┘
                          │
┌─────────────────────────────────────────────────────┐
│        Analyze and Compare Solutions (Pugh and QFD)   │
└─────────────────────────────────────────────────────┘
                          │
┌─────────────────────────────────────────────────────┐
│                 Critical Design Review                │
└─────────────────────────────────────────────────────┘
                          │
┌─────────────────────────────────────────────────────┐
│        Build a Proof of Concept Prototype or Model    │
└─────────────────────────────────────────────────────┘
                          │
┌─────────────────────────────────────────────────────┐
│        Document in Preliminary Design Document        │
└─────────────────────────────────────────────────────┘
```

Figure 1-1. Capstone Engineering Preliminary Design Process.

1.2   The Engineering Design Process

The engineering design process is a structured and iterative approach to solving complex engineering problems. It involves several key steps, each critical to developing effective and innovative solutions.

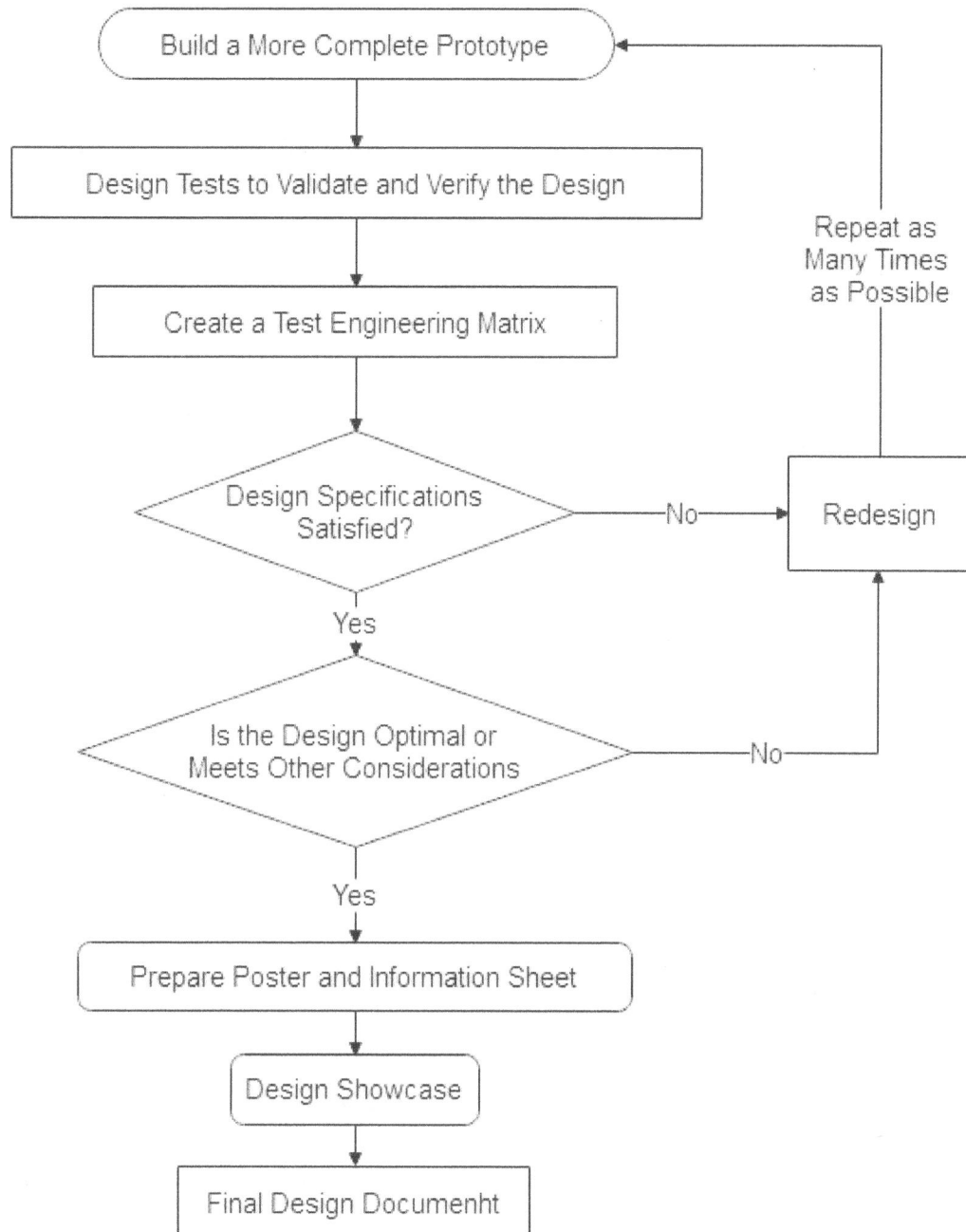

Figure 1-2. Engineering capstone design process final phase.

1.2.1 Defining the Problem

Identifying and clearly articulating the problem is the first and most critical step in the engineering design process. This involves understanding the requirements and constraints, which often need to be completed and require further investigation.

Engineers must ask probing questions and conduct thorough research to ensure they fully grasp the problem's scope and nuances.

Use this workbook to document your problem definition. Write down the questions you need to answer and the research you conduct. This log will serve as a reference throughout your project, helping you stay focused and organized.

1.2.2   Gathering Information

Once a problem has been defined, additional information about the problem area and related topics can be gathered through literature searches, patent reviews, and competitor analysis. This helps build a comprehensive understanding of the problem and informs the development of potential solutions.

Document your findings in this workbook. Include notes from your literature searches, summaries of relevant patents, and any competitor information you find. Keeping this information organized will make it easier to refer back to it as you develop your solutions.

1.2.3   Developing Design Specifications

The problem-solving engineer is responsible for developing the design specifications and identifying relevant constraints. These specifications serve as a blueprint for the design process, guiding the development of solutions that meet the identified needs and constraints.

Use this workbook to outline your design specifications. Detail the constraints you must work within and the requirements your solution must meet. This log will ensure that you stay aligned with your project goals.

1.2.4   Generating Solutions

Solutions are then generated as concepts that satisfy the design specifications and constraints. These concepts may be devices or processes. If the solution is a device, prototyping may be possible. If the concept is a process, it can be modeled, perhaps in a software model. This phase emphasizes creativity and innovation, encouraging engineers to think outside the box and explore multiple potential solutions.

Document your brainstorming sessions and initial concepts in this workbook. Sketch your ideas and write down your thought process. This record will be invaluable as you refine your solutions and track the evolution of your designs.

1.2.5   Prototyping and Modeling

The prototype or model is tested against the design specifications, constraints, and additional considerations such as ethical, environmental, and social factors. This phase is crucial for validating the solution's feasibility and identifying areas for improvement.

Use this workbook to log your prototyping and modeling efforts. Record the materials used, the steps taken, and the results of your tests. Documenting this phase ensures you have a clear understanding of what works and what needs adjustment.

1.2.6   Testing and Iteration

The test results are used to assess the fitness of the solution concept for the problem. Iterations on the concept, prototype, and model improve, optimize, or refine the solution. This iterative process ensures that the final solution is robust, efficient, and effective in addressing the identified problem.

Log each iteration in this workbook. Detail the changes made, the reasons for those changes, and the results of subsequent tests. This iterative log will help you track your progress and ensure that your final design is well-documented.

1.2.7   Documentation and Communication

Once a satisfactory solution is achieved, it is documented, communicated, and delivered to the sponsor of the problem or project. Proper documentation ensures that the design process is transparent and replicable, and effective communication ensures that all stakeholders understand and support the final solution.

This workbook is your primary tool for documentation. Use it to compile your final report, summarizing all aspects of your design process, from problem definition to final testing. This comprehensive record will be essential for presenting your work to sponsors, mentors, and peers.

1.3   Engineering Design vs. Scientific Method

Engineering design distinguishes itself from the scientific method, which is fundamental to scientific discovery. Scientists develop hypotheses based on observations of natural phenomena and test these hypotheses through experiments. These experiments either confirm or refine the hypotheses, leading to widely accepted theories when validated through peer review.

Engineering design, on the other hand, involves applying scientific principles and tools to create solutions that meet human needs. It is iterative and considers practical constraints such as cost, time, and resource availability. The outcome is a tangible product or process that addresses a specific problem. Figure 1-3 shows the flow chart for the scientific method, highlighting the differences between the two approaches.

The engineering design process is focused on creating functional solutions to practical problems. Unlike the scientific method, which seeks to understand natural phenomena, engineering design aims to develop products and processes that solve specific human problems. This distinction is crucial in understanding the role of engineers in society.

As you use this workbook, remember that it is designed to guide you through the engineering design process while also serving as a logbook for documenting your work. This dual-purpose approach ensures that you are not only learning about engineering design but also actively engaging with it and creating a comprehensive record of your project.

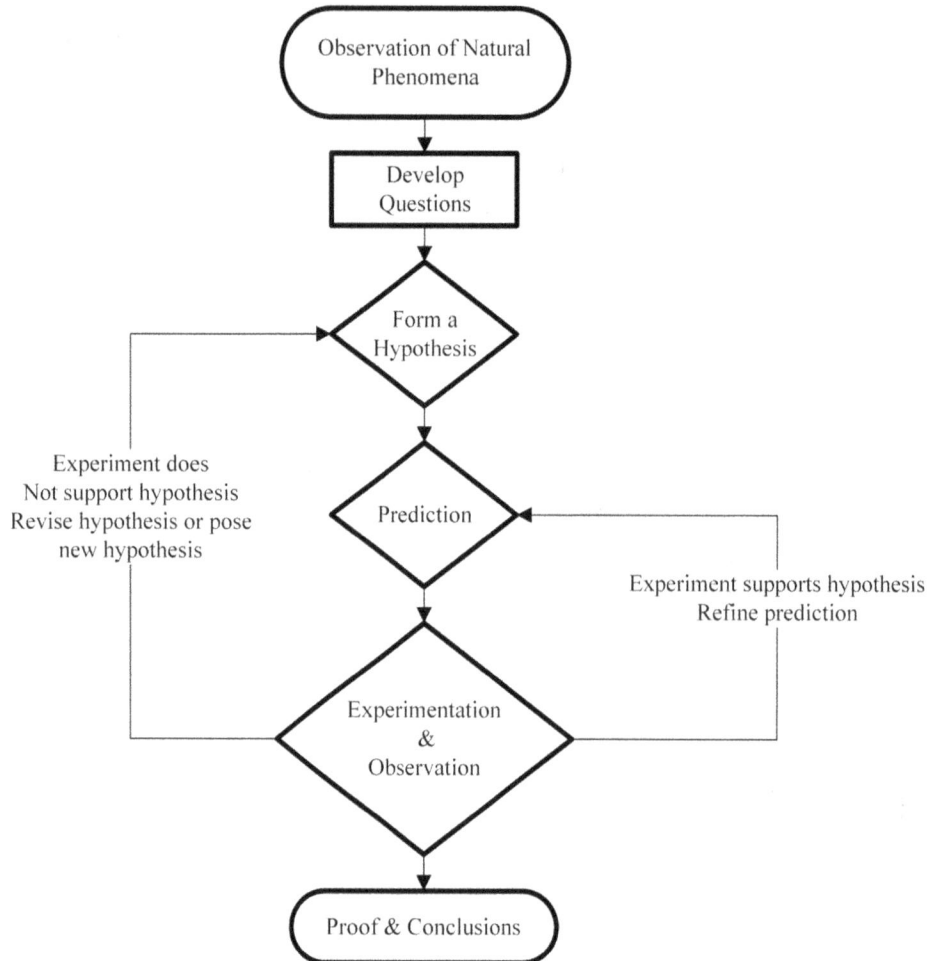

Figure 1-3. The scientific method.

1.4   Incorporating Cost and Schedule in Design

In professional settings, engineering design must account for cost and schedule constraints. These factors often drive decision-making, with longer or costlier concepts potentially being shelved. Accurate cost estimation and timeline prediction are crucial, often forming part of a contractual agreement between design engineers and sponsors. These constraints influence the design process, encouraging more conservative approaches to ensure on-time and within-budget delivery.

1.4.1   Cost Estimation

Accurate cost estimation involves considering all aspects of the design process, including materials, labor, manufacturing, testing, and distribution. Engineers must also account for potential risks and uncertainties that could impact costs. Practical cost estimation helps in making informed decisions about which concepts to pursue and how to allocate resources efficiently.

Use this workbook to document your cost estimates. Record the costs associated with different aspects of your project and any assumptions you make. This log will help you keep track of your budget and make adjustments as needed.

1.4.2   Scheduling

Scheduling involves creating a detailed timeline for the design process, including milestones, deadlines, and deliverables. Engineers must consider the time required for each phase and ensure that the project stays on track. Effective scheduling helps manage expectations and ensures that the project is completed within the agreed-upon timeframe.

Document your project schedule in this workbook. Outline the key milestones and deadlines for each phase of your project. This schedule will serve as a roadmap for your work and help you stay organized and on track.

1.4.3   Balancing Cost and Schedule

Balancing cost and schedule is a critical aspect of engineering design. Engineers must make trade-offs between cost, quality, and time to deliver a solution that meets the sponsor's requirements. This often involves prioritizing certain aspects of the design while compromising on others to achieve an optimal balance.

Use this workbook to record the decisions you make regarding cost and schedule trade-offs. Document the rationale behind your choices and any adjustments you make along the way. This log will provide valuable insights into your decision-making process and help you manage your project effectively.

1.5   Accreditation and Design Education

Design is central to engineering practice, as reflected in the accreditation requirements by ABET (Accreditation Board for Engineering and Technology). ABET accreditation ensures engineering programs provide the technical and professional skills needed in the industry. Programs must establish educational objectives, student learning outcomes, and assessment processes to demonstrate continuous improvement. Engineering students should be familiar with these criteria, as they are integral to their education and professional development.

1.5.1   History and Role of ABET

Engineering accreditation in the United States started with the Engineer's Council for Professional Development (EPCD), which was later renamed the Accreditation Board for Engineering and Technology (ABET). ABET's primary role is to ensure engineering programs meet the standards necessary to produce graduates ready to enter the global workforce. ABET accreditation is recognized worldwide as a mark of quality and excellence in engineering education.

1.5.2   Criteria for Accreditation

ABET's criteria for accrediting engineering programs include several categories, such as students, program educational objectives, student outcomes, continuous improvement, curriculum, faculty, facilities, and institutional support. Table 1.1 shows the engineering accreditation criteria for 2020-2021, providing a detailed overview of what is expected from accredited programs.

Table 1-1. Engineering Accreditation Criteria 2024-2025.

Criterion 1: Students
Criterion 2: Program Educational Objectives
Criterion 3: Student Outcomes
Criterion 4: Continuous Improvement
Criterion 5: Curriculum
Criterion 6: Faculty
Criterion 7: Facilities
Criterion 8: Institutional Support
Program Specific Criteria (established by lead society)

1.5.3   Importance of Accreditation

Accreditation ensures that engineering programs provide a high-quality education that meets the needs of students and employers. It promotes continuous improvement in engineering education by encouraging programs to review and update their curricula, teaching methods, and assessment processes regularly. Accreditation also assures employers that graduates have the necessary knowledge and skills to succeed in the engineering profession.

1.5.4   Student Outcomes

ABET's criteria emphasize student outcomes, which define what students should know and be able to do by the time they graduate. Table 1.2 outlines the student outcomes specified under Criterion 3 of the engineering accreditation criteria for 2020-2021. These outcomes are critical for preparing graduates to enter the professional practice of engineering.

Table 1-2. ABET 2024-2025 Engineering Accreditation Criterion 3 Student Outcomes.

1.	an ability to identify, formulate, and solve complex engineering problems by applying principles of engineering, science, and mathematics.
2.	an ability to apply engineering design to produce solutions that meet specified needs with consideration of public health, safety, and welfare, as well as global, cultural, social, environmental, and economic factors.
3.	an ability to communicate effectively with a range of audiences.
4.	an ability to recognize ethical and professional responsibilities in engineering situations and make informed judgments, which must consider the impact of engineering solutions in global, economic, environmental, and societal contexts.
5.	an ability to function effectively on a team whose members together provide leadership, create a collaborative and inclusive environment, establish goals, plan tasks, and meet objectives.
6.	an ability to develop and conduct appropriate experimentation, analyze and interpret data, and use engineering judgment to draw conclusions.
7.	an ability to acquire and apply new knowledge as needed, using appropriate learning strategies.

1.6   Open-Ended Design Problems

Engineering programs do an excellent job of educating students to become problem solvers. Traditional textbooks reinforce this through exercise problems that have one correct answer. However, real-world engineering problems are often open-ended, lacking clear definitions and having multiple possible solutions. These problems require an investigative approach to gather necessary information and develop a complete problem definition.

1.6.1   Characteristics of Open-Ended Problems

Open-ended problems are inherently more complex and satisfying. They encourage active participation in problem-solving and design, requiring students to ask many questions and seek additional information independently. These problems are typically new and have not been solved before, posing a unique challenge to the design engineer.

Use this workbook to document your problem-solving process. Write down the questions you ask, the information you find, and your thought process. This log will help you track your progress and ensure that you are systematically addressing the problem.

1.6.2   Investigative Approach

To tackle open-ended problems, students need to take an investigative approach. This involves identifying what information is needed, where to find it, and how to interpret it. Students must also consider non-technical aspects, such as ethical implications and stakeholder interests, to develop comprehensive solutions.

Document your investigative approach in this workbook. Keep track of the sources you consult, the experts you talk to, and the insights you gain. This record will be invaluable as you develop your solution and ensure that you have thoroughly explored all aspects of the problem.

1.6.3   Documentation and Iteration

Documentation of the problem-solving process is crucial. It helps track the steps taken, sources consulted, and decisions made. Iteration is also a key component, allowing for continuous improvement of the solution based on testing and feedback.

Use this workbook to document each iteration of your design. Record the changes you make, the reasons for those changes, and the results of subsequent tests. This iterative log will help you track your progress and ensure that your final design is well-documented.

1.7   Collaborative Nature of Capstone Design

Capstone design projects are inherently collaborative, involving multiple stakeholders, including students, faculty, industry mentors, and sponsors. This collaboration ensures that students not only apply what they have learned but also gain valuable insights and feedback from experienced professionals.

1.7.1   Role of Students

Students are the primary drivers of capstone design projects. They apply their technical knowledge, creativity, and problem-solving skills to develop innovative solutions. They also gain experience in project management, teamwork, and communication, which are essential for their future careers.

Use this workbook to document your role in the project. Record your contributions, the tasks you complete, and any challenges you face. This log will help you reflect on your work and identify areas for improvement.

1.7.2   Role of Faculty

Faculty mentors guide students throughout the capstone design process. They provide technical expertise, feedback, and support, helping students navigate challenges and refine their solutions. Faculty also play a crucial role in ensuring that projects meet academic standards and learning outcomes.

Document the feedback and guidance you receive from faculty in this workbook. Keep track of the advice given and how you implement it in your project. This record will help you ensure that you are meeting academic expectations and making the most of your mentor's expertise.

1.7.3   Role of Industry Mentors and Sponsors

Industry mentors and sponsors provide real-world context and practical insights. They offer guidance based on their industry experience and help students understand the practical constraints and considerations that impact engineering design. Sponsors also provide resources and support, such as funding and access to facilities and equipment.

Record the interactions and feedback you receive from industry mentors and sponsors in this workbook. Document the resources provided and how they impact your project. This log will help you track the support you receive and ensure that you are leveraging it effectively.

1.8   Importance of Effective Communication

Effective communication is critical to the success of capstone design projects. Students must communicate their ideas, progress, and results to various stakeholders, including faculty, mentors, sponsors, and peers. This involves presenting technical information clearly and concisely, both in written and oral formats.

1.8.1   Written Communication

Written communication includes reports, documentation, and correspondence. Students must document their design process, including problem definition, research, design specifications, concept development, testing, and iteration. Clear and detailed documentation ensures that the project is transparent and replicable.

Use this workbook as your primary tool for written communication. Record your design process in detail, including all relevant documentation. This comprehensive record will be essential for presenting your work to sponsors, mentors, and peers.

1.8.2   Oral Communication

Oral communication includes presentations, meetings, and discussions. Students must present their ideas and progress to faculty, mentors, sponsors, and peers, effectively conveying technical information and responding to questions and feedback. Practical oral communication skills are essential for successful collaboration and project management.

Prepare your presentations and meeting notes in this workbook. Practice your presentations and record any feedback you receive. This log will help you refine your communication skills and ensure that you are effectively conveying your ideas.

1.9   Ethical and Professional Responsibilities

Ethical and professional responsibilities are integral to engineering practice. Engineers must consider the ethical implications of their designs and make informed decisions that prioritize public health, safety, and welfare. They must also adhere to professional standards and practices, ensuring the integrity and quality of their work.

1.9.1   Ethical Considerations

Ethical considerations include issues such as environmental impact, sustainability, social responsibility, and equity. Engineers must evaluate the potential consequences of their designs and strive to create solutions that benefit society and minimize harm.

Use this workbook to document your ethical considerations. Record the potential impacts of your design and the steps you take to address them. This log will help you ensure that your design is responsible and aligns with ethical standards.

1.9.2   Professional Standards

Professional standards include adhering to codes of conduct, best practices, and industry standards. Engineers must maintain the highest standards of professionalism, ensuring that their work is reliable, safe, and effective. Continuous learning and professional development are also essential to stay current with advancements in the field.

Document your adherence to professional standards in this workbook. Record any professional development activities you undertake and how they impact your project. This log will help you demonstrate your commitment to professionalism and continuous improvement.

1.10   Conclusion

This workbook aims to guide you through the engineering design process, from defining the problem to delivering the final solution. By understanding the nuances of engineering design, incorporating cost and schedule considerations, and tackling open-ended problems, you will develop the skills necessary to succeed in your capstone project and beyond. The chapters that follow will delve into each aspect of the design process, providing detailed guidance and practical examples to help you navigate your capstone design journey.

In summary, this workbook is both a guide and a logbook for your capstone design project. It is designed to help you learn about the engineering design process while also providing a structured format for documenting your work. As you use this workbook, you will create a comprehensive record of your design process, from problem definition to final testing. This dual-purpose approach ensures that you are actively engaging with the material and creating a valuable resource for your future career.

The capstone design experience is a transformative journey that prepares you for the professional world. It challenges you to apply your knowledge, collaborate with others, and develop innovative solutions. Embrace this opportunity to learn, grow, and make a meaningful impact in your chosen field of engineering. As you embark on your capstone design project, use this workbook as your companion, providing the tools and insights needed to achieve success.

1.11 Preparing for Capstone Design

The capstone design experience stands out from any other lecture or laboratory course in the engineering curriculum. Its primary purpose is to prepare engineering graduates for the challenges and dynamics of working in the engineering or technology industry. Unlike traditional courses, the capstone design course engages students in the process of engineering design by tackling open-ended, real-world problems. This experience leverages the knowledge and skills accumulated through coursework, internships, co-op experiences, study abroad programs, professional engineering society activities, hobbies, and personal projects.

1.11.1 Importance of Prerequisites

Capstone courses typically have several prerequisites to ensure that students are adequately prepared for this culminating experience. These prerequisites guarantee that students possess the foundational knowledge necessary to complete their capstone projects. As such, students should be at least in their senior year (4th year or beyond) of their engineering program when enrolling in capstone courses.

1.11.2 Components of Engineering Curriculum

Engineering curricula are structured around five core components: science, mathematics, engineering science, discipline-specific engineering, and general education. These components collectively equip students with the essential skills and knowledge required for capstone design projects.

1.11.2.1 Science and Mathematics

The science and mathematics components provide the theoretical foundation necessary for understanding and applying engineering principles. Courses in physics, chemistry, calculus, and differential equations, among others, develop problem-solving skills and analytical thinking. These courses are crucial for understanding the technical aspects of capstone projects, such as calculating stresses in materials, analyzing fluid flows, or solving complex equations related to system behavior.

1.11.2.2   Engineering Science

Engineering science courses, such as statics, dynamics, strength of materials, fluid mechanics, and thermodynamics, bridge the gap between pure science and practical engineering. These courses teach students to apply scientific principles to real-world engineering problems, preparing them for the analytical and modeling tasks they will encounter in their capstone projects.

1.11.2.3   Discipline-Specific Engineering

Discipline-specific courses delve into the specialized knowledge and techniques of a particular engineering field. For example, mechanical engineering students might study machine design and heat transfer, while electrical engineering students might focus on circuit design and electromagnetics. These courses are essential for developing the specific skills needed to tackle capstone projects related to each discipline.

1.11.2.4   General Education

General education courses provide a well-rounded education, fostering critical thinking and analysis skills that are essential for addressing the broader aspects of engineering projects. These courses cover topics such as environmental impact, ethical issues, political considerations, safety/risk management, financial analysis, project management, and aesthetics evaluation. The skills gained in these courses are invaluable for considering the societal implications of engineering designs and making informed decisions.

1.11.3   Applying Knowledge to Capstone Projects

In capstone design projects, students must apply the comprehensive knowledge they have gained throughout their engineering education. This requires an interdisciplinary approach, as projects often span multiple fields and require a broad skill set.

1.11.3.1   Flexibility in Applying Knowledge

The flexibility to apply knowledge outside one's specific field is a crucial aspect of the capstone design experience. For instance, a mechanical engineering student may work on a project involving laser technology, or an electrical engineering student might focus on cooling systems for electronic components. The ability to draw on knowledge from various disciplines enhances the problem-solving process and leads to more innovative solutions.

1.11.3.2   Laboratory Courses

Laboratory courses play a vital role in preparing students for the practical aspects of their capstone projects. These courses teach students how to design and conduct experiments, analyze data, and draw conclusions based on empirical evidence. Skills developed in laboratory courses are directly applicable to testing and validating design concepts in capstone projects.

1.11.4   Resume Preparation

As part of the capstone design preparation, students may need to create or update their resumes. Professors and project sponsors often use resumes to assign students to specific projects based on their skills, experience, and knowledge. A well-crafted resume that highlights relevant coursework, projects, and extracurricular activities can significantly enhance a student's chances of being selected for a preferred capstone project.

1.11.4.1   Tips for Creating an Effective Resume

- **Highlight Relevant Skills:** Focus on skills directly applicable to potential capstone projects, such as technical proficiency, problem-solving abilities, and project management experience.
- **Include Practical Experience:** List internships, co-op positions, research projects, and any hands-on experience that demonstrates your ability to apply engineering principles in real-world settings.
- **Emphasize Teamwork and Leadership:** Capstone projects often require collaboration and leadership. Highlight experiences where you have worked effectively in teams or taken on leadership roles.
- **Showcase Academic Achievements:** Include details about relevant coursework, academic honors, and any specialized training or certifications.
- **Tailor to the Project:** If possible, tailor your resume to align with the specific requirements and goals of the capstone projects you are interested in.

1.11.4.2   Utilizing Capstone Resources

The capstone design experience often provides access to various resources that can enhance the quality and scope of student projects. These resources may include funding, specialized equipment, software, and access to industry mentors.

1.11.5   Funding and Material Purchases

In your planning, you must account for the administrative work required to process materials and service purchase requests, as well as availability and shipping times. Consult your professor to understand the procedures established by your engineering department or institution for purchasing. If you need to acquire software or plug-ins for existing software, follow the specific procedures for those types of purchases, as they may differ from those for buying physical materials.

1.11.6   Skills and Assistance

Assess your team's skills in manufacturing, modeling, and software development. Determine if you have the necessary expertise to build your design. If not, identify individuals or resources that can assist your team. Ensure you have access to required resources such as funding, tools, software, and machines. If you cannot create the design as planned, consider redesigning it to fit your available resources.

1.11.7   Verification and Validation

Once you have built your design or model, verify that it meets your design specifications. Create experiments or tests to assess your design against the specified parameters. If there are gaps in your design specifications, revisit and establish values for them. If specific parameters remain undefined, consider eliminating them from your specifications. Conduct tests to determine or refine the numerical values of your design specifications.

1.11.8   Planning Test Engineering Activities

Carefully plan your test engineering activities, focusing only on tests that will yield helpful information for redesign and optimization. Avoid conducting tests solely based on familiarity with previous courses. Instead, analyze what needs to be tested and how to perform those tests effectively. Consider whether your design must comply with any regulations or codes and follow standard testing methods published by national professional engineering societies when applicable.

1.11.9   Data Collection and Analysis

From your test engineering procedures, collect data to analyze the performance of your design. Use the analysis results to guide optimization and error correction. The build-test-redesign cycle may need to be repeated multiple times to achieve a robust design. The number of iterations can vary widely depending on the scale and complexity of the design.

Preparing for the capstone design experience requires a comprehensive understanding of your engineering curriculum, effective planning, and utilization of available resources. By leveraging your accumulated knowledge, collaborating with team members, and adhering to established procedures, you can successfully navigate the capstone design process and develop innovative solutions to real-world problems.

2 Regulations, Codes, and Standards

Design projects that result in products or processes used by people are subject to various regulations, codes, standards, or specifications. These guidelines are essential to protecting people from harm, creating norms for compatibility, and establishing safety and quality measures. Understanding and adhering to these requirements is crucial for engineers to ensure that their designs are safe, effective, and legally compliant.

2.1.1 1.5.1 Regulations

Governments create laws and regulations to protect the health and welfare of their citizens, maintain societal structures, and defend rights and territories. Regulatory bodies and agencies further define rules consistent with the laws of their respective countries, governing the creation, sale, service, delivery, and utilization of devices, processes, and services.

2.1.1.1 Federal Regulatory Agencies in the United States

In the United States, numerous federal agencies regulate and enforce laws for various industries and services. Some key agencies include:
- **Consumer Product Safety Commission (CPSC):** Enforces federal safety standards for consumer products.
- **Environmental Protection Agency (EPA):** Establishes and enforces pollution standards to protect the environment.
- **Equal Employment Opportunity Commission (EEOC):** This agency administers and enforces fair employment practices under Title VII of the Civil Rights Act of 1964.
- **Federal Aviation Administration (FAA):** This agency regulates and promotes air transportation safety, including airports and pilot licensing.
- **Federal Communications Commission (FCC):** This agency regulates interstate

and international communications by radio, television, wire, satellite, and cable.

- **Federal Deposit Insurance Corporation (FDIC):** This agency insures bank deposits, approves mergers, and audits banking practices.
- **Federal Reserve System (the FED):** Regulates banking and manages the money supply.
- **Federal Trade Commission (FTC):** This agency ensures free and fair competition and protects consumers from unfair or deceptive practices.
- **Food and Drug Administration (FDA):** The FDA administers federal food purity laws, drug testing and safety regulations, and cosmetics regulation.
- **Interstate Commerce Commission (ICC):** Enforces federal laws concerning transportation across state lines.
- **National Institute of Standards and Technology (NIST):** Establishes and maintains standards in measurement and technological developments.
- **National Labor Relations Board (NLRB):** Prevents or corrects unfair labor practices by employers or unions.
- **Nuclear Regulatory Commission (NRC):** Licenses and regulates non-military nuclear facilities.
- **Occupational Safety and Health Administration (OSHA):** OSHA develops and enforces federal standards and regulations to ensure safe working conditions.
- **Securities and Exchange Commission (SEC):** Administers federal laws concerning the buying and selling of securities.

These agencies provide certification or licensing processes to enforce laws and regulate conduct, ensuring compliance with regulations. They publish their regulations, usually in the Code of Federal Regulations, and solicit public comment and feedback during the regulatory process. Your design project may be directly subject to some of these regulations or be part of a more extensive system or process regulated by these agencies.

2.1.1.2   State Regulatory Agencies

State governments also have regulatory bodies or authorities supervising activities within their territories. These agencies regulate and enforce laws to ensure safety, compliance with standards, and consumer protection. They focus on complex areas requiring supervision without entanglement in the normal political process of lawmaking, such as:

- **Advertising regulation**
- **Alcoholic beverages**
- **Bank regulation**
- **Cable and Internet services**
- **Child safety and wellbeing**

- **Communication services**
- **Consumer protection**
- **Cybersecurity regulation**
- **Economic regulation**
- **Electricity generation, distribution, and pricing**
- **Environmental regulation**
- **Financial regulation**
- **Food safety and security**
- **Noise regulation**
- **Radiation safety**
- **Occupational safety and health**
- **Public health**
- **Pollution regulation and monitoring**
- **Regulation of therapeutic goods**
- **Vehicle regulation**
- **Roads and transportation**
- **Wage regulation**

Your design projects may be regulated by state agencies in addition to federal agencies. State regulatory authorities publish their requirements and enforce their areas of responsibility through inspections, registrations, certifications, and licensing.

2.1.1.3   International Regulations

If your design project involves products or services sold internationally, you must consider the regulations in each country where they will be marketed. Different countries may have varying regulations, requiring compliance to ensure acceptance and avoid legal issues. An internet search can help you find the relevant regulations for your design project.

2.1.2   Codes and Standards

Codes and standards are developed by professional organizations and government representatives to ensure safety, quality, and compatibility in the industry. These guidelines help establish best practices and ensure that products and services meet the necessary criteria for performance and safety.

2.1.2.1   Professional Organizations and Their Codes

Various professional organizations develop codes and standards specific to their fields. Some key organizations include:
- **American Association for the Advancement of Science (AAAS):** Advances science, engineering, and innovation globally.
- **American Academy of Environmental Engineers and Scientists (AAES):**

Focuses on environmental engineering and science.

- **American Institute of Chemical Engineers (AIChE):** Represents chemical engineering professionals.
- **American Institute of Mining, Metallurgical, and Petroleum Engineers (AIME):** Professional association for mining and metallurgy.
- **American Ceramics Society (ACERS):** Focuses on ceramics science and engineering.
- **Acoustical Society of America (ASA):** Covers all branches of acoustics, both theoretical and applied.
- **American Institute of Aeronautics and Astronautics (AIAA):** Represents professionals in aeronautics and astronautics.
- **American Nuclear Society (ANS):** Unifies activities in nuclear science and technology.
- **American Society of Agricultural and Biological Engineers (ASABE):** Focuses on agricultural and biological engineering.
- **American Society of Civil Engineers (ASCE):** Represents civil engineering professionals worldwide.
- **American Society for Engineering Education (ASEE):** Promotes and improves engineering education.
- **American Society of Heating, Refrigerating, and Air-Conditioning Engineers (ASHRAE):** Focuses on building systems, energy efficiency, and indoor air quality.
- **American Society of Mechanical Engineers (ASME):** Collaborates on knowledge sharing and skills development across engineering disciplines.
- **American Society for Testing and Materials (ASTM):** Develops standards for testing and materials.
- **Biomedical Engineering Society (BMES):** Represents biomedical engineering professionals.
- **Computing Sciences Accreditation Board (CSAB):** Focuses on education quality in computing disciplines.
- **International Council on Systems Engineering (INCOSE):** Professional society for systems engineering.
- **Institution of Engineering and Technology (IET):** Supports the global engineering community and technology innovation.
- **Institute of Electrical and Electronics Engineers (IEEE):** Advances technology through professional association.
- **Institute of Industrial and Systems Engineers (IISE):** Represents industrial engineers.
- **Institution of Mechanical Engineers (IMechE):** Focuses on improving the world through engineering.

- **Institute of Transportation Engineers (ITE):** Meets mobility and safety needs in transportation.
- **National Fire Protection Association (NFPA):** Focuses on fire protection.
- **NSF International:** Protects and improves global human health.
- **National Society of Professional Engineers (NSPE):** Focuses on professional engineering registration and licensure.
- **International Society for Optical Engineering (SPIE):** Advances the science and application of light.
- **Society of Automotive Engineers (SAE):** Represents aerospace, automotive, and commercial-vehicle engineers.
- **Society of Manufacturing Engineers (SME):** Focuses on manufacturing aspects.
- **Society of Naval Architects and Marine Engineers (SNAME):** Advances marine engineering.
- **The Minerals, Metals & Materials Society (TMS):** Encompasses materials scientists and engineers.
- **Society of Photo-Optical Instrumentation Engineers (SPIE):** Focuses on optics and photonics technology.

Each professional engineering society publishes its codes and standards on its website. Sometimes, there is a fee to obtain these documents. Many standard documents can also be found in university libraries. Joining a professional society related to your discipline is highly beneficial for networking, resources, and professional development. Student memberships are usually affordable and offer numerous advantages, including access to industry standards and codes.

2.1.2.2   Importance of Codes

Codes are generally accepted guidelines or requirements for industries. They ensure safety, quality, and standardization. For example, building codes ensure the safety, reliability, and durability of structures. While not laws, codes can become mandatory requirements through registration and licensing, such as electrical and plumbing codes. Codes are intended to be widely applicable, serving as best practices or for compatibility and standardization. It is crucial to check your design project for compliance with applicable codes.

2.1.3   Specifications

Industry groups, consortia, or professional organizations typically create specifications. They allow competing industries to agree on standards, ensuring compatibility and interoperability. For example, IEEE develops specifications for hardware and software used in internet communication.

While specifications may not be regulated or subject to codes and standards, non-compliance can result in products being unacceptable to consumers. Specifications are usually published and accessible, often through a simple internet search.

2.1.3.1   Developing and Using Specifications

Specifications are critical in design projects to ensure that products meet industry standards and are compatible with existing systems. They provide detailed descriptions of the requirements for materials, products, or processes. Specifications can include performance criteria, materials standards, and testing methods.

When developing a design project, it is essential to identify and adhere to relevant specifications. This ensures that your design will be accepted in the market and perform as expected. Documenting how your design meets these specifications is also crucial, as it provides a clear record of compliance and can be used to address any issues that arise during testing or production.

2.1.4   Summary of Regulations, Codes, Standards, and Specifications

Regulations, codes, standards, and specifications are fundamental to engineering design. They ensure that products and processes are safe, reliable, and meet the necessary quality standards. Understanding and adhering to these requirements is critical for engineers to ensure that their designs are legally compliant and acceptable to consumers.

This chapter has provided an overview of the key regulatory bodies and professional organizations involved in developing and enforcing these guidelines. It has also highlighted the importance of documentation and compliance in the design process.

As you work on your design project, use this workbook to document your compliance with relevant regulations, codes, standards, and specifications. This documentation will not only help you stay organized and on track but also provide a valuable record of your design process, ensuring that your project meets all requirements and is ready for implementation.

3 Capstone Design Process

The capstone design course(s) in the curriculum were included by your engineering program to respond directly to ABET (Accreditation Board for Engineering and Technology) requirements for accreditation. These courses are designed to prepare you for engineering practice by immersing you in open-ended, real-world design projects that mirror the challenges you would face in industry. Under the guidance of your professors, you will navigate a structured yet flexible process aimed at fostering innovation, teamwork, and practical problem-solving skills.

Capstone design projects are unique in two key aspects: timeline and resources. The fixed academic calendar imposes a strict timeline on projects, creating management challenges distinct from those in the industry. Additionally, the availability of funding and resources varies widely across institutions and projects. This section outlines a capstone design process refined over the past decade, adaptable to different academic schedules and resource constraints.

3.1 Fixed Timeline Challenges

The project timeline is constrained by the academic calendar, which can span a single semester (10-14 weeks) or a full academic year. This fixed timeline requires careful planning and efficient time management. Unlike industry projects, which can extend over several years, capstone projects must achieve significant milestones within a relatively short period. This constraint necessitates a focused approach to project management, with clear goals and regular progress evaluations.

3.1.1 Academic Periods

Different engineering programs have varying structures for capstone design courses. Some offer a single course over one academic period, while others provide a two-part course spanning an entire academic year. The process described here is divided into two parts to accommodate both models. Part I covers the initial stages of defining the problem, generating and evaluating design concepts, and developing prototypes. Part II focuses on building, testing, and refining the design.

3.2   Funding and Resources

Funding and resources for capstone projects come from various sources, including industry sponsors, government labs, faculty research, design competitions, and in-class exercises. Regardless of the source, projects must meet the requirements set by the capstone design program, aligning with ABET accreditation criteria. These criteria ensure that projects are realistic and feasible and provide meaningful learning experiences.

1. **Sources of Funding**
- **Industry Sponsors:** Provide real-world problems and often fund materials and resources.
- **Government Labs:** Offer projects with public sector applications and sometimes provide access to specialized facilities.
- **Faculty Research:** Aligns capstone projects with ongoing research, offering access to academic expertise and equipment.
- **Design Competitions:** Provide predefined problems and sometimes funding or prizes.
- **In-Class Exercises:** Designed by professors to meet specific educational goals within the capstone framework.

3.3   The Capstone Design Process

The capstone design process is a structured yet flexible framework that guides students through the complexities of open-ended design problems. It is divided into two parts, each mapping to the academic calendar and ensuring thorough exploration and development of design concepts.

3.4   Part I: Conceptual Design

Figure 3-1. illustrates the process flow for Part I of the capstone design. Part I involves defining the problem, generating design concepts, and selecting the most promising solutions through decision-making methods, expert reviews, and prototyping. This phase is critical for laying a solid foundation for the project.

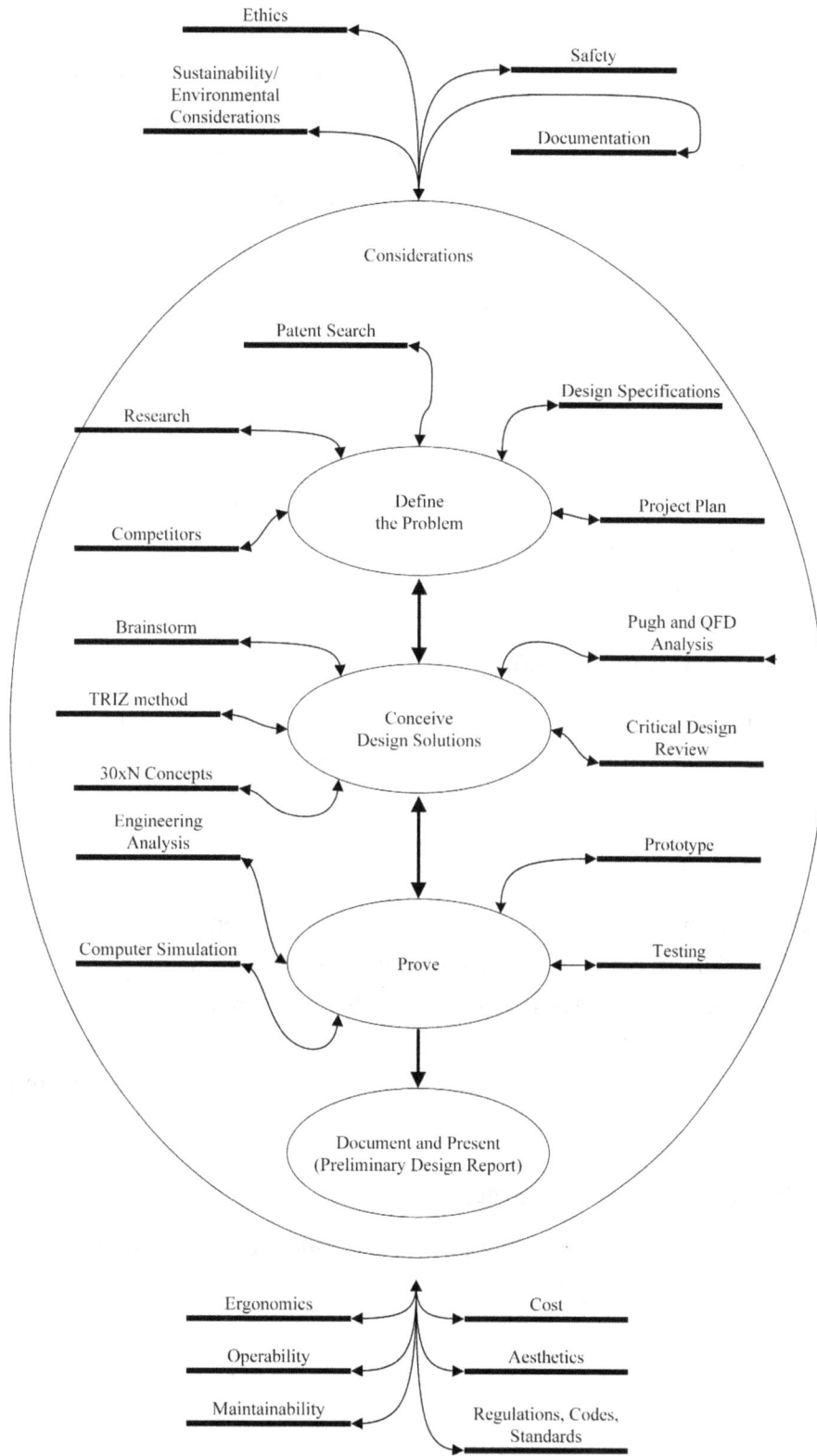

Figure 3-1. Engineering Capstone Design Process Part I diagram.

3.4.1   Defining the Problem

Capstone design projects are inherently open-ended. The problem statement, provided by the sponsor or professor, outlines the project's scope and objectives but leaves room for exploration and innovation. This statement includes realistic constraints such as aesthetics, safety, cost, ergonomics, environmental impact, operability, and maintainability.

Steps to Define the Problem:
1. **Research:** Gather additional information from online sources, technical publications, patent databases, experts, consumer surveys, and competitor analysis.
2. **Design Specifications:** Develop a comprehensive set of design specifications. If some parameters are initially unknown, mark them as TBD (to be determined) and update them as more information becomes available.
3. **Team Collaboration:** Work collaboratively to define the problem and establish a clear understanding of the project's goals and constraints.

3.4.2   Generating Design Concepts

Once the problem is defined, the next step is to generate as many viable design concepts as possible. This phase encourages creativity and collaboration among team members.

3.4.2.1   Steps to Generate Concepts:

- **Brainstorming Sessions:** Conduct team meetings to brainstorm different design concepts. Aim to generate at least 30 concepts per team member.
- **Drawings and Schematics:** Use visual aids to capture and communicate ideas effectively.
- **Evaluation:** Discuss the pros and cons of each concept and select the most promising ones for further development.

3.4.2.2   Concept Evaluation and Selection

The next phase involves evaluating and selecting the best design concepts using formal decision-making methods and reviews.

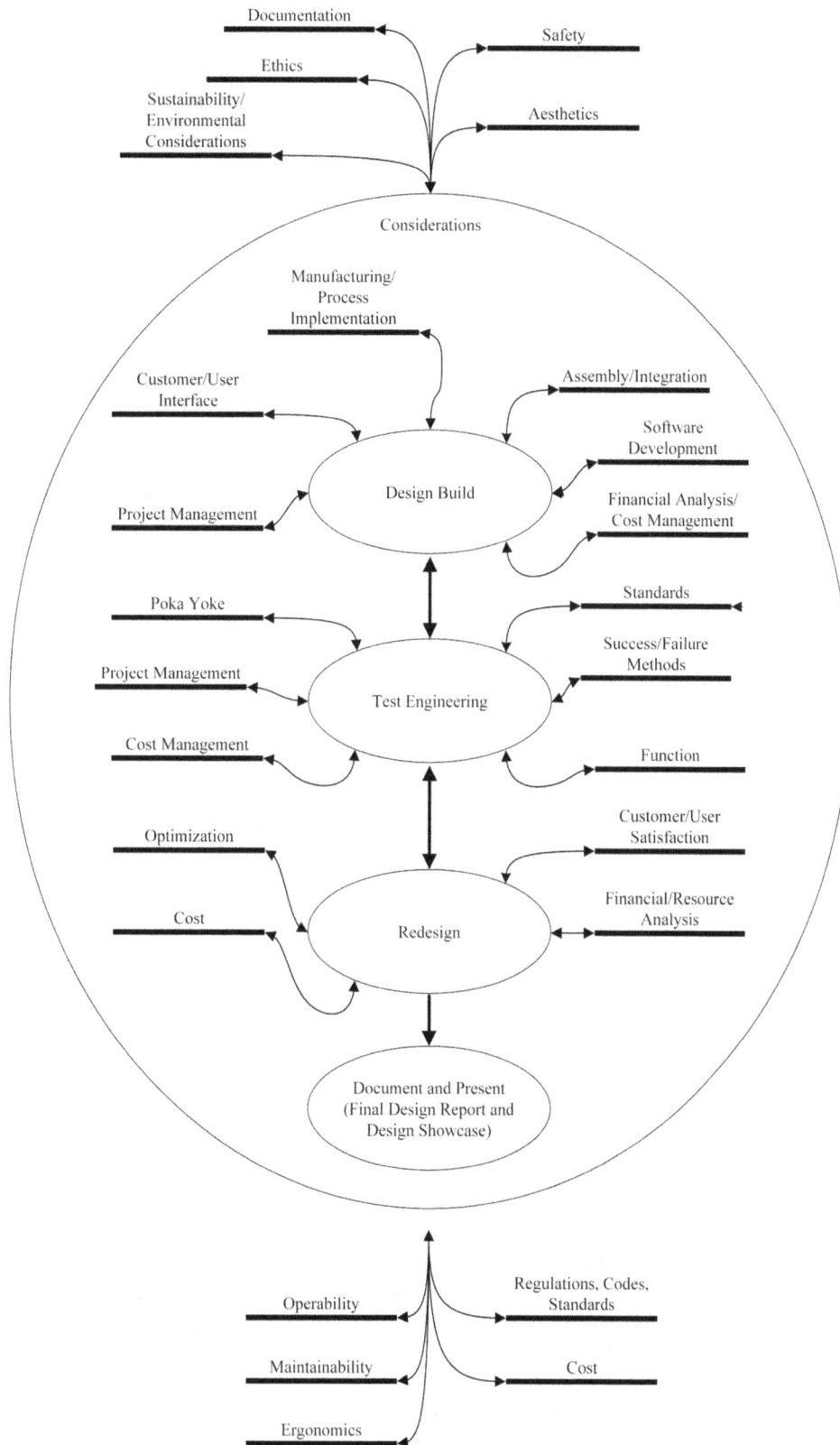

Figure 3-2. Engineering Capstone Design Process Part II diagram.

3.4.2.2.1   Steps to Evaluate and Select Concepts:

1. **Critical Design Review (CDR):** Present selected concepts to peers, experts, sponsors, mentors, and faculty. Gather feedback and critique.
2. **Prototyping:** Develop prototypes or simulations of the selected concepts.
3. **Testing:** Assess prototypes against design specifications and document the results.
4. **Preliminary Design Report (PDR):** Compile findings and feedback into a comprehensive design report. Submit the PDR to professors and sponsors for further comments.

3.5   Part II: Detailed Design and Implementation

Figure 3-2. illustrates the process flow for Part II of the capstone design. Part II involves building, testing, and refining the design, culminating in a final presentation and report.

3.5.1   Building the Design

This phase involves translating the conceptual design into a physical or virtual prototype.

3.5.1.1   Steps to Build the Design:

- **Project Planning:** Develop a detailed project plan outlining tasks, timelines, and resource requirements.
- **Materials and Parts:** Procure materials and parts necessary for building the prototype.
- **Construction:** Assemble the prototype according to the design specifications.

3.5.1.2   Testing and Refinement

Testing is a critical step to ensure that the design meets the required specifications and performs as expected.

3.5.1.3   Steps to Test and Refine the Design:

1. **Testing:** Conduct thorough tests to evaluate the performance of the prototype.
2. **Data Analysis:** Analyze test data to identify areas for improvement.
3. **Iteration:** Make necessary adjustments and refinements to the design based on test results.
4. **Documentation:** Keep detailed records of the testing process and iterations.

3.6   Final Presentation and Report

The final phase involves presenting the completed design and documenting the entire process.

3.6.1   Steps to Finalize the Project:

- **Final Report:** Compile a comprehensive final design report detailing all aspects of the project, from problem definition to final testing.
- **Presentation:** Prepare and deliver a presentation to peers, professors, and sponsors, showcasing the design process and final prototype.
- **Feedback:** Gather feedback from the audience and incorporate it into the final report if necessary.

3.7   Open-Ended Design Problems

Open-ended design problems are a hallmark of capstone projects. These problems mimic real-world challenges by providing an initial problem statement with inherent uncertainties and multiple possible solutions.

3.7.1   Characteristics of Open-Ended Problems

- **Incomplete Information:** The problem statement may lack critical details, requiring students to conduct research and gather additional information.
- **Multiple Solutions:** There are often several viable solutions, each with its pros and cons.
- **Realistic Constraints:** Problems include constraints such as cost, safety, and environmental impact, mirroring real-world considerations.

3.7.2   Importance of Design Specifications

Design specifications are central to the capstone design process. They define the requirements and constraints for the project, guiding all subsequent design activities.

3.7.2.1   Developing Design Specifications

- **Comprehensive:** Ensure specifications are as complete as possible, with parameters marked as TBD if unknown.
- **Iterative:** Update specifications as more information becomes available through research and testing.
- **Collaborative:** Work as a team to develop and refine specifications, ensuring a shared understanding of the project's goals.

3.7.3   Team Collaboration

Effective teamwork is crucial for the success of capstone projects. Team members must collaborate closely, leveraging each other's strengths and expertise.

3.7.3.1   Steps to Enhance Team Collaboration

- **Regular Meetings:** Schedule regular team meetings to discuss progress, brainstorm ideas, and resolve issues.
- **Clear Roles:** Define clear roles and responsibilities for each team member.
- **Communication:** Maintain open and effective communication within the team and with external stakeholders.

3.7.4   Critical Design Review (CDR)

The CDR is a formal part of the capstone design process, where teams present their design concepts to an audience for feedback.

3.7.4.1   Preparing for CDR

- **Presentation:** Develop a clear and concise presentation, highlighting key design concepts and justifications.
- **Feedback:** Be prepared to receive and incorporate feedback from peers, experts, and sponsors.
- **Documentation:** Document the feedback and use it to refine the design concepts.

3.7.5   Prototyping and Testing

Prototyping and testing are crucial phases that validate the design and ensure it meets the required specifications.

3.7.5.1   Steps to Prototype and Test

- **Develop Prototypes:** Create physical or virtual prototypes of the selected design concepts.
- **Conduct Tests:** Perform thorough tests to evaluate the performance of the prototypes.
- **Analyze Data:** Analyze test data to identify strengths and weaknesses.
- **Iterate:** Make necessary adjustments and improvements based on test results.

3.7.6   Preliminary Design Report (PDR)

The PDR is a comprehensive document that summarizes the design process and findings from Part I.

3.7.6.1   Contents of PDR

- **Problem Definition:** Detailed description of the problem and its constraints.
- **Design Specifications:** Comprehensive list of design specifications.
- **Design Concepts:** Description and evaluation of the generated design concepts.
- **Prototyping and Testing:** Summary of prototyping and testing activities and

results.

- **Feedback:** Documentation of feedback received and how it was addressed.

3.7.7 Building and Refining the Design

Part II of the capstone design process focuses on building, testing, and refining the design.

3.7.7.1 Steps to Build and Refine

- **Project Plan:** Develop a detailed plan outlining tasks, timelines, and resources.
- **Construction:** Assemble the prototype according to specifications.
- **Testing:** Conduct tests to evaluate performance.
- **Iteration:** Make necessary adjustments based on test results.
- **Documentation:** Keep detailed records of the building and refinement process.

3.7.8 Final Presentation and Report

The final presentation and report mark the culmination of the capstone design project.

3.7.8.1 Preparing the Final Presentation

- **Clear Communication:** Develop a clear and engaging presentation that highlights key aspects of the design process and final prototype.
- **Audience Engagement:** Be prepared to engage with the audience, answer questions, and receive feedback.
- **Documentation:** Ensure that the final report is comprehensive and accurately reflects the entire design process.

The capstone design process is a rigorous yet rewarding experience that prepares students for real-world engineering challenges. By following a structured process, collaborating effectively, and adhering to design specifications and constraints, students can develop innovative solutions to complex problems.

In your planning, it's crucial to account for the administrative work required to process materials and service purchase requests, as well as availability and shipping times. Consult your professor to follow your engineering department's procedures or your institution's established purchasing guidelines. If you need to buy software or plug-ins for existing software to create your model or software, these purchases may follow different procedures than buying physical materials.

Consider your team's skills in manufacturing, modeling, and software development. Assess whether you have the necessary skills to build your design. If not, identify individuals who can assist your team or determine if additional resources are required, such as funding, tools, software, or machines. If you cannot create the design as planned, you may need to redesign it.

Once you have built your design or model, you need to verify that it meets your design specifications. Validate your design by creating experiments or tests to assess it against the parameters outlined in your design specifications. Revisit any gaps in your design specifications and establish values for them. If specific parameters cannot be defined, consider eliminating them from your specifications. Some tests may be required to determine or refine the numerical values of your design specifications.

Carefully plan your test engineering activities, focusing only on tests that will yield helpful information for your redesign and optimization work. Avoid conducting tests simply because you are familiar with them from previous courses. Instead, analyze what needs to be tested and how to conduct those tests effectively. Consider whether your design needs to comply with any regulations or codes and ensure you follow standard testing methods published by national professional engineering societies when applicable.

From your test engineering procedures, collect data to analyze the performance of your design. Use the analysis results to guide you in optimizing your design or correcting errors. The build-test-redesign cycle must be repeated as many times as necessary to achieve a robust design. Past projects have ranged from two iterations to 150 iterations, depending on the scale and complexity of the design and its build.

This chapter has provided an overview of the capstone design process, highlighting the importance of a fixed timeline, funding and resources, open-ended problems, and effective teamwork. As you embark on your capstone project, use this workbook to document each step of your journey, ensuring a thorough and organized approach to engineering design.

3.8   Advice to Students in Capstone Design

3.8.1   Background Research on the Project

Conduct thorough background research when preparing for your capstone design project. The sources of design projects can vary widely, depending on the approach and philosophy of your engineering program, department, and college. Most engineering programs seek to maintain close relationships with their students' employers, looking for design project opportunities with these companies. If you are interested in working on projects with companies where your friends or family members are employed, start by learning about the projects from your professor or company presentations.

Here are some steps to conduct adequate background research:

3.8.1.1   Consult with Your Professor:

- o **Initial Discussion**: Start by talking with your professor about the project. Professors often have valuable insights and background knowledge.
- o **Taking Notes**: Take detailed notes during your conversation to capture all

critical points for further research.

3.8.1.2   Engage with the Project Sponsor:

o   **Direct Contact**: Reach out to the project sponsor to ask detailed questions. The sponsor is usually the most knowledgeable person about the project's specifics.

o   **Follow-up**: If directed to other individuals within the company, follow up to gather more information.

3.8.1.3   Explore Patent Databases:

o   **Patent Research**: Search through patent databases for similar or related inventions. This can provide insights into existing solutions and potential innovation areas.

o   **Company Patents**: Look for patents filed by the sponsoring company to understand their technological focus and previous work.

3.8.1.4   Review Scholarly Works:

o   **Academic Sources**: Use resources like Google Scholar, ResearchGate, and Academia.edu to find scholarly articles related to the project.

o   **University Library**: Utilize your university library's database to access academic papers.

3.8.1.5   Analyze Competitor Products and Processes:

o   **Competitor Research**: Identify and study competitor products or processes. This can provide a benchmark and highlight areas for improvement or differentiation.

o   **Sponsor's Input**: Ask the sponsor about their competitors to better understand the market landscape and specific challenges.

Be flexible in your project preferences. Interest in a broad range of technical areas can serve you well in your professional career and increase your chances of being assigned to a project that matches your skills and interests.

3.8.2   Educational Preparation

Capstone design sequences are intended to be the culminating experience of your engineering education, utilizing all the knowledge you've gained throughout your coursework. The curriculum includes four primary components: mathematics, sciences, engineering sciences, discipline-specific engineering courses, and general education. Engineering professional electives can also be crucial if they align with your capstone project.

To maximize your educational preparation:

1. **Core Components**:
   o **Mathematics**: Strengthen your understanding of calculus, differential equations, linear algebra, and probability and statistics. These tools are essential for modeling and solving engineering problems.
   o **Sciences**: Ensure a solid grasp of physics and chemistry, as these principles underpin many engineering phenomena.
   o **Engineering Sciences**: Focus on mechanics, thermodynamics, fluid dynamics, materials science, and electrical circuits, as these are directly applicable to designing and optimizing engineering systems.
   o **Discipline-Specific Courses**: Pay attention to courses specific to your engineering discipline (mechanical, electrical, civil, chemical, etc.), as they provide the specialized knowledge needed for your project.
   o **General Education**: Develop your analytical, ethical, and communication skills through humanities, social sciences, and communication courses.

2. **Professional Electives and Specializations**:
   o **Professional Electives**: Choose electives that align with your interests and the requirements of your capstone project.
   o **Minors and Additional Degrees**: Consider pursuing a minor or second degree for additional skills and perspectives, which can give you a competitive edge in multidisciplinary projects.

3.8.3   Professional Preparation

The capstone design experience is a form of practice for becoming a professional engineer. Sponsors of capstone design projects look for prospective future engineers to hire from these teams. They expect team members to be qualified and capable of contributing positively and professionally to the design solution.

To prepare professionally:

1. **Complete Coursework**:
   o Ensure you have completed all required courses in your curriculum to have a solid educational foundation.

2. **Professional Electives**:
   o Select electives that align with your capstone project to gain more profound knowledge and practical skills relevant to your design problem.

3. **Internships and Co-ops**:
   o Gain practical, hands-on experience through internships or co-op programs. These experiences are invaluable for applying theoretical knowledge to real-world problems.
   o Internships with the project's sponsor can facilitate better communication and understanding of the design problem.

4. **Industry Work Experience**:
   - o Any prior or concurrent industry work experience enhances your ability to work on real-world design problems and develop practical solutions.

3.8.4   Qualifications

When applying for a capstone project, highlight the qualifications that are relevant to the project's needs. Consider the following:

3.8.4.1   Academic Performance:

   - o Your transcript is a good indicator of your performance in required engineering, science, mathematics, and other relevant courses.

3.8.4.2   Professional and Elective Courses:

   - o Highlight any professional elective courses that are relevant to the design problem.

3.8.4.3   Internship and Co-op Experience:

   - o Emphasize any internships or co-ops, especially those related to the project's sponsor or industry.

3.8.4.4   Research Experience:

   - o Include any experience working in a research lab, as it demonstrates detailed investigation and problem-solving skills.

3.8.4.5   Teamwork Experience:

   - o Mention any experience working in a team, whether in athletic, professional, military, or other settings.

3.8.4.6   Special Skills and Certifications:

   - o Highlight any special skills or certifications relevant to the design problem.

3.8.4.7   Hobbies and Extracurricular Activities:

   - o Include relevant hobbies or extracurricular activities that contribute to the project.

3.8.4.8   Personal Statement

   - o Clearly explain why you are interested in the project and why you should be assigned to it. Highlight your passion, commitment, and relevant qualifications.

Treat your application for a design project as practice when applying for a job. List and explain all your qualifications and clearly articulate why you are the best candidate for the specific project. Given the competitive nature of capstone project assignments, a well-prepared and thorough application is essential for securing a desired position.

3.9 Assignments

Assignment 3-1: Prepare a tailored resume for capstone.

Prepare a comprehensive and tailored resume that showcases your qualifications, strengths, and experiences relevant to your capstone project. This resume will inform the professor in charge of the class about your background and the unique value you can bring to the project.

Instructions:
1. Format and Layout
   - Use a professional and transparent format for your resume.
   - Choose a readable font (e.g., Arial, Times New Roman) and maintain consistent font sizes throughout the document.
   - Use bullet points to present information concisely and make it easy to read.
   - Ensure proper spacing and margins for a visually appealing layout.
2. Contact Information
   - Include your full name, email address, phone number, and location at the top of the resume.
   - Use a professional email address and ensure that your voicemail message is appropriate.
3. Educational Background
   - List your current degree program, university, and expected graduation date.
   - Include your major, minor (if applicable), and any relevant coursework or projects.
   - Highlight your academic achievements, such as a high GPA, scholarships, or awards.
   - Mention any honors societies or academic distinctions you have received.
4. Relevant Coursework and Certifications
   - List any elective courses you have taken that are relevant to your capstone project or demonstrate your knowledge in a specific area.
   - Include any certifications or training programs you have completed that are applicable to your field of study or the project.
5. Internship and Work Experience
   - Describe any internships or work experience you have had, especially those related to your field of study or the project.
   - Use action verbs to highlight your responsibilities, achievements, and the skills you developed.
   - Quantify your accomplishments whenever possible (e.g., "Implemented a new inventory system that reduced waste by 15%").

6. Research Experience
   - If you have any research experience, include a section describing your role, the project, and any significant findings or contributions.
   - Highlight any publications, presentations, or conferences you have participated in.
7. Technical Skills
   - List any technical skills you possess that are relevant to your field or the project, such as programming languages, software proficiencies, or laboratory techniques.
   - Include your level of proficiency for each skill (e.g., beginner, intermediate, advanced).
8. Teamwork and Leadership
   - Describe experiences that demonstrate your ability to work effectively in a team environment.
   - Highlight any leadership roles or responsibilities you have held in group projects, student organizations, or extracurricular activities.
9. Non-Academic Skills and Experiences
   - Include any non-academic skills or experiences that showcase your unique qualities and interests.
   - This could include volunteer work, community involvement, or personal projects that demonstrate your passion, creativity, or problem-solving abilities.
10. Hobbies and Interests
    - Include a brief section on your hobbies and interests to give the professor a sense of your personality and well-roundedness.
    - Keep this section concise and relevant to the project or your field of study.
11. Proofreading and Review
    - Carefully proofread your resume for any spelling, grammar, or formatting errors.
    - Ask a friend, family member, or career services professional to review your resume and provide feedback.
12. Submission
    - Save your resume as a PDF file with a clear and professional file name (e.g., "FirstName_LastName_CapstoneResume.pdf").
    - Upload your resume under the designated assignment on Brightspace before the specified deadline.

Remember, your resume should be tailored to highlight the skills, experiences, and qualities that make you a strong candidate for your capstone project. Be concise, specific, and honest in your descriptions, and use action-oriented language to showcase your accomplishments.

If you have any questions or need further guidance, please don't hesitate to contact your professor or the career services office.

4 Capstone Design Project Presentations

4.1 Introduction

Capstone design project presentations are a vital component of the capstone experience. They provide students with the opportunity to learn about the various projects available and make informed decisions about which projects they are most qualified to work on. These presentations, which take place over two class days, offer valuable insights into the problem statements, objectives, and requirements of each project, enabling students to assess their qualifications and interests in relation to the projects.

The presentation schedule is carefully designed to accommodate the large number of projects being presented within the limited class time. Each sponsor is allotted a total of 15 minutes for their presentation, with 8 minutes dedicated to the presentation itself and 7 minutes reserved for questions and answers. This format ensures that all projects receive adequate attention and that students have sufficient time to engage with the project sponsors and gather additional information to make well-informed decisions.

4.2 Presentation Logistics

4.2.1 Presentation Dates and Times

The capstone design project presentations are scheduled to take place on the second and third days of class. This allows students to settle into the course and familiarize themselves with the overall capstone design process before diving into the specifics of each project. The presentations are spread out over two days to accommodate the large number of projects and to provide students with ample time to absorb the information presented.

4.2.2   Classroom Setup and Technology

The classroom setup and technology play a crucial role in promoting engagement and interaction between project sponsors and students, thereby facilitating effective presentations. The seating arrangement, designed to provide clear sightlines to the presentation area, encourages discussion and collaboration among students. Moreover, the room is equipped with the necessary audio-visual tools, including a projector, screen, and sound system, to support multimedia presentations and ensure that all attendees can clearly see and hear the content being presented.

4.2.3   Presentation Order and Flow

The order of presentations is carefully planned to ensure a logical flow and to maintain student engagement throughout the sessions. Projects may be grouped by theme, complexity, or other relevant criteria to create a cohesive presentation sequence. Transitions between presentations are managed efficiently, with the 7-minute Q&A sessions serving as brief breaks to allow for setup and preparation.

4.3 Student Preparation and Engagement

4.3.1   Previewing Presentation Materials

To maximize student engagement and understanding during the presentations, project sponsors are encouraged to provide their presentation materials in advance, whenever possible. These materials are made available to students before the start of the presentations, allowing them to preview the content and familiarize themselves with the projects. This advanced preparation enables students to formulate questions, identify areas of interest, and better contextualize the information presented during the live sessions.

4.3.2   Active Listening and Note-Taking During Presentations

Students are expected to listen and take detailed notes during the project presentations actively. They should focus on critical elements such as the problem statement, project objectives, deliverables, and any specific requirements or constraints mentioned by the sponsors. Effective note-taking strategies, such as using a structured template or digital note-taking tools, can help students capture vital information and organize their thoughts for later review and analysis.

4.3.3   Engaging with Project Sponsors and Presenters

Students are strongly encouraged to make the most of the 7-minute Q&A sessions by asking well-thought-out questions and seeking additional details about the projects. This interaction helps students gain a deeper understanding of the project scope, challenges, and expectations, which is crucial for determining their qualifications and fit for each project. Sponsors and presenters, as the key architects of these projects, should be prepared to address student inquiries and provide further insights into their projects. Their role is not just to present but to foster a collaborative and interactive learning environment where students can actively engage with the projects and gain a richer learning experience.

4.4 Project Sponsor Presentations

4.4.1 Sponsor Presentation Guidelines

Project sponsors are provided with guidelines to ensure that their presentations are informative, concise, and well-structured. The recommended content includes an overview of the problem statement, project objectives, deliverables, and any specific requirements or constraints. Sponsors are encouraged to use visual aids, such as slides or demonstrations, to effectively communicate their ideas and engage the audience within the allotted 8-minute presentation time.

4.4.2 Handling Sponsor Presentation Materials

The course instructors or support staff are responsible for collecting and organizing the sponsor presentation materials. This includes gathering electronic copies of slides, handouts, or any other relevant documents. These materials are then distributed to the students, either electronically through a learning management system or as printed copies, depending on the preferences and resources available.

4.4.3 Facilitating Sponsor-Student Interactions

During the 7-minute Q&A sessions, the course instructors or designated moderators facilitate interactions between sponsors and students. They encourage students to ask questions and engage in meaningful discussions with the sponsors. The moderators ensure that the Q&A sessions remain focused, respectful, and productive while also managing time effectively to allow all students an equal opportunity to participate and gather the information they need to make informed project choices.

4.5 Student Reflection and Analysis

4.5.1 Post-Presentation Reflection

After each presentation, students are given time to reflect on the information presented and capture their initial thoughts and impressions. This reflection process helps students consolidate their understanding of the projects and identify key takeaways and insights. Students may be provided with guiding questions or prompts to facilitate their reflection, encouraging them to consider aspects such as the project's relevance to their qualifications, the skills required, and the potential impact of the project.

4.5.2   Comparative Analysis of Projects

As students attend multiple project presentations, they are encouraged to engage in comparative analysis, identifying common themes, challenges, and opportunities across different projects. This analysis helps students appreciate the breadth and diversity of the capstone design projects and allows them to evaluate the relative complexity and scope of each project. By comparing and contrasting the projects, students can make more informed decisions about which projects align best with their qualifications and skill sets.

4.5.3   Assessing Project Alignment with Qualifications and Skills

Throughout the presentation process, students assess each project's alignment with their qualifications, strengths, and career goals. They consider factors such as the project's technical focus, the opportunity for skill application and development, and the potential for making a meaningful impact. Students are encouraged to reflect on their engineering education, other educational experiences, professional experience, and any other relevant qualifications to determine which projects they are best suited for.

4.6   Project Application Process

4.6.1   Project Application Assignment

After attending all project presentations, students are required to complete a project application assignment. This assignment involves filling out a form for each project they feel qualified to participate in, up to a maximum of eight projects. The application form requires students to articulate their qualifications, skills, and experiences that make them suitable candidates for each project they apply to.

4.6.2   Emphasizing Qualifications over Interest

While student interest in a project is essential, it is not the sole determining factor for project assignment. Students must demonstrate that they possess the necessary qualifications, skills, and experiences to contribute effectively to the projects they apply for. The project application process encourages students to critically evaluate their strengths and match them with the requirements and demands of each project.

4.6.3   Preparing Strong Applications

To prepare robust project applications, students should draw upon their engineering education, other relevant coursework, professional experiences (such as internships or co-op positions), and any additional qualifications or skills they have acquired. Students should provide specific examples and evidence to support their claims of qualification, showcasing how their background aligns with the needs of each project. Well-crafted and compelling applications will increase the likelihood of students being assigned to their preferred projects.

4.7 Project Assignment Process

4.7.1 Reviewing Project Applications

The course instructors carefully review the project applications submitted by students, evaluating the qualifications and fit of each student for the projects they have applied to. The review process takes into account the project requirements, team composition needs, and the overall distribution of skills and backgrounds across the project teams.

4.7.2 Assigning Students to Projects

The project assignment process is designed to create well-balanced teams with a diverse range of skills and qualifications. This careful team composition ensures that each team has the necessary expertise to tackle their assigned project successfully. Based on the review of project applications, the course instructors assign students to projects. Priority may be given to students in the honors section or those seeking a specific minor, such as nuclear engineering, to ensure that their educational goals are met.

4.7.3 Communicating Project Assignments

Once the project assignments are finalized, students are notified of their assigned projects through an official communication channel, such as email or the learning management system. The notification includes details about the project, the team composition, and any necessary documentation or resources. Students are encouraged to reach out to their assigned teammates and project sponsors to begin the process of team formation and project planning.

Capstone design project presentations are a key element in helping students make informed decisions about which projects they are most qualified to work on. The 15-minute presentation format, with 8 minutes for the presentation and 7 minutes for Q&A, allows students to engage with project sponsors, ask questions, and gather valuable insights. This process enhances their understanding of the projects and their fit, thereby empowering them to make the best project choices.

The project application process is a critical component of the capstone experience, as it requires students to critically evaluate their qualifications and articulate how their skills and experiences align with the needs of each project. By emphasizing qualifications over interest alone, the application process ensures that students are assigned to projects where they can make the most meaningful contributions and maximize their learning outcomes.

As students embark on their capstone design journey, they are encouraged to leverage their engineering education, additional qualifications, and professional experiences to prepare robust project applications. By actively participating in the presentation and application process, students will be well-positioned to join project teams that align with their strengths and provide opportunities for growth and success.

4.8 Guidelines for Effective Project Applications

When applying for capstone design projects, it is essential to submit well-crafted and compelling applications that showcase your qualifications, skills, and experiences. The following guidelines will help you prepare practical project applications:

4.8.1 Review Project Requirements

- o Carefully review the project descriptions, objectives, and any specific requirements or skills mentioned during the project presentations or in the provided materials.
- o Make note of the essential qualifications and expertise sought by each project sponsor.
- o Identify projects that align with your educational background, technical skills, and professional experiences.

4.8.2 Highlight Relevant Coursework

- o Emphasize relevant coursework you have completed that directly relates to the project you are applying for.
- o Mention specific courses, projects, or assignments that have provided you with knowledge and skills applicable to the project.
- o Explain how your coursework has prepared you to contribute effectively to the project.

4.8.3 Showcase Technical Skills

- o Identify the technical skills required for each project and assess your proficiency in those areas.
- o Highlight any programming languages, software tools, or laboratory techniques you have mastered that are relevant to the project.
- o Provide examples of how you have applied these skills in previous projects or assignments.

4.8.4 Describe Professional Experiences

- o If you have relevant professional experiences, such as internships, co-op positions, or part-time jobs, highlight them in your application.
- o Explain how these experiences have provided you with practical skills and knowledge that are applicable to the project.
- o Use specific examples to demonstrate your ability to work in a team, solve problems, or deliver results.

4.8.5 Emphasize Transferable Skills

- o In addition to technical skills, highlight any transferable skills you possess that can contribute to the success of the project.
- o These may include skills such as communication, leadership, project management, or problem-solving.
- o Provide examples of how you have demonstrated these skills in previous academic or professional settings.

4.8.6 Tailor Your Application

- o Customize your application for each project you are applying to rather than submitting a generic application.
- o Demonstrate your understanding of the project's specific requirements and explain how your qualifications align with them.
- o Use language and terminology that resonates with the project sponsor and shows your familiarity with the project domain.

4.8.7 Provide Supporting Evidence

- o Where possible, provide evidence to support your claims of qualification and expertise.
- o This may include links to relevant projects, portfolios, GitHub repositories, or other examples of your work.
- o Ensure that any supporting evidence is easily accessible and clearly demonstrates your skills and accomplishments.

4.8.8 Be Concise and Clear

- o While it is crucial to provide sufficient detail, aim to keep your application concise and to the point.
- o Use clear and precise language to convey your qualifications and experiences effectively.
- o Organize your application logically, using headings and bullet points to improve readability.

4.8.9   Proofread and Review

- Before submitting your application, proofread it carefully to identify and correct any spelling, grammar, or formatting errors.
- Review your application to ensure that it is complete, coherent, and effectively communicates your qualifications and fit for the project.
- Consider seeking feedback from peers, mentors, or the writing center to further refine your application.

4.8.10   Demonstrate Enthusiasm and Professionalism

- Express your enthusiasm for the project and explain why you are particularly interested in working on it.
- Maintain a professional tone throughout your application, demonstrating respect for the project sponsor and the application process.
- Close your application by thanking the reviewer for their consideration and expressing your eagerness to contribute to the project.

By following these guidelines, you can prepare practical project applications that showcase your qualifications, skills, and experiences. Remember, the goal is to demonstrate your fit for the project and convince the reviewers that you have the necessary expertise to contribute to its success. Take the time to craft compelling applications that highlight your strengths and set you apart from other applicants.

4.9 Assignments

Assignment 4-1: Preparing and Asking Questions for Sponsor Project Presentations

Objective: Actively participate in the sponsor project presentations by preparing thoughtful and relevant questions to gather additional information about the proposed projects. Asking well-crafted questions will help you gain a deeper understanding of the project scope, requirements, and expectations, enabling you to make informed decisions when applying for projects.

Instructions:

1. Review Project Summaries
   - Prior to the sponsor presentations, carefully review the project summaries or any provided materials to familiarize yourself with each project.
   - Identify critical aspects of the projects, such as the problem statement, objectives, deliverables, and any mentioned requirements or constraints.

2. Identify Areas for Clarification
   - As you review the project summaries, note down any areas that require further clarification or elaboration.
   - Consider aspects of the project that may not be fully explained or that you would like to know more about.

3. Prepare Specific Questions
   - Based on your review and identified areas for clarification, prepare a list of specific questions for each project you are interested in.
   - Aim to create questions that will elicit valuable information and help you gain a deeper understanding of the project.
   - Consider the following categories of questions: a. Project Scope and Objectives b. Technical Requirements and Challenges c. Required Skills and Expertise d. Sponsor Expectations and Involvement e. Impact and Significance of the Project

4. Prioritize Your Questions
   - Review your list of questions and prioritize them based on their relevance and importance to your understanding of the project.
   - Select the top 2-3 questions for each project that you believe will provide the most valuable insights.

5. Ask Questions During the Presentations
   - During the sponsor presentations, actively listen to the information being shared and take notes.
   - When the opportunity arises, raise your hand or use the designated method to ask your prepared questions.
   - Be concise and clear when asking your questions, ensuring that they are easily understandable to the presenter and the audience.

6. Engage in Follow-up Discussion

      o   After asking your initial question, be prepared to engage in a follow-up discussion if the presenter seeks further clarification or provides an opportunity for additional queries.

      o   Use this opportunity to delve deeper into the project details and gain a more comprehensive understanding of the project requirements and expectations.

7. Record Responses and Insights

      o   As the presenter responds to your questions, take detailed notes to capture the critical information and insights provided.

      o   These notes will be valuable when reflecting on the projects and making informed decisions during the application process.

Sample Questions: Here are 15 sample questions that you can consider asking during the sponsor project presentations:

1. What are the primary objectives and goals of this project?
2. What are the key deliverables expected at the end of the project?
3. What specific technical skills or knowledge would be most valuable for students working on this project?
4. What are the significant challenges or obstacles anticipated in completing this project?
5. Are there any particular engineering disciplines or backgrounds that would be especially well-suited for this project?
6. What resources or support will be provided by the sponsor throughout the project?
7. Are there any specific tools, software, or methodologies that the sponsor recommends or expects to be used in this project?
8. How will the success of the project be measured, and what metrics will be used?
9. Is there any prior work or research that the project will build upon?
10. What is the potential impact or significance of this project in the broader context of the sponsor's organization or industry?
11. Are there any opportunities for student innovation or creativity within the project scope?
12. How frequently does the sponsor expect to meet with the project team, and what is the expected level of communication?
13. What are the sponsor's expectations for the quality and depth of the project outcomes?
14. Are there any specific skills or experiences that would make a student particularly well-suited for this project?
15. How does this project align with the sponsor's long-term goals or vision?

Remember, the key is to ask questions that demonstrate your engagement, curiosity, and desire to understand the project in greater depth. By asking thoughtful and targeted questions, you will be better equipped to make informed decisions when applying for projects. You will show your potential as a valuable contributor to the project team. Keep in mind that the class guidelines predetermine the project timeline, documentation requirements, and team selection process, so focus your questions on aspects that are specific to the project itself and the sponsor's expectations.

Project #
Sponsor

Title

Presenter

Questions:

Notes:

Project #
Sponsor

Title

Presenter

Questions:

Notes:

Project #
Sponsor

Title

Presenter

Questions:

Notes:

Project #
Sponsor

Title

Presenter

Questions:

Notes:

Project #
Sponsor

Title

Presenter

Questions:

Notes:

Project #
Sponsor

Title

Presenter

Questions:

Notes:

Project #
Sponsor

Title

Presenter

Questions:

Notes:

Project #
Sponsor

Title

Presenter

Questions:

Notes:

Project #
Sponsor

Title

Presenter

Questions:

Notes:

Project #
Sponsor

Title

Presenter

Questions:

Notes:

Project #
Sponsor

Title

Presenter

Questions:

Notes:

Project #
Sponsor

Title

Presenter

Questions:

Notes:

Project #
Sponsor

Title

Presenter

Questions:

Notes:

Project #
Sponsor

Title

Presenter

Questions:

Notes:

Project #
Sponsor

Title

Presenter

Questions:

Notes:

Project #
Sponsor

Title

Presenter

Questions:

Notes:

Project #
Sponsor

Title

Presenter

Questions:

Notes:

Project #
Sponsor

Title

Presenter

Questions:

Notes:

Project #
Sponsor

Title

Presenter

Questions:

Notes:

Project #
Sponsor

Title

Presenter

Questions:

Notes:

Project #
Sponsor

Title

Presenter

Questions:

Notes:

Project #
Sponsor

Title

Presenter

Questions:

Notes:

Project #
Sponsor

Title

Presenter

Questions:

Notes:

Project #
Sponsor

Title

Presenter

Questions:

Notes:

Project #
Sponsor

Title

Presenter

Questions:

Notes:

Project #
Sponsor

Title

Presenter

Questions:

Notes:

Project #
Sponsor

Title

Presenter

Questions:

Notes:

.

5 Design Project Team

During the past three decades, there has been a profound change in industry practices in performing engineering design work. Previously, extensive engineering companies organized their engineers into technical groupings within design divisions. Mechanical engineers worked in one group, electrical engineers in another, and so on. Over time, companies realized that this approach was limiting and that diversity in backgrounds and expertise produced better project and design outcomes. They discovered that teams composed of individuals from various educational backgrounds could produce more robust design results. It also became evident that technically, excellent engineers who graduated from elite engineering programs but worked alone were less effective than teams of engineers who collaborated effectively.

5.1 Selection of Team Members for Design Projects

Success in capstone design projects results from both a structured approach and a careful selection of team members. The primary goal of capstone design is to prepare students for professional engineering practice by simulating real-world environments. In industry, team assignments are based on the skills and qualifications best suited for the project. Following a similar process in capstone courses will create more capable teams and lead to better design project outcomes.

Several methods can be used to assign students to capstone design project teams:

1. **Random Assignment**:

    At the start of the year, each student is randomly assigned to a project. While this ensures every student is assigned a project, the results are unpredictable, leading to potentially imbalanced teams.

2. **Interest-Based Assignment**:

    Students express their interest in various projects, and assignments are made based on these preferences. This method can lead to students working only with friends, which may not reflect typical workplace environments.

3. **Student-Selected Projects**:

    Students propose their projects, subject to the professor's approval. This approach can lead to high engagement but requires careful review to ensure appropriate project complexity and difficulty.

    4. **Qualification-Based Assignment**:

Students are "hired" into teams based on their qualifications, including academic performance, internship or co-op experience, diversity, and resumes summarizing their experience. This approach simulates the hiring process in the industry and has proven to produce the best outcomes.

Regardless of the approach used, students must be assigned to project teams quickly at the start of the academic year.

5.1.1   Background Research on the Project

When you learn about design projects from your professor or company presentations, it is crucial to conduct thorough background research on the problems that interest you. This research not only enhances your understanding but also strengthens your qualifications for the project.

Here are steps to effectively conduct background research:

1. **Consult with Your Professor**:
   - Start by discussing the project with your professor, who often has valuable insights and background knowledge.
   - Take detailed notes during your conversation to capture all critical points for further research.

2. **Engage with the Project Sponsor**:
   - Contact the project sponsor to ask detailed questions. Sponsors are usually the most knowledgeable about the project's specifics.
   - If directed to other individuals within the company, follow up with them to gather more information.

3. **Explore Patent Databases**:
   - Search through patent databases for similar or related inventions. This research can provide insights into existing solutions and potential innovation areas.
   - Look for patents filed by the sponsoring company to understand their technological focus and previous work.

4. **Review Scholarly Works**:
   - Use resources like Google Scholar, ResearchGate, and Academia.edu to find scholarly articles related to the project.
   - Utilize your university library's database to access academic papers.

5. **Analyze Competitor Products and Processes**:
   - Identify and study competitor products or processes to provide a benchmark and highlight areas for improvement or differentiation.
   - Ask the sponsor about their competitors to better understand the market landscape and specific challenges.

Flexibility in project preferences can be beneficial. Being open to a broad range of technical areas can serve you well in your professional career and increase your chances of being assigned to a project that matches your skills and interests.

5.1.2   Educational Preparation

Capstone design sequences are intended to be the culminating experience of your engineering education, utilizing all the knowledge you've gained throughout your coursework. The curriculum typically includes four primary components: mathematics, sciences, engineering sciences, discipline-specific engineering courses, and general education. Engineering professional electives can also play a significant role if their content aligns with your capstone project.

To maximize your educational preparation:

1. **Core Components**:
   - **Mathematics**: Strengthen your understanding of calculus, differential equations, linear algebra, and probability and statistics. These tools are essential for modeling and solving engineering problems.
   - **Sciences**: Ensure a solid grasp of physics and chemistry, as these principles underpin many engineering phenomena.
   - **Engineering Sciences**: Focus on mechanics, thermodynamics, fluid dynamics, materials science, and electrical circuits, as these are directly applicable to designing and optimizing engineering systems.
   - **Discipline-Specific Courses**: Pay attention to courses specific to your engineering discipline (mechanical, electrical, civil, chemical, etc.), as they provide the specialized knowledge needed for your project.
   - **General Education**: Develop your analytical, ethical, and communication skills through humanities, social sciences, and communication courses.
2. **Professional Electives and Specializations**:
   - **Professional Electives**: Choose electives that align with your interests and the requirements of your capstone project.
   - **Minors and Additional Degrees**: Consider pursuing a minor or second degree for additional skills and perspectives, which can give you a competitive edge in multidisciplinary projects.

5.1.3   Professional Preparation

The capstone design experience is a form of practice for becoming a professional engineer. Sponsors of capstone design projects look for prospective future engineers to hire from these teams. They expect team members to be qualified and capable of contributing positively and professionally to the design solution.

To prepare professionally:

1. **Complete Coursework**:
   - o Ensure you have completed all required courses in your curriculum to have a solid educational foundation.
2. **Professional Electives**:
   - o Select electives that align with your capstone project to gain more profound knowledge and practical skills relevant to your design problem.
3. **Internships and Co-ops**:
   - o Gain practical, hands-on experience through internships or co-op programs. These experiences are invaluable for applying theoretical knowledge to real-world problems.
   - o Internships with the project's sponsor can facilitate better communication and understanding of the design problem.
4. **Industry Work Experience**:
   - o Any prior or concurrent industry work experience enhances your ability to work on real-world design problems and develop practical solutions.

5.1.4   Qualifications

When applying for a capstone project, highlight the qualifications that are relevant to the project's needs. Consider the following:

1. **Academic Performance**:
   - o Your transcript is a good indicator of your performance in required engineering, science, mathematics, and other relevant courses.
2. **Professional and Elective Courses**:
   - o Highlight any professional elective courses that are relevant to the design problem.
3. **Internship and Co-op Experience**:
   - o Emphasize any internships or co-ops, especially those related to the project's sponsor or industry.
4. **Research Experience**:
   - o Include any experience working in a research lab, as it demonstrates detailed investigation and problem-solving skills.
5. **Teamwork Experience**:
   - o Mention any experience working in a team, whether in athletic, professional, military, or other settings.
6. **Special Skills and Certifications**:
   - o Highlight any special skills or certifications relevant to the design problem.
7. **Hobbies and Extracurricular Activities**:
   - o Include relevant hobbies or extracurricular activities that contribute to the project.

8. **Personal Statement**:
    - Clearly explain why you are interested in the project and why you should be assigned to it. Highlight your passion, commitment, and relevant qualifications.

Treat your application for a design project as practice when applying for a job. List and explain all your qualifications and clearly articulate why you are the best candidate for the specific project. Given the competitive nature of capstone project assignments, a well-prepared and thorough application is essential for securing a desired position.

5.2 Professionalism in Interactions

Engineers are expected to be truthful, discreet, reliable, respectful of others' opinions, and effective communicators. As you prepare for your profession after graduation, it is crucial to behave professionally in class and your interactions with sponsors, mentors, advisors, teaching assistants, and professors.

Professionalism encompasses self-presentation, attitude, and communication style. Here are some key traits of professionalism:

1. **Attention to Detail**:
    - Pay close attention to sponsor needs, requirements, and communications. Be responsive to clients, sponsors, mentors, and team members.
    - Proactivity is essential for anticipating issues and solving problems. Show interest in the project and pay attention to details.

2. **Reliability and Dependability**:
    - Honor your commitments and demonstrate responsibility for assigned tasks. Be punctual and prepared for scheduled meetings.
    - Commitment and reliability show respect for team members and stakeholders.

3. **Commitment to Project Goals**:
    - View yourself as a dedicated team member with a commitment to the project's success. Your dedication makes you a respected and valued team member.

4. **Professional Appearance**:
    - Dress according to the norms of your work environment. Respect others and observe proper etiquette in interactions with team members and stakeholders.
    - Keep personal matters from interfering with team productivity and project success.

5. **Confidentiality and Non-Disclosure**:
    - Strictly observe sponsor policies on confidentiality and non-disclosure of project information. Read and understand any non-disclosure agreements

(NDAs) you sign and honor their terms throughout the project.

By adhering to these principles of professionalism, you will enhance your reputation, increase your chances of being assigned to top-choice projects, and advance your career in engineering.

5.3 Team Dynamics

The development of effective team dynamics is crucial for the success of capstone design projects. B.W. Tuchman proposed a four-stage developmental process for project teams to transform into productive units: Forming, Storming, Norming, and Performing. Understanding and navigating these stages can help capstone teams achieve their goals and deliver successful projects.

5.3.1 Forming Stage

The forming stage is the initial phase where team members come together, often not knowing each other well. In extensive engineering programs, students may be strangers, and it takes time for them to become a cohesive unit. During this stage:

- **Positive Attitude**: Students are generally excited to be part of a project team and exhibit a positive attitude.
- **Politeness and Caution**: Team members are polite and cautious as they try to understand the group dynamics.
- **Anxiety and Disappointment**: Some students may feel anxious about their roles and responsibilities or disgruntled if they are not assigned to their preferred team.
- **Effort to Connect**: Team members must make an effort to get to know each other and understand the project goals.

5.3.2 Storming Stage

The storming stage is characterized by the emergence of conflicts as team members begin to push the boundaries established during the forming phase. This phase can be challenging and may lead to the failure of some teams if not managed properly:

- **Conflicts and Frustration**: Team members may experience conflicts and frustration due to differing opinions and approaches.
- **Overwhelming Tasks**: The complexity of the design problem and the amount of work required can overwhelm students.
- **Questioning Abilities**: Some students may doubt their qualifications and the value of the project.
- **Stress and Isolation**: Those who remain focused on the tasks may experience stress, especially if they lack strong relationships with peers or feel unprepared.

5.3.3   Norming Stage

As teams progress, they enter the norming stage, where they start to resolve differences and work more collaboratively:

- **Resolution of Differences**: Team members work through their differences and begin to respect each other and their roles.
- **Socialization**: Increased social interaction helps in building a sense of camaraderie and trust.
- **Collaborative Effort**: The team starts functioning more cohesively, sharing responsibilities, and supporting each other.
- **Progress in Project**: There is noticeable progress towards project goals, though occasional relapses into the storming stage may occur.

5.3.4   Performing Stage

The performing stage is when the team reaches its peak productivity and effectiveness:

- **High Performance**: The team's hard work starts yielding tangible results, such as defining the design problem, creating credible design specifications, and finding valuable information.
- **Positive Feedback**: Positive feedback from professors, sponsors, and peers boosts the team's morale.
- **Rewarding Experience**: The structure and process of the capstone design become rewarding, and team members feel accomplished and relaxed.
- **Adjourning Stage**
  The final phase is adjourning, where the team disbands upon project completion. This often coincides with graduation, marking a transition to professional careers or further education.

5.3.5   Characteristics of Effective Capstone Design Teams

Capstone design teams are typically self-managing and self-organizing, centered around peer leadership rather than top-down management. Key characteristics of successful teams include:

1. **Member Leadership**:
   o Team members coach each other, remove obstacles, and mitigate distractions.
   o Leadership is distributed among team members rather than being centralized.
2. **Collaborative Learning**:
   o Team members continuously learn from and teach each other new skills and knowledge.

      o   There is a commitment to mutual growth and project success.
3. **Consensus Decision-Making**:
   - Teams strive to reach agreements and make decisions collaboratively.
   - Members work to remove bottlenecks and are willing to take on different tasks as needed.
4. **Communication and Relationship Building**:
   - Effective communication within the team and with sponsors and mentors is crucial.
   - Building solid relationships with stakeholders enhances project outcomes.
5. **Handling Dysfunction**:
   - Teams proactively address dysfunction and work to reengage members who are distracted or unproductive.
   - Collaborative problem-solving is essential for maintaining team cohesion and productivity.

5.3.5.1   Team Size and Composition

The optimal team size for capstone projects is typically three to five members, allowing for efficient use of human talent. Teams should aim for a balance of skills and backgrounds to tackle the multifaceted nature of design problems.

5.3.5.2   Commitment to Team and Project

Successful capstone design teams exhibit a solid commitment to both the project and each other. This involves:
- **Responsibility**: Members take ownership of their roles and tasks.
- **Punctuality and Preparation**: Being on time and well-prepared for meetings and milestones.
- **Respect and Etiquette**: Maintaining a professional demeanor and respecting each other's contributions.
- **Flexibility**: Being willing to adapt to changing project needs and support team members in various capacities.

By understanding and applying these principles of team dynamics, capstone design teams can navigate the challenges of their projects and achieve successful outcomes.

5.4   Effective Team Membership

A team is a collaborative effort. Working effectively in a team is not straightforward; it involves navigating agreements and disagreements with your team members during interactions and discussions. It is paramount not to lose sight of the team's goal: to create a successful design solution. Effective team membership requires core qualities that team members must have or develop during the capstone design project. These qualities include a positive attitude, problem-solving skills, curiosity, dedication, creativity, responsiveness, and an appreciation of diversity.

5.4.1 Positive Attitude

Attitude is a critical factor that can make or break a design project. Having a positive attitude towards the project and your team members is essential for a successful design outcome. A positive attitude:

- **Fosters Collaboration**: Encourages open communication and collaboration among team members.
- **Build resilience**: Helps the team navigate challenges and setbacks with a constructive mindset.
- **Enhances Morale**: Contributes to a supportive and motivating team environment.

5.4.2 Problem Solver

Being a problem solver is at the heart of engineering. As an engineering student, you have been educated in solving technical problems, and these skills are essential for the design aspects of your project. However, problem-solving extends beyond technical challenges:

- **Team Dynamics**: Addressing conflicts and facilitating effective teamwork.
- **Administrative Challenges**: Navigating organizational and bureaucratic hurdles.
- **Sponsor Interactions**: Effectively communicating and meeting the expectations of project sponsors.
- **Course Requirements**: Ensuring all academic and project milestones are met.

5.4.3 Curiosity

Curiosity is crucial for tackling complex capstone design projects. These projects often involve open-ended problems with no straightforward solutions. Curiosity drives:

- **In-Depth Exploration**: Investigating all aspects of the design problem to uncover hidden issues and potential solutions.
- **Continuous Learning**: Seeking out new information, techniques, and technologies that can contribute to the project.
- **Innovative Thinking**: Challenging assumptions and exploring alternative approaches to design problems.

5.4.4 Dedication

Dedication to the project and team is essential for success. Open-ended design problems require persistence and commitment. Dedication manifests in several ways:

- **Persistence**: Staying focused on the problem despite obstacles and setbacks.

- **Commitment to Team**: Supporting fellow team members and contributing consistently to the team's efforts.
- **Time Management**: Allocating sufficient time and effort to meet project deadlines and milestones.

5.4.5   Creativity

Creativity is highly valued in capstone design projects. These projects often require new and innovative solutions to previously unsolved problems. Creativity involves:

- **Thinking Outside the Box**: Developing unique and unconventional solutions.
- **Applying Knowledge Innovatively**: Using existing knowledge and skills in novel ways to address design challenges.
- **Encouraging Innovation**: Creating an environment where team members feel free to propose and explore new ideas.

5.4.6   Responsiveness

Responsiveness is critical given the fixed timeline of capstone projects dictated by the academic calendar. Being responsive ensures:

- **Effective Communication**: Timely responses to team members, sponsors, and other stakeholders.
- **Prompt Action**: Quick resolution of issues and timely execution of tasks.
- **Maintaining Momentum**: Keeping the project on track and avoiding delays.

5.4.7   Appreciation of Diversity

Appreciation of diversity involves recognizing and valuing the different perspectives and backgrounds of team members. Working in a diverse technical and cultural environment is essential for productive teamwork. This appreciation includes:

- **Listening**: Actively listening to and considering the viewpoints of all team members.
- **Understanding**: Making an effort to understand the diverse experiences and knowledge that team members bring.
- **Adapting**: Being flexible and willing to adjust approaches based on diverse inputs.

Effective team membership is foundational to the success of capstone design projects. By embodying these core qualities, team members can navigate the complexities of their projects and work collaboratively towards innovative and successful design solutions.

5.5 Team Leadership

In a capstone design team, one person typically assumes the role of team leader. The team leader plays a crucial role in guiding the team towards successful project completion. Their responsibilities include scheduling team meetings, handling external communications, maintaining the project plan, and ensuring the timely submission of reports and assignments. Effective team leadership requires strong communication skills, organizational abilities, and a proactive attitude.

5.5.1   Responsibilities of the Team Leader

1. **Scheduling Team Meetings**:
   - **Planning Meetings**: The team leader is responsible for scheduling regular team meetings. This includes finding times that work for all team members and ensuring meetings are frequent enough to maintain progress.
   - **Setting Agendas**: The team leader sets the agenda for each meeting, focusing on current tasks, upcoming deadlines, and any issues that need to be addressed.
   - **Facilitating Meetings**: During meetings, the team leader ensures discussions stay on track, all members have a chance to contribute, and decisions are made efficiently.
2. **Handling External Communications**:
   - **Communication with Sponsors**: The team leader manages communications with project sponsors, ensuring that updates, questions, and issues are communicated clearly and promptly.
   - **Liaising with Professors**: The team leader also communicates with the course professor, providing updates on the team's progress and seeking guidance as needed.
   - **Networking**: Building and maintaining professional relationships with external stakeholders is essential for gaining support and resources for the project.
3. **Maintaining the Project Plan**:
   - **Project Planning**: The team leader helps develop and maintain a detailed project plan, outlining tasks, timelines, milestones, and responsibilities.
   - **Monitoring Progress**: Regularly monitor the team's progress against the project plan and make adjustments as necessary to stay on track.
   - **Resource Management**: Ensuring that all necessary resources are available and utilized effectively to meet project goals.
4. **Ensuring Timely Submission of Reports and Assignments**:
   - **Deadline Management**: Keeping track of all deadlines and ensuring that the team submits reports and assignments on time.
   - **Quality Assurance**: Reviewing submissions for quality and completeness

before they are turned in.

- o **Documentation**: Maintaining thorough documentation of the team's work and decisions to support reports and presentations.

5.5.2   Qualities of an Effective Team Leader

1. **Effective Communicator**:
   - o **Clarity**: Communicating clearly and concisely with team members and external stakeholders.
   - o **Listening Skills**: Actively listening to team members' ideas, concerns, and feedback.
   - o **Diplomacy**: Handling conflicts and disagreements within the team with tact and fairness.
2. **Organizational Skills**:
   - o **Time Management**: Effectively managing time and prioritizing tasks to ensure project milestones are met.
   - o **Detail-Oriented**: Paying attention to details to ensure nothing is overlooked in the project plan or communications.
   - o **Multitasking**: Balancing multiple responsibilities and tasks simultaneously.
3. **Proactive Attitude**:
   - o **Initiative**: Taking the initiative to address issues and challenges before they escalate.
   - o **Problem-solving**: Identify potential problems early and develop solutions.
   - o **Motivation**: Keeping the team motivated and focused on project goals.

5.5.3   Team Leader Election or Assignment

1. **Election by Team**:
   - o **Democratic Process**: The team may elect a leader through a democratic process, ensuring that the chosen leader has the support and confidence of the team members.
   - o **Nomination and Voting**: Team members can nominate candidates, and a vote is held to select the leader.
2. **Assignment by Professor**:
   - o **Professor's Decision**: The course professor may assign a team leader based on their knowledge of the student's skills, experience, and leadership potential.
   - o **Criteria for Selection**: The professor considers factors such as academic performance, previous leadership experience, and interpersonal skills when assigning a leader.

5.5.4   Supporting the Team Leader

While the team leader has significant responsibilities, it is essential for the entire team to support their leader:

1. **Collaboration**:
   - **Active Participation**: Team members should actively participate in meetings and contribute to discussions and decision-making processes.
   - **Sharing Workloads**: Team members should take on their share of tasks and responsibilities to ensure the team functions smoothly.
2. **Communication**:
   - **Open Dialogue**: Maintaining open and honest communication with the team leader about any issues or concerns.
   - **Feedback**: Providing constructive feedback to the team leader to help them improve their effectiveness.
3. **Respect**:
   - **Trust**: Trusting the team leader's decisions and respecting their role in guiding the team.
   - **Support**: Offering support and encouragement to the team leader, particularly during challenging times.

Effective team leadership is crucial for the success of capstone design projects. The team leader's role in organizing meetings, handling communications, maintaining the project plan, and ensuring timely submissions is essential for keeping the project on track. By embodying qualities such as effective communication, organizational skills, and a proactive attitude, a team leader can guide their team to achieve their project goals. Moreover, the support and collaboration of all team members are vital for the team leader to perform their role effectively. Together, a well-led and cohesive team can overcome challenges and deliver a successful design solution.

5.6   Roles and Responsibilities

To maximize efficiency and ensure all aspects of a capstone project are effectively managed, teams should structure themselves in a way that leverages each member's strengths and expertise. By defining distinct roles and responsibilities, team members can concentrate on specific areas, ensuring comprehensive coverage of all project facets. Here are some typical roles and their associated responsibilities, along with tips on how to creatively define and adapt these roles to fit the unique needs of the team and project.

5.6.1   Common Team Roles

1. **Team Leader/Manager**
   - **Responsibilities**:
     - Schedule and facilitate team meetings.
     - Coordinate communication with sponsors, professors, and other stakeholders.
     - Maintain the project plan and ensure timely submission of reports and assignments.
   - **Skills Required**: Leadership, organization, time management, communication.
   - **Creative Titles**: Project Coordinator, Team Captain, Project Facilitator.

2. **Design Engineer**
   - **Responsibilities**:
     - Develop design concepts and solutions based on project requirements.
     - Create detailed engineering drawings and specifications.
     - Conduct simulations and tests to validate designs.
   - **Skills Required**: CAD proficiency, creativity, problem-solving, and attention to detail.
   - **Creative Titles**: Design Specialist, Innovation Engineer, Concept Developer.

3. **Research Engineer**
   - **Responsibilities**:
     - Conduct background research to support the design process.
     - Analyze existing solutions and technologies.
     - Compile and present research findings to the team.
   - **Skills Required**: Analytical skills, attention to detail, technical writing, data analysis.
   - **Creative Titles**: Knowledge Engineer, Research Analyst, Technology Scout.

4. **Financial Analyst**
   - **Responsibilities**:
     - Develop and manage the project budget.
     - Conduct cost-benefit analyses and financial feasibility studies.
     - Prepare financial reports and documentation.
   - **Skills Required**: Financial modeling, budgeting, analytical skills, attention to detail.
   - **Creative Titles**: Financial Strategist, Budget Analyst, Economic Planner.

5. **Marketing Engineer**
   - **Responsibilities**:

- Develop strategies to market the project or product.
- Conduct market research and analyze consumer needs.
- Create marketing materials and presentations.
  - **Skills Required**: Market research, communication, creativity, and presentation skills.
  - **Creative Titles**: Market Strategist, Product Evangelist, Customer Insights Analyst.

6. **Test Engineer**
   - **Responsibilities**:
     - Develop and implement testing protocols for the design.
     - Analyze test results and recommend design modifications.
     - Ensure the final product meets all specifications and standards.
   - **Skills Required**: Analytical skills, attention to detail, problem-solving, technical expertise.
   - **Creative Titles**: Quality Assurance Engineer, Validation Specialist, Test Coordinator.

7. **Documentation Specialist**
   - **Responsibilities**:
     - Compile and maintain all project documentation.
     - Ensure all reports, manuals, and presentations are professionally formatted and error-free.
     - Facilitate knowledge sharing within the team.
   - **Skills Required**: Technical writing, organization, attention to detail, communication.
   - **Creative Titles**: Information Manager, Documentation Engineer, Technical Writer.

8. **Logistics Coordinator**
   - **Responsibilities**:
     - Manage the procurement of materials and equipment.
     - Coordinate project timelines and schedules.
     - Ensure timely delivery of project components.
   - **Skills Required**: Organizational skills, time management, problem-solving, negotiation.
   - **Creative Titles**: Operations Manager, Supply Chain Coordinator, Resource Manager.

5.6.2   Defining Custom Roles

Depending on the specific needs of the project, team members may need to define custom roles. This flexibility allows the team to adapt to unique challenges and ensure all areas are covered. Here are some steps to define custom roles:

1. **Assess Project Requirements**:
   o Analyze the project's scope to identify all necessary tasks and responsibilities.
   o Consider any unique challenges or requirements that may arise.
2. **Identify Team Member Strengths**:
   o Evaluate the skills, experiences, and interests of each team member.
   o Match roles to individuals based on their strengths and preferences.
3. **Create Descriptive Titles**:
   o Develop creative and descriptive titles that reflect the specific focus and responsibilities of each role.
   o Ensure titles are clear and easily understood by all stakeholders.
4. **Define Responsibilities**:
   o Clearly outline the responsibilities and expectations for each role.
   o Ensure all critical tasks are assigned, and no areas are overlooked.
5. **Flexibility and Adaptation**:
   o Be prepared to adapt roles as the project progresses and new challenges arise.
   o Encourage team members to take on additional responsibilities when needed.

5.6.3   Importance of Defined Roles

Clearly defined roles and responsibilities are essential for several reasons:
1. **Efficiency**:
   o Ensures all tasks are covered without overlap or gaps.
   o Allows team members to focus on their areas of expertise, increasing productivity.
2. **Accountability**:
   o Each team member knows what is expected of them, enhancing accountability.
   o Clear responsibilities help in tracking progress and identifying issues.
3. **Collaboration**:
   o Well-defined roles facilitate better collaboration and communication among team members.
   o Helps prevent conflicts and misunderstandings.
4. **Professional Development**:
   o Allows team members to develop and showcase their skills in specific areas.
   o Provides valuable experience relevant to their future careers.

By thoughtfully defining and assigning roles, capstone design teams can optimize their performance and work more effectively towards their project goals. Each member's contributions become integral to the team's success, ensuring a well-rounded and comprehensive approach to the design problem.

5.7 Effective Team Management

Effective team management is crucial to the success of capstone design projects. Team meetings are a critical component of this management, as they provide a structured platform for discussing progress, addressing issues, and making decisions. To ensure that meetings are productive and time is used efficiently, teams must adhere to best practices in scheduling, planning, and documentation.

5.7.1 Importance of Effective Team Management

Effective team management ensures that:
- **Resources are utilized efficiently**.
- **Team members remain aligned with project goals**.
- **Communication is clear and compelling**.
- **Deadlines are met**.
- **Problems are identified and addressed promptly**.

5.7.2 Key Elements of Productive Team Meetings

1. **Regular Schedule**:
   - **Consistency**: Establish a regular meeting schedule, such as once or twice a week, to maintain momentum and ensure consistent progress.
   - **Availability**: Choose meeting times that are convenient for all team members to maximize attendance and participation.
   - **Flexibility**: Be willing to adjust the schedule as needed to accommodate essential deadlines or unexpected challenges.
2. **Meeting Agenda**:
   - **Preparation**: Prepare an agenda before each meeting to outline the topics to be discussed. This ensures that meetings are focused and productive.
   - **Relevance**: Ensure the agenda items are relevant to all team members and the project. Include updates on progress, upcoming tasks, and any issues that need addressing.
   - **Prioritization**: Prioritize agenda items to address the most critical topics first, ensuring that essential discussions are not rushed or overlooked.
3. **Minutes of the Meetings**:
   - **Documentation**: Assign a team member to take minutes during each meeting. These minutes should capture essential discussions, decisions made, and action items.
   - **Accessibility**: Distribute the minutes to all team members promptly after

the meeting. This ensures everyone is informed and can refer back to the discussions as needed.

- o **Accountability**: Use the minutes to track action items and ensure that responsibilities are clear. Review previous minutes at the start of each meeting to follow up on outstanding tasks.

5.7.3   Strategies for Effective Team Management

1. **Clear Communication**:
   - o **Transparency**: Maintain open and honest communication within the team. Ensure that all members are aware of project updates, changes, and any challenges that arise.
   - o **Feedback**: Encourage team members to provide feedback on the project and team dynamics. Address any concerns promptly to maintain a positive working environment.
2. **Defined Roles and Responsibilities**:
   - o **Role Clarity**: Clearly define each team member's roles and responsibilities to avoid overlap and ensure all tasks are covered.
   - o **Accountability**: Hold team members accountable for their assigned tasks. Regularly review progress and provide support as needed.
3. **Effective Time Management**:
   - o **Prioritization**: Prioritize tasks based on their importance and deadlines. Focus on high-impact activities that drive the project forward.
   - o **Efficiency**: Encourage efficient work practices and discourage time-wasting activities. Utilize tools and techniques to streamline processes and improve productivity.
4. **Conflict Resolution**:
   - o **Proactive Approach**: Address conflicts early before they escalate. Foster a team culture that encourages open discussion and resolution of issues.
   - o **Mediation**: If conflicts persist, consider mediation by a neutral team member or the course professor to find a fair and amicable solution.
5. **Use of Technology**:
   - o **Collaboration Tools**: Utilize collaboration tools such as project management software, shared document platforms, and communication apps to facilitate teamwork and organization.
   - o **Meeting Tools**: Video conferencing tools can be used for virtual meetings if team members cannot meet in person. Record meetings, if necessary, for later reference.
6. **Motivation and Morale**:
   - o **Recognition**: Recognize and celebrate individual and team achievements. This boosts morale and motivates team members to continue performing

well.

- o **Support**: Provide support and resources to team members to help them succeed. This includes access to necessary materials, mentorship, and constructive feedback.

5.7.4   Best Practices for Conducting Meetings

1. **Start and End on Time**:
   - o Respect everyone's time by starting and ending meetings punctually. This demonstrates professionalism and ensures that meetings do not overrun and affect other commitments.
2. **Encourage Participation**:
   - o Ensure that all team members have the opportunity to speak and contribute. This fosters a sense of ownership and encourages diverse perspectives.
3. **Focus on Solutions**:
   - o Keep discussions solution-oriented. Avoid dwelling on problems without addressing potential solutions. Encourage creative thinking and problem-solving.
4. **Summarize and Review**:
   - o At the end of each meeting, summarize the key points discussed and review the action items. This ensures clarity and reinforces responsibilities.
5. **Follow-Up**:
   - o Follow up on action items between meetings to ensure progress is being made. This can be done through regular check-ins or updates via communication tools.

Effective team management is vital for the success of capstone design projects. By adhering to best practices in meeting scheduling, planning, and documentation, teams can ensure that their time is used efficiently and that project goals are met. Clear communication, defined roles, effective time management, conflict resolution, use of technology, and maintaining motivation and morale are all essential components of effective team management. Through these strategies, teams can navigate challenges, foster collaboration, and achieve successful project outcomes.

5.8 Scheduling

Effective scheduling is a critical component of team management, especially in capstone design projects where many engineering seniors juggle part-time jobs, academic commitments, and personal responsibilities. Allocating time for regular team meetings and project work is essential for maintaining progress and ensuring the successful completion of the project. This section explores strategies for scheduling team meetings, including considerations for part-time employment, optimizing meeting times, and integrating meetings into project plans.

5.8.1   Challenges of Scheduling

1. **Part-Time Jobs**:
   o Many engineering seniors work part-time jobs, which can complicate finding expected meeting times.
   o Flexibility and understanding are required from all team members to accommodate each other's schedules.

2. **Academic Commitments**:
   o Students have varying class schedules, lab sessions, and other academic obligations.
   o Balancing these commitments with project work requires careful planning.

3. **Personal Responsibilities**:
   o Personal commitments and extracurricular activities add another layer of complexity to scheduling.

5.8.2   Strategies for Effective Scheduling

1. **Establishing Regular Meeting Times**:
   o **Frequency**: Aim to hold at least two team meetings per week to maintain momentum and ensure regular progress updates.
   o **Consistency**: Establish regular meeting times that remain consistent throughout the project duration. This helps team members plan their other activities around the meetings.

2. **Utilizing Common Periods**:
   o **Course Allocation**: Capstone design courses often allocate standard periods specifically for team meetings and design work. Make full use of these periods.
   o **Synchronization**: Coordinate with the course professor to identify these standard periods and schedule meetings during these times.

3. **Creating a Scheduling Plan**:
   o **Initial Survey**: Conduct an initial survey of all team members' availability, including class schedules, work hours, and other commitments.
   o **Optimal Times**: Identify overlapping free times that are most convenient for all team members. These are the optimal times for scheduling regular meetings.
   o **Flexibility**: Build flexibility into the schedule to accommodate unexpected changes in availability.

4. **Integrating Meetings into Project Plans**:
   o **Project Timeline**: Include scheduled meetings in the project timeline. This ensures that meeting times are considered part of the required project activities.
   o **Calendar Integration**: Use digital calendars (e.g., Google Calendar, Outlook) to schedule meetings and send invites to all team members. This helps everyone keep track of meeting times and receive reminders.
5. **Handling Conflicts and Adjustments**:
   o **Conflict Resolution**: Be prepared to adjust meeting times if conflicts arise. Use doodle polls or similar tools to find new expected times when necessary.
   o **Emergency Meetings**: Schedule additional meetings as needed to address urgent issues or milestones.

5.8.3   Best Practices for Scheduling

1. **Early and Clear Communication**:
   o Communicate meeting times well in advance to ensure all team members can adjust their schedules accordingly.
   o Use clear and concise communication to avoid misunderstandings about meeting times and locations.
2. **Prioritizing Attendance**:
   o Emphasize the importance of attending scheduled meetings to all team members. Regular attendance is crucial for maintaining team cohesion and progress.
   o Encourage team members to inform the team in advance if they are unable to attend a meeting and arrange for them to catch up on missed discussions.
3. **Maximizing Meeting Efficiency**:
   o Plan meetings to be as efficient as possible to respect everyone's time. Start and end meetings on time and stick to the agenda.
   o Assign a timekeeper if necessary to ensure the meeting stays on track.
4. **Using Technology**:
   o Utilize technology to facilitate scheduling and communication. Tools such as Doodle polls can help find the best times for everyone.
   o Use virtual meeting platforms (e.g., Zoom, Microsoft Teams) for remote team members or when in-person meetings are not feasible.
5. **Reviewing and Adjusting Schedules**:
   o Regularly review the meeting schedule to ensure it continues to meet the team's needs. Adjust as necessary to accommodate changing circumstances.

    ○   Periodically check in with team members to ensure that the meeting times are still convenient and make adjustments as needed.

5.8.4   Benefits of Effective Scheduling

1. **Improved Coordination**:
   - ○ Consistent scheduling improves coordination among team members, ensuring everyone is on the same page and working towards common goals.
2. **Enhanced Productivity**:
   - ○ Regular and well-planned meetings enhance productivity by providing a structured environment for discussing progress, addressing issues, and planning the next steps.
3. **Reduced Stress**:
   - ○ Clear and consistent scheduling reduces stress by providing predictability and allowing team members to plan their other activities around the meetings.
4. **Greater Accountability**:
   - ○ Scheduled meetings create a sense of accountability, as team members are expected to report on their progress and contribute to discussions regularly.

Effective scheduling is a cornerstone of successful team management in capstone design projects. By establishing regular meeting times, utilizing standard periods, creating a scheduling plan, and integrating meetings into project plans, teams can ensure that they make the most of their time together. Clear communication, prioritizing attendance, maximizing meeting efficiency, using technology, and regularly reviewing schedules are best practices that can help teams navigate the complexities of scheduling. With these strategies, teams can improve coordination, enhance productivity, reduce stress, and maintain accountability, ultimately contributing to the successful completion of their capstone design projects.

5.9 Agenda and Time Management

Effective agenda planning and time management are essential for productive team meetings in capstone design projects. An agenda provides structure, ensures all relevant topics are covered, and helps keep the meeting on track. This section provides guidelines for creating and managing agendas, engaging team members, and optimizing meeting efficiency.

5.9.1   Guidelines for Preparing an Agenda

1. **Assign a Responsible Member**:
   - o Typically, the team leader prepares the agenda and shares it with team members before the meeting. This ensures that everyone knows what to expect and can prepare accordingly.
2. **Solicit Input from Team Members**:
   - o Invite team members to submit agenda items to the team leader. If time permits, include these items in the upcoming meeting; otherwise, schedule them for future meetings.
3. **Pose Agenda Items as Questions**:
   - o Frame agenda items as questions that team members can answer. This approach engages all members and encourages participation.
4. **Allocate Time for Each Topic**:
   - o Assign a specific duration for each agenda item. Be precise and concise, allowing enough time for meaningful discussion without overrunning the schedule. Be careful not to underestimate the time required for each item.
5. **Distribute Documents and Handouts**:
   - o Include any necessary documents or handouts with the meeting agenda. Send the agenda and materials well in advance, allowing team members sufficient time to review them before the meeting.
6. **Review Meeting Effectiveness**:
   - o Allocate time at the end of each meeting to assess its effectiveness. Discuss what worked well and identify areas for improvement.

5.9.2   Example Agenda Table

Table 5-1 shows an example agenda format that you can use to create your meeting agendas. It includes columns for the topic, duration, purpose, lead, preparation, and process.

5.9.3   Best Practices for Agenda and Time Management

1. **Clear Communication**:
   - o Share the agenda and relevant documents well before the meeting. This gives team members time to prepare and come ready to contribute.
2. **Strict Adherence to the Agenda**:
   - o Follow the agenda closely and adhere to the allotted time for each topic. This ensures that all items are covered and prevents the meeting from overrunning.
3. **Engage All Team Members**:
   - o Pose agenda items as questions to encourage participation. Assign

different team members to lead various topics, fostering a sense of
ownership and involvement.

4. **Time Management**:
   - Use a timekeeper to monitor the discussion and ensure each item stays
     within its allocated time. If a topic requires more time than planned,
     consider scheduling a follow-up discussion rather than extending the
     current meeting.

5. **Meeting Evaluation**:
   - At the end of the meeting, review its effectiveness. Discuss what went
     well and identify improvements for future meetings. This practice helps
     continuously enhance meeting productivity.

Table 5-1. Creating a Practical Meeting Agenda.

Topic	Duration	Purpose	Lead	Preparation	Process
What do we need to add or change in our project plan?	15 min	Decision	Alex	Review the project plan	Alex presents some options
What questions can we ask about the problem so we can define it better?	20 min	Decision	Taylor	Read problem definition	Taylor presents some ideas and questions
What design specifications should we include?	20 min	Decision	Jordan	Review design specifications development template	Jordan will guide the team through the template questions
Was this a productive meeting? What should we do better for the next meeting?	10 min	Discussion	Casey	None	Team members identify pluses and minuses for this meeting. Agree to make changes to correct minuses.

5.9.5 Example Meeting Agenda and Minutes

Here is an example of a complete meeting agenda with corresponding minutes to illustrate effective time management and documentation.

Meeting Agenda:
Date: [Insert Date]
Time: [Insert Time]
Location: [Insert Location or Virtual Meeting Link]

Topic	Duration	Purpose	Lead	Preparation	Process
Welcome and Review Agenda	5 min	Information	Alex	None	Alex reviews the agenda and objectives for the meeting
What do we need to add or change in our project plan?	15 min	Decision	Alex	Review the project plan	Alex presents some options
What questions can we ask about the problem so we can define it better?	20 min	Decision	Taylor	Read problem definition	Taylor presents some ideas and questions
What design specifications should we include?	20 min	Decision	Jordan	Review design specifications development template	Jordan will guide the team through the template questions
Was this a productive meeting? What should we do better for the next meeting?	10 min	Discussion	Casey	None	Team members identify pluses and minuses for this meeting. Agree to make changes to correct minuses.

Meeting Minutes:
    Date: [Insert Date]
Time: [Insert Time]
Location: [Insert Location or Virtual Meeting Link]

Topic	Discussion Points	Action Items	Responsible	Due Date
Welcome and Review Agenda	Alex reviewed the agenda and meeting objectives	N/A	N/A	N/A
What do we need to add or change in our project plan?	Discussed potential changes to the project plan; agreed on adjustments	Update project plan with agreed changes	Alex	[Insert Date]
What questions can we ask about the problem so we can define it better?	Identified key questions to refine the problem definition	Compile and review the list of questions	Taylor	[Insert Date]
What design specifications should we include?	Reviewed the design specifications development template; discussed key specifications	Complete the design specifications template	Jordan	[Insert Date]
Was this a productive meeting? What should we do better for the next meeting?	Identified compelling aspects of the meeting and areas for improvement	Implement identified improvements for the next meeting	Casey	[Insert Date]

By following these guidelines and using the provided example table, capstone design teams can ensure their meetings are well-organized, efficient, and productive. This approach helps teams stay on track, meet deadlines, and achieve their project goals.

5.10 Minutes

Meeting minutes are an official record of what was discussed, decided, and assigned during a team meeting. They serve as a valuable reference for team members and other stakeholders, such as professors and sponsors, to review and evaluate the team's progress or note exceptional accomplishments and deficiencies. Well-documented minutes ensure transparency, accountability, and continuity within the team. This section outlines the essential elements of meeting minutes and provides guidelines for creating effective minutes.

5.10.1 Essential Elements of Meeting Minutes

1. **Date, Time, and Duration**:
   - o Record the exact date, start time, end time, and total duration of the meeting.
2. **Location**:
   - o Indicate where the meeting took place, whether it was in person or virtual.
3. **Attendance**:
   - o List the names and emails of all attendees. This ensures that everyone's participation is documented and provides a point of contact for follow-up.
4. **Team Information**:
   - o Include the name of the team and the title of the project to identify the context of the meeting clearly.
5. **Project Sponsor**:
   - o Note the name of the project sponsor. This helps in tracking involvement and accountability for project outcomes.
6. **Agenda Items**:
   - o List the agenda items discussed during the meeting. For each item, provide a summary of the discussion, decisions made, and action items assigned.

5.10.2 Guidelines for Creating Effective Meeting Minutes

1. **Conciseness**:
   - o Keep the minutes concise and focused. Aim for no more than one page per meeting, summarizing key points and decisions without unnecessary detail.
2. **Clarity**:
   - o Use clear and straightforward language. Avoid jargon and ensure that the minutes are understandable to all team members and stakeholders.
3. **Accuracy**:
   - o Ensure that the minutes accurately reflect the discussions and decisions made during the meeting. Double-check for any errors or omissions.
4. **Timeliness**:
   - o Distribute the minutes promptly after the meeting, ideally within 24-48 hours. This ensures that the information is fresh and actionable.
5. **Attachments**:
   - o Attach any supporting documents, such as the project plan or other relevant materials, or provide links to these documents. This provides additional context and resources for team members.

5.10.3   Example Meeting Minutes Template

Below is an example template for meeting minutes, illustrating how to structure and document the essential elements.

Meeting Minutes

Team Name:

Date: [Insert Date]

Time: [Start Time] - [End Time]

Duration: [Total Duration]

Location: [Insert Location or Virtual Meeting Link]

Attendees:
- Alex Smith (alex.smith@example.com)
- Taylor Johnson (taylor.johnson@example.com)
- Jordan Lee (jordan.lee@example.com)
- Casey White (casey.white@example.com)

Team Name: **Mechanix Mavericks**

Project Title: Solar-Powered Water Purification System

Project Sponsor: Clean Water Initiative

Agenda Items:

1. **Welcome and Review Agenda** (5 min)
   - **Discussion**: Alex reviewed the agenda and meeting objectives.
   - **Decisions**: N/A
   - **Action Items**: N/A

2. **What do we need to add or change in our project plan?** (15 min)
   - **Discussion**: Discussed potential changes to the project plan; agreed on adjustments.
   - **Decisions**: Update the project plan with agreed changes.
   - **Action Items**: Alex will update the project plan by [Insert Due Date].

3. **What questions can we ask about the problem so we can define it better?** (20 min)
   - **Discussion**: Identified key questions to refine the problem definition.
   - **Decisions**: Compile and review the list of questions.
   - **Action Items**: Taylor is to compile and review questions by [Insert Due Date].

4. **What design specifications should we include?** (20 min)
   - **Discussion**: Reviewed the design specifications development template; discussed vital specifications.
   - **Decisions**: Complete the design specifications template.
   - **Action Items**: Jordan is to complete the design specifications template by

[Insert Due Date].
5. **Was this a productive meeting? What should we do better for the next meeting?** (10 min)
   - **Discussion**: Identified compelling aspects of the meeting and areas for improvement.
   - **Decisions**: Implement identified improvements for the next meeting.
   - **Action Items**: Casey will document meeting feedback and improvements by [Insert Due Date].

   Supporting Documents:
   - [Project Plan]
   - [Design Specifications Table]

By following these guidelines and using the provided template, capstone design teams can create effective meeting minutes that enhance communication, accountability, and project management. Well-documented minutes help ensure that all team members are aligned and informed, contributing to the overall success of the project.

5.11 Action Items

Action items are a crucial component of effective project management in capstone design projects. They represent a list of tasks or actions that the team needs to complete to advance the project. Managing action items effectively ensures that the team remains organized, focused, and accountable for the various tasks required to achieve project goals. This section outlines the importance of action items, how to manage them, and best practices for integrating them into the project plan.

5.11.1 Importance of Action Items

1. **Clarity and Focus**:
   - Action items provide clarity on what needs to be done, by whom, and by when.
   - They help the team stay focused on specific tasks and milestones, ensuring steady progress toward the project objectives.
2. **Accountability**:
   - Assigning action items to specific team members establishes clear responsibility and accountability.
   - This promotes a sense of ownership and commitment to completing assigned tasks.
3. **Progress Tracking**:
   - Action items serve as a tool for tracking progress and identifying any delays or obstacles.
   - Regularly reviewing action items helps the team stay on track and make

timely adjustments as needed.

5.11.2   Managing Action Items

1. **Identification**:
   - o During meetings, discussions, and planning sessions, identify tasks that need to be completed.
   - o Ensure that each action item is relevant to the project's progress and success.

2. **Documentation**:
   - o Record each action item in a shared document or project management tool.
   - o Include details such as the task description, responsible team member, due date, and any necessary resources or steps.

3. **Assignment**:
   - o Assign each action item to a specific team member who will be responsible for its completion.
   - o Ensure that the assigned tasks are distributed evenly and appropriately based on team members' strengths and availability.

4. **Integration with Project Plan**:
   - o Add action items to the overall project plan, ensuring they are tracked alongside other project tasks and milestones.
   - o Update the project plan regularly to reflect the status of each action item.

5. **Tracking and Review**:
   - o Regularly review the list of action items during team meetings to monitor progress.
   - o Update the status of each action item, noting which tasks are completed, in progress, or pending.
   - o Discuss any challenges or delays and adjust the plan as necessary.

5.11.3   Best Practices for Action Items

1. **Specific and Clear Descriptions**:
   - o Ensure that each action item has a clear and specific description so that team members understand what is required.
   - o Avoid vague or ambiguous language.

2. **Realistic Deadlines**:
   - o Set realistic and achievable deadlines for each action item.
   - o Consider the complexity of the task and the availability of resources and team members.

3. **Prioritization**:
   - o Prioritize action items based on their importance and urgency.
   - o Focus on high-impact tasks that significantly contribute to the project's

progress.

4. **Regular Updates**:
   o Update the action item list regularly to reflect the current status of each task.
   o Communicate any changes to the team promptly.
5. **Documentation of Rationale**:
   o Document the rationale behind each action item to provide context and justification.
   o This helps in understanding the importance of each task and facilitates informed decision-making.
6. **Example Action Item List**
   Below is an example template for documenting and tracking action items.

Table 5-2. Action Item List.

Action Item	Description	Assigned To	Due Date	Status	Notes
Update project plan	Incorporate changes discussed in the last meeting	Alex	[Insert Date]	In Progress	Review changes with the team
Compile a list of problem-definition questions.	Develop questions to refine the problem definition	Taylor	[Insert Date]	Completed	Presented in the last meeting
Complete design specifications template	Fill out the template with key design specifications	Jordan	[Insert Date]	In Progress	Review with team in next meeting
Conduct literature review	Research existing solutions and technologies related to the project	Casey	[Insert Date]	Not Started	Use university library resources
Schedule a meeting with the project sponsor	Arrange a meeting to discuss project progress and get feedback	Alex	[Insert Date]	Scheduled	Meeting set for [Insert Date]
Test prototype functionality	Perform initial testing on the prototype and document results	Taylor	[Insert Date]	Not Started	Prepare testing protocols

Action items are a fundamental tool for effective team management in capstone design projects. By carefully planning, documenting, and tracking action items, teams can ensure that all necessary tasks are completed efficiently and on time. Clear descriptions, realistic deadlines, and regular updates are crucial to managing action items effectively. Integrating action items into the project plan and documenting their rationale further enhances transparency and accountability. By following these guidelines and best practices, capstone design teams can stay organized, maintain steady progress, and achieve their project goals successfully.

5.12 Progress Reports

Progress reporting is a crucial activity in engineering projects, ensuring that all stakeholders are informed about the project's status and progress toward its goals and objectives. Regular progress reports are instrumental in updating the project plan, monitoring activities, and making necessary adjustments to keep the project on track. In the context of capstone design projects, progress reports are typically submitted on a weekly basis to the project sponsor and the capstone professor.

5.12.1   Purpose of Progress Reports

1. **Information Dissemination**:
    o   Progress reports keep all stakeholders, including team members, professors, and sponsors, informed about the current status of the project.
    o   They provide a comprehensive update on the activities undertaken, milestones achieved, and any deviations from the plan.
2. **Project Monitoring**:
    o   Progress reports facilitate continuous monitoring of the project, allowing for timely identification and resolution of issues.
    o   In the industry, such monitoring is critical for managing project costs, avoiding cost overruns, and ensuring that customers receive the best product at the lowest price.
3. **Plan Updates**:
    o   Progress reports should be used to update the project plan, reflecting the current status of tasks and any changes in timelines or priorities.
    o   They help maintain an accurate and up-to-date project plan, which is essential for effective project management.
4. **Performance Assessment**:
    o   Progress reports allow the assessment of the impact of team members' activities on the project timeline.
    o   They provide a basis for evaluating individual and team performance, ensuring accountability and continuous improvement.

5.12.2   Format and Content of Progress Reports

A helpful format for a progress report is a memorandum from the team to their supervisors, such as professors and sponsors. The following information should be included as a minimum in the progress report:

1. **Date of Progress Report**:
   - The specific date when the report is prepared and submitted.
2. **Recipients**:
   - Names of individuals the progress report is addressed to (e.g., sponsor, professor).
3. **Project Title and Sponsor**:
   - The title of the project and the name of the project sponsor.
4. **Duration Covered**:
   - The specific period covered by the progress report (e.g., the past week).
5. **Team Members**:
   - Names of students who participated in preparing the progress report along with their project role/title.
6. **Project Recap**:
   - A summary of the project, providing context for the progress report.
7. **Progress Made**:
   - Detailed description of the progress made during the past week, including activities of each team member.
   - Specific results accomplished and milestones achieved.
8. **Problems and Issues**:
   - Describe any problems or issues that surfaced during the reporting period.
   - An analysis of the impact of these issues on the project timeline and overall progress.
9. **Action Plan**:
   - A plan outlining the steps the team will take to resolve problems or obstacles.
   - Any adjustments made to the project plan to accommodate these issues.
10. **Plans for Next Period**:
    - Describe what is planned for the next reporting period.
    - Specific tasks to be undertaken and goals to be achieved.
11. **Attachments**:
    - An updated project plan (Gantt chart) to reflect the current status and any changes.
    - Any other relevant documents, such as drawings, articles, or research findings.

5.12.3 Example Progress Report Template

Below is an example template for a weekly progress report:

Memorandum
Date: [Insert Date]

To: [Professor Name, Sponsor Name]
From: [Team Name]
Subject: Weekly Progress Report for [Project Title]

---

1. Project Title and Sponsor:
- **Project Title**: Solar-Powered Water Purification System
- **Project Sponsor**: Clean Water Initiative

2. Duration Covered:
- **Reporting Period**: [Insert Start Date] to [Insert End Date]

3. Team Members and Roles:
- Alex Smith - Team Leader
- Taylor Johnson - Research Engineer
- Jordan Lee - Design Engineer
- Casey White - Financial Analyst

4. Project Recap:
- The Solar-Powered Water Purification System aims to develop a cost-effective and sustainable solution for providing clean drinking water in remote areas.

5. Progress Made:
- **Alex Smith**: Updated the project plan with new milestones and coordinated the team's activities.
- **Taylor Johnson**: Completed the literature review and compiled a list of potential technologies.
- **Jordan Lee**: Developed initial design sketches and began CAD modeling.
- **Casey White**: Conducted a cost analysis for the proposed design and identified potential funding sources.

6. Problems and Issues:
- **Supply Chain Delays**: Encountered delays in obtaining critical components, which may impact the prototype development timeline.
- **Design Challenges**: Identified issues with integrating the solar panels into the purification system.

7. Action Plan:
- **Supply Chain**: Jordan will contact alternative suppliers to expedite component delivery.
- **Design Challenges**: The team will hold a design review meeting to brainstorm solutions for the integration issues.

8. Plans for Next Period:
- **Alex Smith**: Finalize the updated project plan and submit it to the professor and sponsor.
- **Taylor Johnson**: Begin testing potential technologies in the lab.

- **Jordan Lee**: Complete the CAD model and prepare for prototype construction.
- **Casey White**: Develop a detailed budget proposal for the project.

9. Attachments:
- Updated Project Plan (Gantt chart)
- Design Sketches
- Literature Review Summary

5.12.4   Best Practices for Progress Reporting

1. **Regular Submission**:
   o Submit progress reports consistently on the designated weekly deadline, usually by email.
   o Timely submissions ensure continuous monitoring and prompt feedback.
2. **Accuracy and Honesty**:
   o Ensure that the information in the progress report is accurate and honestly reflects the current status of the project.
   o Transparency about challenges and delays is crucial for effective problem-solving.
3. **Conciseness**:
   o Keep the report concise while including all essential information. Avoid unnecessary details that do not contribute to understanding the project's status.
4. **Engagement**:
   o Engage all team members in preparing the progress report. This promotes accountability and ensures that all perspectives are considered.
5. **Analysis and Adjustment**:
   o Use the progress report to analyze ongoing tasks and look for opportunities to achieve parallelism, reducing the number of sequential tasks where possible.
   o Continuously adjust the project plan based on the insights gained from the progress reports.

By adhering to these guidelines and using the provided template, capstone design teams can create effective progress reports that enhance communication, accountability, and project management. Regular progress reporting helps ensure that the project stays on track, issues are addressed promptly, and all stakeholders are kept informed, ultimately contributing to the successful completion of the project.

5.13 Communications

Effective communication is an essential skill for success in life and is critical for capstone design projects. The members of an engineering design team must communicate effectively within the team and with all of their project stakeholders, including professors, sponsors, and peers. Effective communication ensures that everyone is aligned with the project's goals, progress is shared, issues are addressed promptly, and the project runs smoothly.

5.13.1   Methods of Communication

Current methods or venues for capstone design communication include:

1. **Verbal/Oral Face-to-Face Meetings**:
   - o **Benefits**: Face-to-face meetings provide rich nonverbal content, such as body language and eye contact, enhancing the context and clarity of the communication. They are effective for brainstorming, discussing complex ideas, and building rapport.
   - o **Challenges**: Scheduling can be difficult if team members have conflicting schedules or are geographically dispersed.
2. **Verbal/Oral by Phone or Software**:
   - o **Benefits**: Phone calls and software tools like FaceTime, Skype, WebEx, and Zoom allow for real-time communication, which is effective for urgent issues or when face-to-face meetings are not possible.
   - o **Challenges**: Lack of visual cues can sometimes lead to misunderstandings.
3. **Email**:
   - o **Benefits**: Email provides a written record of communication, which is helpful in tracking decisions, sharing detailed information, and attaching documents. It is location-independent and can be sent and read at any time.
   - o **Challenges**: Emails can be misinterpreted without nonverbal cues, and the response time may vary. There are also concerns about security and confidentiality.

5.13.2   Communication Effectiveness

The method of communication you use for your project will depend on the software and facilities available to you at your university and the sponsor's preferences. Some sponsors may have confidentiality and security concerns, requiring the use of secure certified software for communication. Always check with your sponsor regarding any security requirements for your interactions.

1. **Communication Method Comparison**

Effective communication methods vary in clarity and information content. Table 5-3. illustrates a comparison of different communication methods, showing the balance between clarity and information richness.

Table 5-3. Comparison of Different Communication Methods.

Method	Clarity	Information Content
Face-to-Face	High	High
Video Conference	High	High
Phone Call	Medium	Medium
Email	Low	High

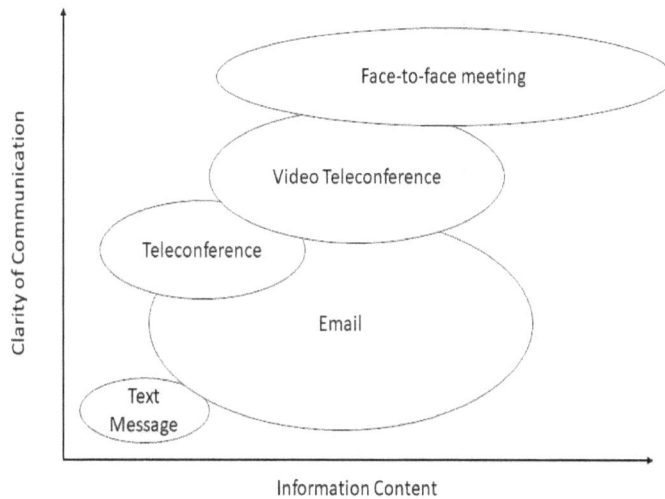

Figure 5-1. Comparison of different communication methods.

2. **Email Communication**

Email will likely be a primary mode of communication for your project. Table 5-4. shows some strengths and weaknesses of email communication.

Table 5-4. Email Strengths and Weaknesses

Strengths	Weaknesses
Quick creation	Can be created too quickly, leading to hasty or unclear messages
Fast delivery	An overload of emails can lead to important messages being missed.
Location independent	Lacks nonverbal context (e.g., eye contact, body language)
Audit trail of communication	Lengthy message chains can become cumbersome.
Facility for providing feedback	There is no guarantee that the email will be read or acted upon
Allows for attachments	Attachments can be overlooked or mishandled.
Environmentally friendly (reduces paper use)	Generally lower quality of communication
The sender assumes the recipient will take action	Low security and confidentiality

5.13.3   Anatomy of an Effective Email

1. **Address Field**:
   - o   Include "to:" and "cc:" fields correctly, addressing the primary recipients and those who should be copied for their information.
2. **Subject Line**:
   - o   The subject line should be informative and allow the recipient to identify the email's purpose quickly. Include the team number, sponsor, and project title for easy reference.
3. **Greetings**:
   - o   Use a professional greeting, including the recipient's title or salutation (e.g., Ms., Mr., Dr., etc.).
4. **Content**:
   - o   Be thoughtful, clear, and concise in your message. Keep it brief but comprehensive.
5. **Signature**:
   - o   Include your contact information (email and phone number) and team information in your signature.
6. **Attachments**:
   - o   Use attachments for essential documents. Clearly explain the content of the attachments in the email body. Name files for easy identification and later reference.

5.13.4   Online Communication

In light of situations like the COVID-19 pandemic, online communication has become more prevalent. Capstone design projects now often require remote collaboration tools for file sharing and video conferencing.

5.13.4.1   Key Considerations for Online Communication

1. **File Sharing**:
   - Cloud storage solutions (e.g., Google Drive, Dropbox, OneDrive) can be used to store and manage project files. Ensure all team members have access and the ability to contribute.
2. **Video Conferencing**:
   - Utilize tools like Zoom, Microsoft Teams, or WebEx for remote meetings. These platforms often provide additional features like screen sharing, which can enhance collaborative efforts.
3. **Security**:
   - Be mindful of security requirements, mainly when dealing with proprietary or sensitive information. Use secure, certified software as required by your sponsor.

Effective communication is the backbone of successful capstone design projects. By leveraging various communication methods appropriately and ensuring clarity, relevance, and timeliness, teams can enhance collaboration and achieve their project goals efficiently. Proper email etiquette and the strategic use of online communication tools will further support the team's efforts to maintain transparency, accountability, and continuous progress.

5.14 Online File Management

Effective online file management is crucial for capstone design teams, mainly when members are operating from remote locations. This involves storing, retrieving, syncing, and sharing files across various devices, including desktop computers, laptops, tablets, and smartphones. Another essential function is the ability to search and find content quickly and accurately. Numerous file-sharing services and tools are available to facilitate these needs. This section discusses some of the most effective tools for online file management.

5.14.1   Key Requirements for Online File Management

1. **Storage and Retrieval**:
   - Files must be stored securely and organized systematically for easy retrieval.

    o The storage system should support large file sizes and various file types (e.g., CAD files, documents, spreadsheets).

2. **Syncing Across Devices**:
   - Files should sync automatically across all devices used by team members to ensure everyone has the most up-to-date information.
   - This includes compatibility with desktop computers, laptops, tablets, and smartphones.

3. **File Sharing and Collaboration**:
   - The system should allow easy sharing of files with team members, professors, and sponsors.
   - Collaborative features, such as real-time editing and commenting, enhance teamwork and productivity.

4. **Search Functionality**:
   - The ability to search the content stored by the team to quickly and accurately find items is essential.
   - Advanced search options, such as keyword search within documents, tags, and filters, improve efficiency.

5.14.2 Popular Tools for Online File Management

1. **Microsoft Office 365**:
   - **Overview**: Available at most universities for students and faculty and widely used by companies.
   - **Key Features**:
     - Integration with Microsoft Word, Excel, PowerPoint, OneNote, and Outlook.
     - Cloud storage and sharing through OneDrive and SharePoint.
     - Real-time collaboration and editing.
     - Additional services like Skype and Exchange for communication.
   - **Advantages**: Familiar interface, robust security features, seamless integration with Microsoft applications.

2. **Google Workspace (formerly G Suite)**:
   - **Overview**: An excellent alternative to Microsoft Office 365, offering a suite of cloud-based tools.
   - **Key Features**:
     - Applications include Gmail, Docs, Sheets, Slides, Sites, Drive, Calendar, and Hangouts.
     - Real-time collaboration and editing.
     - Works across all devices capable of running the Google Chrome browser or operating system.
   - **Advantages**: User-friendly interface, robust collaboration features,

extensive integration with other Google services.

3. **Box**:
   - **Overview**: Secure file storage, sharing, and collaboration tool accessible on multiple platforms.
   - **Key Features**:
     - Supports Windows PCs, Macs, iOS, and Android devices.
     - Advanced security and compliance features.
     - Integrates with various third-party applications.
   - **Advantages**: High security, ease of use, robust collaboration features.

4. **Dropbox**:
   - **Overview**: Another popular file storage, sharing, and collaboration tool.
   - **Key Features**:
     - Supports multiple platforms, including Windows, macOS, iOS, and Android.
     - Easy file sharing with simple link generation.
     - Real-time collaboration with Dropbox Paper.
   - **Advantages**: User-friendly interface, reliable syncing, wide acceptance and use.

5. **Google Drive**:
   - **Overview**: Available at many universities or through personal Google accounts.
   - **Key Features**:
     - Seamless integration with Google Workspace applications.
     - Real-time collaboration and editing.
     - Extensive storage options.
   - **Advantages**: User-friendly, robust collaboration tools, extensive integration with Google services.

6. **Microsoft OneDrive**:
   - **Overview**: A file storage and sharing service that integrates with Microsoft Office.
   - **Key Features**:
     - Real-time collaboration and editing.
     - Works seamlessly with Microsoft Office applications.
     - Advanced security and compliance features.
   - **Advantages**: Strong integration with Microsoft Office, robust security, and ease of use.

5.14.3   Implementing Effective File Management

1. **Choosing the Right Tool**:
   - Select a file management tool that best fits the team's needs, considering factors like ease of use, integration with existing tools, security features, and cost.
   - Ensure the chosen tool is supported by both the university and the project sponsor, especially if there are specific security or confidentiality requirements.

2. **Organizing Files**:
   - Create a clear and consistent folder structure that all team members follow.
   - Use descriptive file names and organize files by project phase, document type, or team member responsibilities.

3. **Setting Up Permissions and Access Controls**:
   - Define and assign appropriate access levels for each team member to ensure data security and prevent unauthorized changes.
   - Regularly review and update permissions as team roles and responsibilities evolve.

4. **Regular Backups**:
   - Ensure that files are regularly backed up to prevent data loss.
   - Use the file management tool's built-in backup features or set up automated backups to an external drive or additional cloud storage.

5. **Training and Onboarding**:
   - Provide training for all team members on how to use the chosen file management tool effectively.
   - Develop a user guide or conduct workshops to ensure everyone understands the best practices for file management, including naming conventions, version control, and collaborative editing.

6. **Example Usage of Google Drive**

   Scenario: A capstone design team decides to use Google Drive for their online file management. Here's how they set it up and use it effectively:

   1. **Initial Setup**:
      - The team leader creates a shared Google Drive folder titled "Capstone Design Project."
      - Within this folder, subfolders are created for each phase of the project: "Research," "Design," "Prototyping," "Testing," and "Final Report."
      - Each team member is given access to the shared folder with editing permissions.

   2. **Organizing Files**:
      - Research documents and literature reviews are stored in the "Research"

folder.

- o CAD drawings and design specifications are stored in the "Design" folder.
- o Prototyping plans and materials lists are stored in the "Prototyping" folder.
- o Test results and analysis are stored in the "Testing" folder.
- o The final report and presentation slides are stored in the "Final Report" folder.

3. **Collaboration and Editing**:
- o Team members use Google Docs for collaborative writing and editing of project reports.
- o Google Sheets are used for managing project budgets and timelines.
- o Comments and suggestions are used to discuss changes and improvements directly within the documents.

4. **Version Control**:
- o Google Drive's version history feature is used to track changes and revert to previous versions if needed.
- o Important milestones and document versions are labeled and archived to maintain a clear project history.

5. **Communication**:
- o The team uses Google Meet for regular video conferences and integrates Google Calendar to schedule meetings and set deadlines.
- o Meeting agendas, minutes, and action items are stored and shared in the "Meeting Documents" subfolder.

By implementing these strategies and using practical file management tools, capstone design teams can ensure that their files are well-organized, easily accessible, and secure. This facilitates efficient collaboration, enhances productivity, and contributes to the overall success of the project.

5.15 Real-Time Conferencing Tools

In the context of capstone design projects, real-time conferencing tools are essential for maintaining effective communication and collaboration, especially when team members are working remotely. While phone meetings can suffice, video teleconferencing offers a much more productive environment by enabling visual communication, screen sharing, and other interactive features. This section explores various real-time conferencing tools, highlighting their features, benefits, and integration capabilities.

5.15.1   Benefits of Video-Conferencing Tools

1. **Visual Communication**:
- o Enhances understanding through visual cues such as facial expressions and body language.

o Builds rapport and strengthens team relationships.
2. **Screen Sharing**:
   o Allows participants to share presentations, documents, and other materials in real time.
   o Facilitates collaborative review and editing of documents.
3. **Session Sharing**:
   o Enables one participant to control another participant's keyboard and mouse, allowing for hands-on collaboration and troubleshooting.
4. **Integration with Other Tools**:
   o Many video-conferencing tools integrate seamlessly with project management, file storage, and productivity tools, enhancing overall efficiency.

5.15.2   Popular Real-Time Conferencing Tools

1. **Cisco WebEx**:
   o **Overview**: WebEx is a secure video-conferencing environment offered by Cisco and is widely used in the government, industry, and medical fields for sensitive communications.
   o **Key Features**:
      ▪ Secure, encrypted communications.
      ▪ Screen sharing and session sharing capabilities.
      ▪ Cisco WebEx Event Center for scheduling meetings with automated notifications.
      ▪ Integration with Cisco's other communication and collaboration tools.
   o **Advantages**: High security, robust feature set, reliable performance.
2. **Zoom**:
   o **Overview**: Zoom is a popular, user-friendly video conferencing system that saw widespread adoption during the COVID-19 pandemic.
   o **Key Features**:
      ▪ High-quality video and audio.
      ▪ Screen sharing and breakout rooms for small group discussions.
      ▪ Recording capabilities for meeting archiving.
      ▪ Integration with various productivity tools and learning management systems.
   o **Advantages**: Ease of use, scalability, extensive features for large and small meetings.
3. **Skype**:
   o **Overview**: Skype is a video-conferencing tool offered by Microsoft, known for its integration with the MS Office environment.

- o **Key Features**:
  - Video and voice calls, instant messaging.
  - Screen sharing and file sharing.
  - Integration with Microsoft Office and Outlook.
- o **Advantages**: Familiar interface, seamless integration with Microsoft products, ease of use.

4. **Google Hangouts**:
  - o **Overview**: Google Hangouts, part of Google Workspace (formerly G Suite), offers video conferencing capabilities that integrate well with Google's suite of productivity tools.
  - o **Key Features**:
    - Video calls, instant messaging, and group chats.
    - Screen sharing and document collaboration through Google Drive.
    - Integration with Gmail and Google Calendar.
  - o **Advantages**: Strong integration with Google Workspace, ease of use, reliable performance.

5. **Microsoft Teams**:
  - o **Overview**: Microsoft Teams is a comprehensive collaboration platform that integrates with Office 365 and offers robust video conferencing capabilities.
  - o **Key Features**:
    - Video calls, chat, and file sharing within channels.
    - Integration with OneDrive and SharePoint for document storage and collaboration.
    - Screen sharing, meeting recording, and real-time document collaboration.
  - o **Advantages**: Centralized platform for all communication and collaboration needs, robust integration with Office 365, supports remote teamwork effectively.

5.15.3   Implementation and Best Practices

1. **Choosing the Right Tool**:
  - o Select a video-conferencing tool that meets the team's needs, considering factors such as ease of use, integration with existing tools, security features, and cost.
  - o Ensure the chosen tool is supported by both the university and the project sponsor, especially if there are specific security or confidentiality requirements.
2. **Setting Up Meetings**:
  - o Schedule regular meetings using the video-conferencing tool's scheduling

features.

- o Send automated notifications to participants, including links to the meeting and any necessary preparation materials.

3. **Conducting Effective Meetings**:
   - o **Preparation**: Ensure all participants have the necessary equipment and software installed and tested before the meeting.
   - o **Agenda**: Prepare and share a clear agenda before the meeting to keep discussions focused and on track.
   - o **Participation**: Encourage active participation from all team members, using features such as screen sharing and breakout rooms to facilitate collaboration.
   - o **Recording and Notes**: Record meetings and take detailed notes to capture critical decisions and action items. Share these with the team after the meeting.

4. **Security Considerations**:
   - o Ensure that meetings discussing sensitive information are conducted using secure, encrypted communication tools.
   - o Use password protection and waiting rooms to control access to meetings.
   - o Regularly update software to protect against security vulnerabilities.

5. **Example Use Case: Using Zoom for Capstone Projects**

   Scenario: A capstone design team decides to use Zoom for their real-time conferencing needs. Here's how they set it up and use it effectively:

   1. **Initial Setup**:
      - o The team leader creates a Zoom account and sets up recurring weekly meetings.
      - o Invitations are sent to all team members, professors, and sponsors, including links to join the meetings.

   2. **Conducting Meetings**:
      - o **Preparation**: Before each meeting, the team leader tests the audio and video setup and ensures all necessary documents are ready for screen sharing.
      - o **Agenda**: An agenda is prepared and shared with all participants ahead of the meeting.
      - o **Engagement**: During the meeting, the team leader facilitates the discussion, using breakout rooms for small group work and screen sharing for presentations.
      - o **Recording**: The meeting is recorded, and the recording link is shared with all participants afterward.

   3. **Follow-Up**:
      - o Meeting notes and action items are documented and shared with the team

via email and a shared Google Drive folder.
- o The team leader schedules follow-up meetings and ensures that any issues or tasks identified during the meeting are addressed.

By leveraging these real-time conferencing tools and following best practices for their implementation, capstone design teams can enhance their communication, collaboration, and overall project management. These tools not only facilitate efficient remote teamwork but also help maintain continuity and productivity throughout the project lifecycle.

5.16 Assignments

Assignment 5-1: Initial Team Meeting

Objective: After being assigned to your capstone project team, participate in the initial team meeting during the designated class time. This meeting will serve as an opportunity to establish communication, discuss individual backgrounds and skills, and plan for future meetings and responsibilities. The initial team meeting will set the foundation for effective collaboration and project success.

Instructions:

- Attend the Meeting
  - The initial team meeting will be held during the designated class time.
  - Ensure that you arrive on time and are prepared to participate actively in the meeting.
- Prepare for the Meeting
  - Review the project description and any available materials to familiarize yourself with the project scope and requirements.
  - Make a list of questions or topics you would like to discuss with your team members.
- Meeting Activities During the meeting, complete the following activities: a. Introductions
  - Take turns introducing yourselves, sharing your background, minors, professional elective courses, and any relevant experience or skills.
  - Discuss what each team member can contribute to the project based on their strengths and expertise.
- Exchange contact information
  - Share your preferred method of contact (e.g., email, phone number) with your team members.
  - Create a team contact list or directory for accessible communication.
- Discuss Schedules and Availability
  - Share your schedules, including classes, work commitments, and other obligations.
  - Identify expected free times for regular team meetings and collaboration sessions, keeping in mind that Thursdays are designated for team project meetings or meetings with the sponsor.
- Assign Roles and Responsibilities
  - Discuss the various roles and responsibilities needed for the project (e.g., project manager, technical lead, documentation specialist).
  - Assign roles based on each team member's strengths, interests, and experience.
- Plan Sponsor Meeting

- o Discuss the need to arrange a meeting with the project sponsor to gather more information and clarify project requirements.
        - o Assign a team member to reach out to the sponsor and schedule the meeting, preferably on a Thursday.
    - Set Next Steps and Action Items
        - o Identify immediate next steps and action items for each team member.
        - o Set deadlines and establish a system for tracking progress and accountability.
- **Meeting Agenda**
- Use the agenda template on Brightspace for your team meetings:
    - o Introductions
    - o Project Overview
    - o Team Member Backgrounds and Skills
    - o Contact Information Exchange
    - o Schedule and Availability Discussion (Thursdays for team meetings and sponsor meetings)
    - o Role and Responsibility Assignment
    - o Sponsor Meeting Planning
    - o Next Steps and Action Items
    - o Open Discussion and Questions
- **Meeting Minutes Template**
- Use the minutes template on Brightspace to record the minutes of your team meeting:

Initial Team Meeting Minutes
Date:
Time:
Location:
Attendees:

Agenda Items:
1. Introductions
   - Team member backgrounds and skills
2. Project Overview
   - Discussion of project scope and requirements
3. Contact Information Exchange
   - Team contact list created
4. Schedule and Availability
   - Regular meeting times and dates established (Thursdays for team meetings and sponsor meetings)

5. Role and Responsibility Assignment
  - Roles assigned to each team member
6. Sponsor Meeting Planning
  - Team member assigned to schedule sponsor meeting (preferably on a Thursday)
7. Next Steps and Action Items
  - List of next steps and action items for each team member
8. Open Discussion and Questions
  - Any additional topics or concerns raised by team members

Action Items:
- [Action Item 1]
  - Assigned to:
  - Deadline:
- [Action Item 2]
  - Assigned to:
  - Deadline:

Next Meeting:
Date:
Time:
Location:

Fill in the template with the relevant information discussed during your meeting, including key points, decisions made, action items assigned, and the details for the next meeting.

- **Follow-up**
  - o Share the meeting minutes with all team members after the meeting.
  - o Ensure that everyone understands their assigned action items and deadlines.
  - o Begin working on your assigned tasks and responsibilities.
  - o Communicate regularly with your team members to provide updates and maintain progress.

Remember, you are expected to attend lectures on Tuesdays and utilize Thursdays for team project meetings and meetings with the sponsor. By actively participating in this initial team meeting and establishing clear roles, responsibilities, and communication channels, you will set your team up for success throughout the capstone project.

6 Project Management

Capstone design problems are inherently open-ended, complex, and multifaceted, necessitating the involvement of multiple people and resources to achieve successful outcomes. To manage the numerous individual and group tasks required to complete such projects within the fixed and rigid time frame of the capstone course—and within any provided budget or cost guidelines—a structured method known as project management is employed.

Project management is an essential discipline in engineering and industry at large. It involves decomposing, estimating, planning, organizing, and managing tasks and resources to accomplish a defined objective within given constraints. These resources can include people, facilities, equipment, money, and time. A robust project management approach ensures that all aspects of the project are coordinated effectively, enabling the team to achieve its goals.

6.1 The Role of a Project Plan

A project plan, or roadmap, is fundamental to managing a project successfully. It is specific to the team and the project, serving multiple critical purposes:

- **Coordination**: It helps coordinate project resources and activities among team members.
- **Communication**: It facilitates communication with sponsors, professors, mentors, and other stakeholders about the project's status and progress.
- **Responsibility Distribution**: It clearly shows the distribution of responsibilities and resources.

6.1.1 Objectives of Project Management

The purposes of project management in the context of a capstone design project are multifaceted:

- **Creating a Realistic Plan**: Develop a practical and achievable project plan.
- **Motivating the Team**: Establish a roadmap that motivates the team toward the end goal of a successful design solution.

- **Communication**: Clearly communicate the design solution process to sponsors and other stakeholders.
- **Fostering Collaboration**: Enhance collaboration among team members by defining a clear scope and objectives.
- **Aligning Goals**: Ensure all team members are aligned with the common goals of the project and understand how to achieve them.
- **Building Confidence**: Develop confidence among team members and stakeholders in the team's ability to achieve a successful design solution.
- **Identifying Issues**: Recognize and communicate when the team is off track and address problems promptly.
- **Receiving Guidance**: Allow sponsors and mentors to observe the team's progress and offer advice or assistance.
- **Resource Identification**: Identify and acquire necessary tools or resources or adapt the plan if specific resources are unavailable.
- **Anticipating Bottlenecks**: Identify potential bottlenecks or problem areas and plan workarounds proactively.

6.1.2   The Project Management Process

Project management begins with problem decomposition and estimation of project details. For capstone design teams, this can be particularly challenging due to their limited experience in project management and the uniqueness of their design problems. The team may not have all the necessary information at the outset to lay out the entire project with all the required steps. However, by following the clearly defined process described in this book, teams can create an initial project plan and then iteratively fill in the details as they learn more about their project and how to complete it.

6.1.3   Practical Steps for Project Planning

Planning should start with a comprehensive table listing all tasks, the resources needed, time duration, and cost for each step. This structured approach ensures that all aspects of the project are considered and managed effectively. The table provides an example list for a capstone project. 6-1, which serves as a template for teams to develop their project plans.

By adopting a structured project management approach, capstone design teams can navigate the complexities of their projects, align their efforts with defined objectives, and achieve successful outcomes. The following sections of this chapter will delve deeper into specific project management techniques, tools, and best practices tailored to the needs of capstone design projects.

6.2 Planning

Project planning is a vital group activity for capstone projects, requiring the team to collaboratively prepare lists of tasks, the order of events, necessary resources, and time durations. Since most engineering students may lack extensive experience in project planning, initial efforts often involve estimations. These estimations can be refined as the team gains a better understanding of the project and its collaborative dynamics. Planning is an ongoing process throughout the design project, with the goal of producing a comprehensive project plan. This plan will guide the team in monitoring progress, coordinating efforts, and communicating with sponsors and professors.

6.2.1   Objectives of Project Planning

1. **Task Identification**:
   o Identify all necessary tasks required to complete the project.
   o Ensure each task is well-defined and contributes to the overall project goals.
2. **Resource Allocation**:
   o Determine the resources needed for each task, including materials, equipment, and non-human resources.
   o Assign responsibilities to team members based on their strengths and expertise.
3. **Timeline Establishment**:
   o Create a timeline for task completion, ensuring that the project stays on schedule.
   o Incorporate critical milestones and deadlines set by the course and sponsors.

6.2.2   Elements of a Project Plan

Each item in the project plan should be specified with the following attributes:
1. **Descriptive Task Title**:
   o Clearly describe the task to ensure understanding among all team members.
2. **Task Responsibility**:
   o Assign a responsible person or group for each task, ensuring accountability.
3. **Non-People Resources Needed**:
   o Identify the materials, equipment, software, or other resources required for the task.
4. **Task Start Time**:
   o Specify when the task can begin, considering dependencies and resource

availability.

5. **Task Dependencies**:
   - o List other tasks that must be completed before this task can start.
6. **Estimated Duration**:
   - o Estimate how long the task will take based on available information and experience.

6.2.3   Reverse Planning Method

A reverse planning method is particularly effective in capstone design projects due to their fixed end dates, which align with the university's academic calendar. Starting with the final deliverables in mind—a successful design solution and documentation—the team can work backward to identify and schedule all necessary tasks. This approach ensures that each task is planned with the end goal in focus, facilitating the timely completion of the project.

6.2.3.1   Key Steps in Reverse Planning:

1. **Define Final Deliverables**:
   - o Identify the final deliverables required by the course and sponsor.
2. **Identify Checkpoints**:
   - o Establish key milestones and checkpoints, such as critical design reviews and class assignments.
3. **Schedule Tasks Backward**:
   - o Plan tasks in reverse order from the final deliverable to the project's start, ensuring all dependencies are considered.
4. **Example Capstone Project Plan**

5. **Continuous Planning and Adjustment**
   Planning in capstone design is a continuous process. The initial project plan serves as a starting point, but it must be regularly reviewed and adjusted based on the team's progress and any new information or challenges that arise. This dynamic approach ensures that the project remains on track and can adapt to any changes or obstacles encountered.
   Critical Thinking and Problem-Solving:
   - Critical thinking is essential in planning, requiring the team to consider the problem statement, sponsor requirements, course requirements, time constraints, and budget limitations.
   - The capstone project's complexity often exceeds typical engineering problems, providing students with unparalleled experience and substantial benefits to their education and future careers.

Table 6-2. Capstone First List of Activities or Tasks

Task	Estimated Duration	Resources	Dependencies
Organize team meetings	1 day	Team members	None
Meet with sponsor	1 day	Team members and sponsor	None
Review the problem definition and prepare a list of questions	1 week	Team members, sponsors, mentors, experts	None
Search for additional information about the design problem, patent search, article search	2 weeks	Team members	Problem definition
Create design specifications	2 weeks	Team members, sponsors, mentors, experts	Problem definition, information search
Generate design concepts	1 week	Team members	Problem definition, information search, design specifications
Analyze concepts and down-select to a smaller number	1 week	Team members	Information search, competition research
Prepare presentation for critical design review	1 week	Team members	Concept analysis
Critical design review	1 day	Team members, class, sponsors, mentors, professor	All previous tasks
Use feedback from critical design review to decide on a concept to prototype	1 week	Team members	Critical design review
Prove design concept – create a prototype/model of the design solution	4 weeks	Team members	None
Document the design work – create a preliminary design report	2 weeks	Team members	None
Conclude Part I by presenting/meeting with the sponsor	1 week	Team members	None

Effective project planning is the cornerstone of a successful capstone design project. By collaboratively identifying tasks, allocating resources, establishing timelines, and continuously refining the plan, the team can navigate the complexities of their project and achieve their goals. A well-structured project plan not only facilitates internal coordination but also enhances communication with sponsors, professors, and other stakeholders, ensuring a comprehensive and successful design solution.

6.3   Scheduling

Scheduling is a crucial aspect of project management in capstone design projects. It involves mapping the tasks identified during the planning phase onto a timeline, ensuring each task has a start date, end date, and duration. Scheduling also involves managing dependencies between tasks, allocating resources, and making realistic predictions about time requirements. This section delves into the intricacies of scheduling, the tools available for it, and best practices to ensure an effective scheduling process.

6.3.1   The Scheduling Process

1. **Mapping Tasks to a Timeline**:
   o   Each task identified during planning needs to be scheduled with a specific start date, end date, and estimated duration.
   o   Consider task dependencies where specific tasks cannot begin until preceding tasks are completed.
2. **Resource Allocation**:
   o   Assign resources to each task, including human resources (team members) and non-human resources (equipment, materials).
   o   Resources can be allocated on a percentage basis (e.g., a team member may spend 50% of their time on a specific task).
3. **Handling Dependencies**:
   o   Define task dependencies clearly to ensure that the project timeline accounts for these sequential activities.
   o   Use tools to visualize and manage these dependencies effectively.
4. **Resource Utilization**:
   o   Optimize the use of resources, ensuring that they are neither overburdened nor underutilized.
   o   Plan for the availability of specific resources, such as tools or equipment, that may only be needed for a short duration within a broader task.

6.3.2   Importance of Scheduling

1. **Time Management**:
   o   Scheduling helps predict the time required for each task, enabling the team to manage their time effectively.
   o   It ensures that all tasks are completed within the fixed time frame of the capstone course.
2. **Cost Estimation**:
   o   A well-defined schedule helps estimate project costs and perform financial

analyses.
- o Accurate scheduling is essential for creating realistic proposals and securing sponsor approval.

3. **Task Assignment**:
   - o Equitable and appropriate task assignment ensures that team members work on tasks matching their skills and knowledge.
   - o This maximizes productivity and leverages the strengths of each team member.

4. **Handling External Dependencies**:
   - o Scheduling must account for external factors such as purchasing, shipping times, and resource availability.
   - o Proper planning helps mitigate delays caused by these external dependencies.

6.3.3   Steps to Create a Schedule

1. **Define Tasks and Subtasks**:
   - o Break down the project into smaller, manageable tasks using the problem decomposition method (divide-and-conquer).
   - o Ensure that each task is well-defined and contributes to the overall project goals.

2. **Assign Tasks to Team Members**:
   - o Assign tasks based on team members' skills, availability, and workload.
   - o Ensure equitable distribution of tasks to maintain team morale and productivity.

3. **Estimate Task Durations**:
   - o Use best estimates to predict how long each task will take.
   - o Refine these estimates as the project progresses and more information becomes available.

4. **Establish Task Dependencies**:
   - o Identify which tasks depend on the completion of others.
   - o Use tools to visualize these dependencies and plan the sequence of activities.

5. **Load-Level Resources**:
   - o Ensure that resources are allocated efficiently across tasks.
   - o Avoid overloading or underutilizing any single resource.

6. **Integrate Fixed Dates and Milestones**:
   - o Incorporate fixed dates from the capstone design course and any other critical milestones.
   - o Ensure the schedule aligns with these key dates to avoid missing deadlines.

7. **Avoid Common Pitfalls**:
   - Avoid making unreasonable assumptions about the project or team members' skills.
   - Do not underestimate the level of effort required for complex tasks like coding, machining, welding, or learning new software.
   - Make realistic predictions about task durations and success probabilities.

6.3.4   Tools for Scheduling

1. **Microsoft Project**:
   - Widely used in engineering industries for comprehensive project management.
   - Features include task scheduling, resource allocation, Gantt charts, and critical path analysis.

2. **PRIMAVERA**:
   - Often integrated with AutoCAD and other CAD systems.
   - Suitable for large-scale projects with detailed scheduling and resource management needs.

3. **Cloud-Based Project Management Software**:
   - Examples include Asana, Trello, Monday.com, and Smartsheet.
   - These tools allow real-time collaboration, task tracking, and easy access for all team members.
   - Cloud-based solutions enhance team engagement and project transparency.

4. **Integrated CAD Systems**:
   - Some CAD systems like CATIA have built-in project management features.
   - Integration with design tools streamlines the scheduling process for engineering projects.

Effective scheduling is pivotal for the success of capstone design projects. It translates the planning phase into actionable tasks mapped on a timeline, ensuring that all activities are completed within the allotted time frame. By utilizing appropriate tools and methodologies, teams can create realistic schedules, allocate resources efficiently, and navigate dependencies effectively. Continuous refinement of the schedule as the project progresses helps accommodate new information and unforeseen challenges, ensuring that the team remains on track to achieve its project goals.

8. **Example of a Scheduling Table**

Table 6-3. Example Schedule for Capstone Project

Task	Start Date	End Date	Duration	Resources	Dependencies
Organize team meetings	01/01/2024	01/01/2024	1 day	Team members	None
Meet with sponsor	02/01/2024	02/01/2024	1 day	Team members, sponsor	None
Review problem definition and prepare questions	03/01/2024	10/01/2024	1 week	Team members, sponsors, mentors, experts	None
Search for additional information	11/01/2024	25/01/2024	2 weeks	Team members	Problem definition
Create design specifications	26/01/2024	09/02/2024	2 weeks	Team members, sponsors, mentors, experts	Problem definition, information search
Generate design concepts	10/02/2024	16/02/2024	1 week	Team members	Problem definition, information search, design specifications
Analyze concepts and down-select	17/02/2024	23/02/2024	1 week	Team members	Information search, competition research
Prepare presentation for critical design review	24/02/2024	02/03/2024	1 week	Team members	Concept analysis
Critical design review	03/03/2024	03/03/2024	1 day	Team members, class, sponsors, mentors, professor	All previous tasks
Decide on a concept to prototype	04/03/2024	10/03/2024	1 week	Team members	Critical design review
Prove design concept – create prototype/model	11/03/2024	07/04/2024	4 weeks	Team members	None
Document the design work – preliminary design report	08/04/2024	21/04/2024	2 weeks	Team members	None
Conclude Part I by presenting to the sponsor	22/04/2024	28/04/2024	1 week	Team members	None

6.4   Entering Project Information into MS Project

Microsoft Project (MS Project) is a powerful project management tool that allows teams to plan, schedule, and manage their projects effectively. This tutorial will guide you through the process of entering project information into MS Project, using examples relevant to capstone design projects. MS Project is available both as a cloud-based product and as standalone software. The cloud-based version is accessible through Microsoft or can be installed by an enterprise for shared access by team members.

6.4.1   Getting Started with MS Project

1. **Opening a New Project**:
    o   Launch MS Project and open a new project file. You will see a blank entry field, as shown in Figure 6-1..

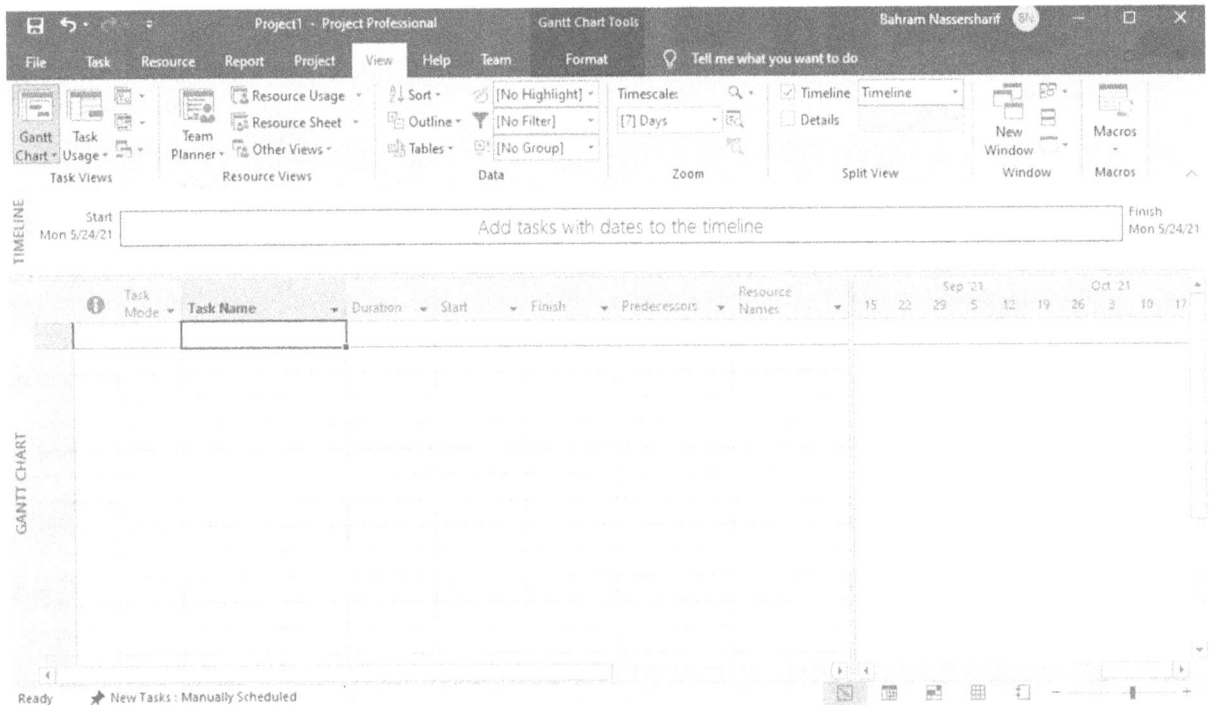

Figure 6-1. Project entry fields in MS Project.Preparing to Enter Project Information:
    o   Before entering data, ensure you have the following information prepared:
        ▪   Descriptive task names.
        ▪   Approximate durations of tasks (in days, by default).
        ▪   Task dependencies (predecessors).
        ▪   Resource assignments (team members and equipment).
        ▪   Fixed start dates for specific tasks (if required by course schedules).

6.4.2   Step-by-Step Guide to Entering Project Information

1. **Set the Project Start Date**:
   - Go to the "Project" tab and select "Project Information."
   - Set the "Start date" to the date when your team begins work on the project.
2. **Enter Task Names**:
   - In the "Task Name" column, enter the descriptive names of each task. This should match your prepared task list.
3. **Set Task Durations**:
   - In the "Duration" column, enter the estimated duration for each task. Use the default unit of days unless a different unit is required.
4. **Define Task Dependencies**:
   - In the "Predecessors" column, enter the task number of the preceding task. This ensures that tasks are scheduled in the correct order.
   - For example, if "Generate Design Concepts" depends on "Review Problem Definition," enter the task number for "Review Problem Definition" as the predecessor for "Generate Design Concepts."
5. **Assign Resources to Tasks**:
   - In the "Resource Names" column, enter the initials of the team members responsible for each task (e.g., TM1, TM2, TM3).
   - Include any additional resources, such as equipment (e.g., 3D printers).
6. **Auto-Schedule Tasks**:
   - Ensure all tasks are set to "Auto-Scheduled" mode. This allows MS Project to automatically calculate start and finish dates based on dependencies and durations.
   - To set tasks to auto-schedule, select the tasks, then click on the "Task" tab and choose "Auto Schedule."
7. **Fixed Start Dates for Specific Tasks**:
   - For tasks with fixed start dates (e.g., project presentations), enter the exact start date in the "Start" column.

6.4.3   Example Project Layout

Figure 6-2. shows an example project layout for a semester-long Phase I capstone course. You can adapt your schedule based on this example.

		Task Mode	Task Name	Duration	Start	Finish	Predecessor	Resource Names
1			Star of Project	0 days	Mon 9/20/21	Mon 9/20/21		
2			Organize and set up team meetings	1 day	Mon 9/20/21	Mon 9/20/21	1	TM1,TM2,TM3,TM4
3			**Weekly Progress Report**	**51 days**	**Fri 9/24/21**	**Fri 12/3/21**		
15			Schedule meeting with sponsor and meet	1 wk	Mon 9/20/21	Fri 9/24/21	1	TM1,TM2,TM3,TM4, Sponsor
16			Review problem definition and prepare a list of questions	1 wk	Mon 9/20/21	Fri 9/24/21	1	TM1,TM2,TM3,TM4
17			Search for additional information about the design problem	1 wk	Mon 9/20/21	Fri 9/24/21	1	TM1,TM2,TM3,TM4
18			Literature search	1 wk	Mon 9/27/21	Fri 10/1/21	17	TM1,TM2,TM3,TM4
19			Patent Search	1 wk	Mon 9/27/21	Fri 10/1/21	17	TM1,TM2,TM3,TM4
20			Create Design Specifications	1 wk	Mon 10/4/21	Fri 10/8/21	19	TM1,TM2,TM3,TM4
21			Generate design concepts	1 wk	Mon 10/11/21	Fri 10/15/21	20	TM1,TM2,TM3,TM4
22			Analyze Concepts and Down Select	1 wk	Mon 10/18/21	Fri 10/22/21	21	TM1,TM2,TM3,TM4
23			Prepare Presentation	1 wk	Mon 10/18/21	Fri 10/22/21	21	TM1,TM2,TM3,TM4
24			Critical Design Review	1 day	Mon 10/25/21	Mon 10/25/21	23	TM1,TM2,TM3,TM4
25			Select concept to model or prototype	3 days	Tue 10/26/21	Thu 10/28/21	24	TM1,TM2,TM3,TM4, Sponsor
26			Create prototype or model	4 wks	Fri 10/29/21	Thu 11/25/21	25	TM1,TM2,TM3,TM4
27			Document Design	2 wks	Fri 11/26/21	Thu 12/9/21	26	TM1,TM2,TM3,TM4
28			Final Presentation and Delivery	1 day	Fri 12/10/21	Fri 12/10/21	27	TM1,TM2,TM3,TM4
29			End of Phase I	1 day?	Mon 12/13/21	Mon 12/13/21	28	

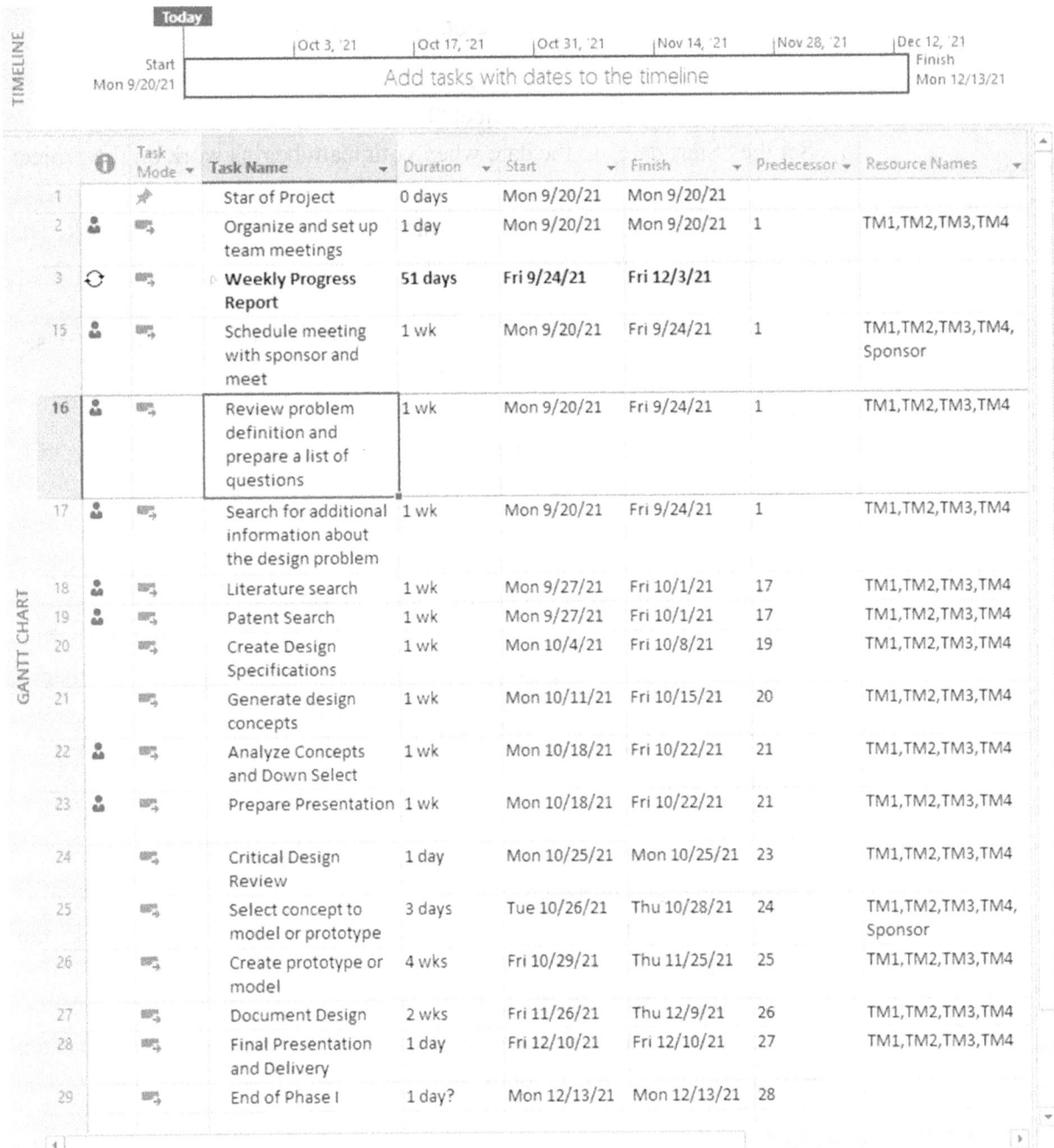

Figure 6-2. Sample the first project plan, including capstone design process steps.Fine-Tuning the Project Schedule

1. **Adjusting Task Durations**:
   o Review and adjust task durations based on team input and progress. Refine the estimates as the project progresses.
2. **Managing Task Dependencies**:
   o Ensure all dependencies are accurately reflected. Modify dependencies if the order of tasks changes.

3. **Resource Leveling**:
   - o Use MS Project's resource leveling feature to balance the workload among team members. This helps avoid over-allocation and ensures optimal resource utilization.
4. **Monitoring and Updating**:
   - o Regularly monitor the project schedule and update it with actual start and finish dates. This keeps the schedule accurate and reflects the project's real-time progress.
5. **Milestones and Deadlines**:
   - o Identify critical milestones and deadlines in the schedule. Ensure these are clearly marked and communicated to the team.

Entering project information into MS Project is a structured process that transforms your planning efforts into a detailed, manageable project schedule. By following the steps outlined in this tutorial, capstone design teams can effectively utilize MS Project to organize their tasks, allocate resources, manage dependencies, and ensure timely completion of their projects. Regular monitoring and adjustments will keep the project on track, helping the team achieve its goals within the fixed academic timelines.

6.5 Independent and Dependent Tasks

In project management, tasks can generally be categorized as either independent or dependent. Understanding the distinction between these types of tasks is crucial for adequate scheduling and resource allocation. This section explores the characteristics of independent and dependent tasks, their implications for project planning, and how they can be visualized and managed using tools like MS Project.

6.5.1 Independent Tasks

Definition: Independent tasks are tasks that can be performed in parallel with one another without any interference. These tasks do not rely on the completion of other tasks and can be scheduled flexibly.

Characteristics:
- **Parallel Execution**: Independent tasks can be executed simultaneously.
- **Flexible Scheduling**: These tasks can be started, paused, and completed in any order relative to one another.
- **Resource Optimization**: Multiple team members can work on different independent tasks simultaneously, optimizing the use of resources.

Example: Consider two tasks, Task 1 and Task 2, which are independent of each other. As shown in Figure 6-3., these tasks can be performed in parallel, meaning Task 1 can be completed at the same time as Task 2.

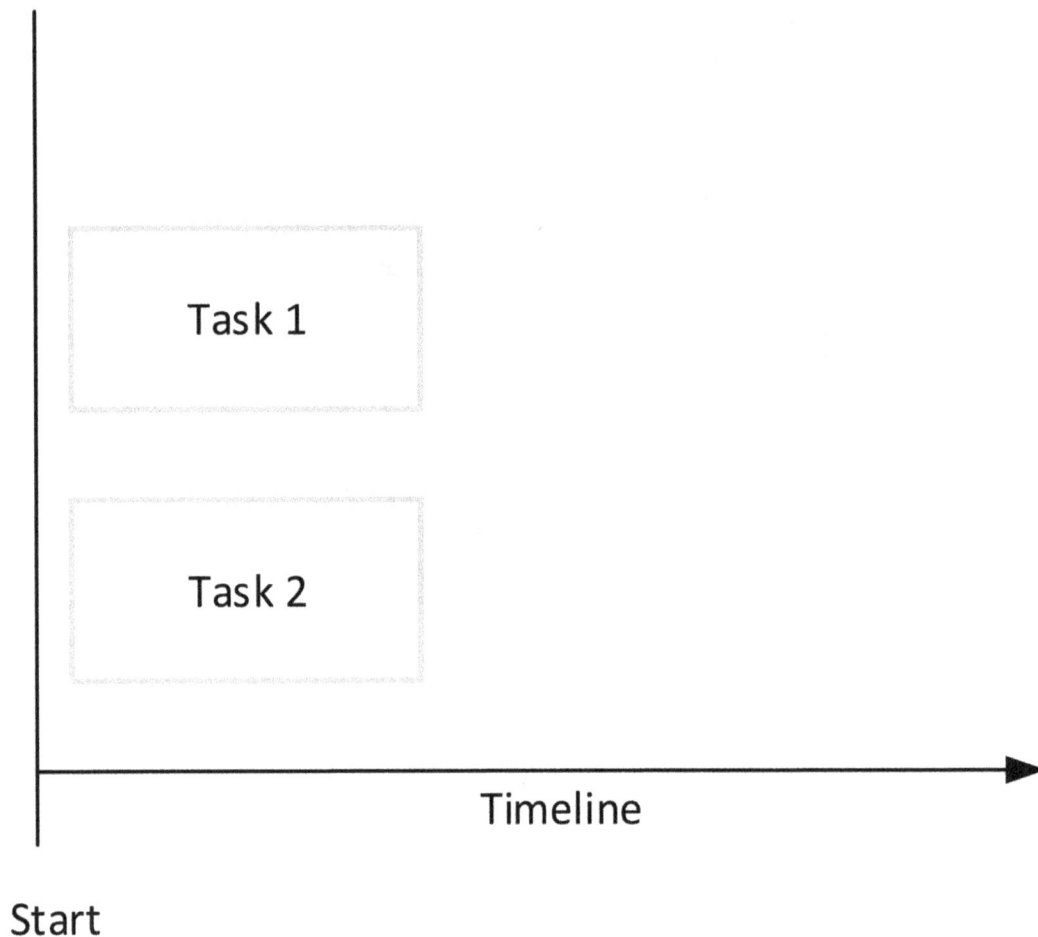

Figure 6-3. Illustration of independent tasks.

6.5.2   Dependent Tasks

Definition: Dependent tasks are tasks that must be carried out in a specific sequence. One task relies on the completion of another task before it can begin.

Characteristics:

- **Sequential Execution**: Dependent tasks must follow a specific order, where one task cannot start until the preceding task is completed.
- **Scheduling Constraints**: The start date of a dependent task is constrained by the end date of its predecessor.
- **Critical Path**: The sequence of dependent tasks forms the critical path, which is crucial for determining the project timeline.

Example: Consider two tasks, Task 1 and Task 2, where Task 2 depends on the completion of Task 1. As illustrated in Figure 3.4, Task 2 cannot start until Task 1 has been completed.

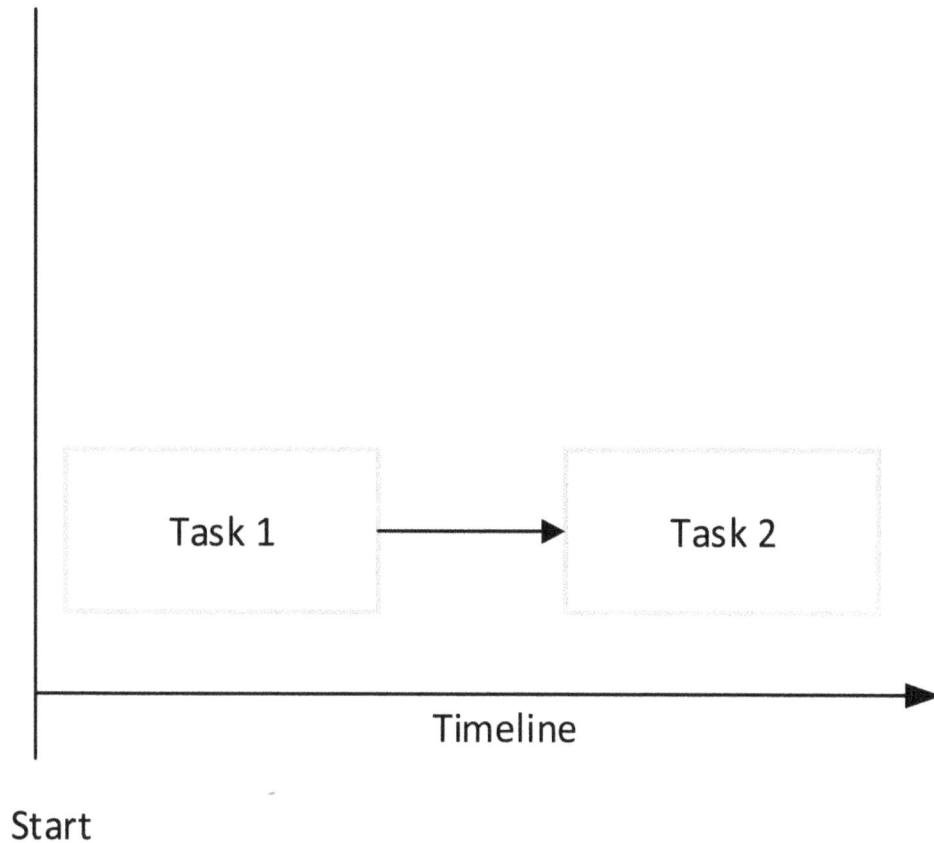

Figure 6-4. Illustration of dependent tasks.

Multiple Dependent Tasks: When multiple tasks are dependent on each other, they form a sequential chain. Figure 6-5. illustrates a sequence of four dependent tasks, where the completion of each subsequent task depends on the completion of the previous one.

6.5.3   Task Networks

Task networks can take various forms based on the relationships between independent and dependent tasks. These configurations are essential for visualizing the project's structure and effectively planning the workflow.

Example Configurations: Figure 6-6. This figure shows some possible configurations of task networks involving four tasks. These configurations help map out the capstone design project tasks into a network relationship, providing a clear visualization of how tasks interact.

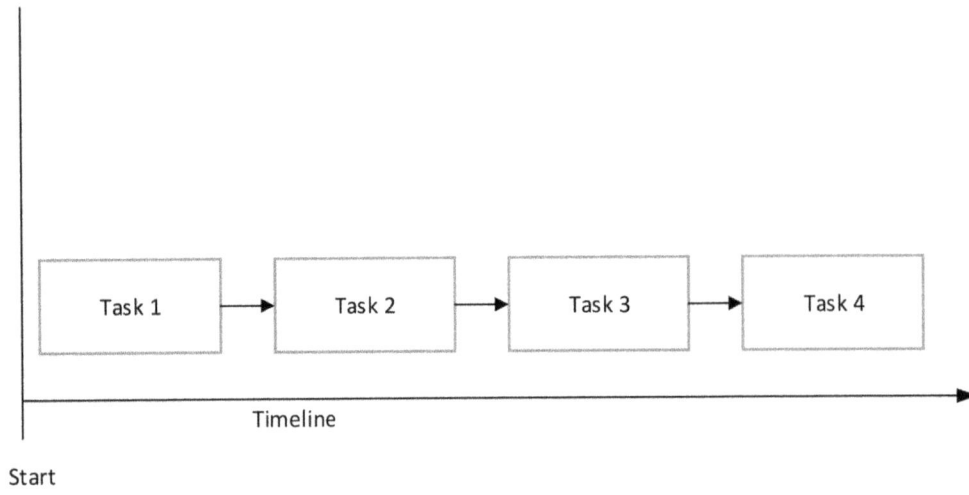

Figure 6-5. Illustration of multiple dependent tasks.

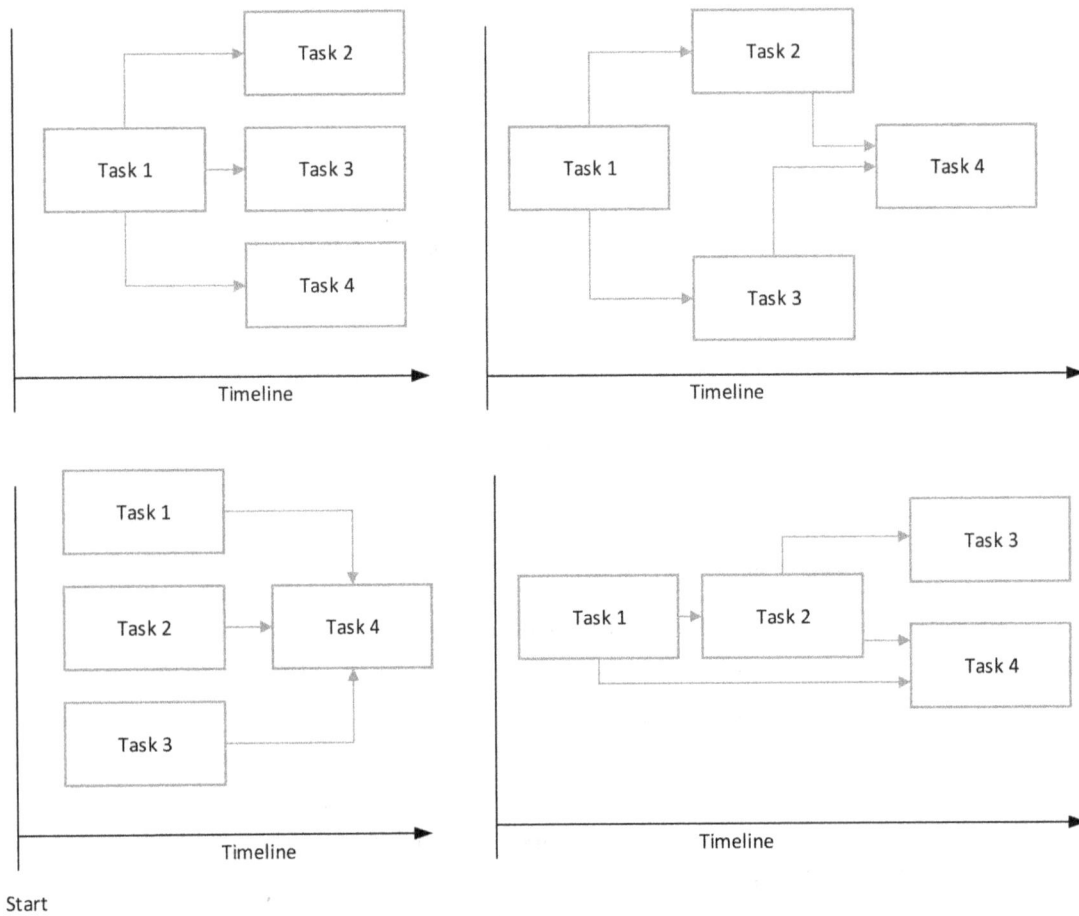

Figure 6-6. Illustration of combinations of independent and dependent tasks.

6.5.4   Time Attributes of Task Dependencies

The arrows connecting the task boxes in task networks can also have time attributes assigned to them for more precise project planning. This allows for better management of task start times, particularly when considering delays or early starts.

Favorable Lag Time: If Task 2 needs to start sometime after Task 1 is completed, a favorable lag time is added. For example, if shipping time for a part is needed between Task 1 and Task 2, Task 2 would start after Task 1 plus the shipping time.

Negative Lag Time: If Task 2 can start before Task 1 is fully completed, a negative lag time is used. This might occur if resources are freed up earlier than expected, allowing Task 2 to begin ahead of the original schedule.

Example: Consider a scenario where Task 2 can start three days after Task 1 is completed (favorable lag time) or two days before Task 1 is finished (negative lag time). These dependencies are illustrated in Figure 6-7..

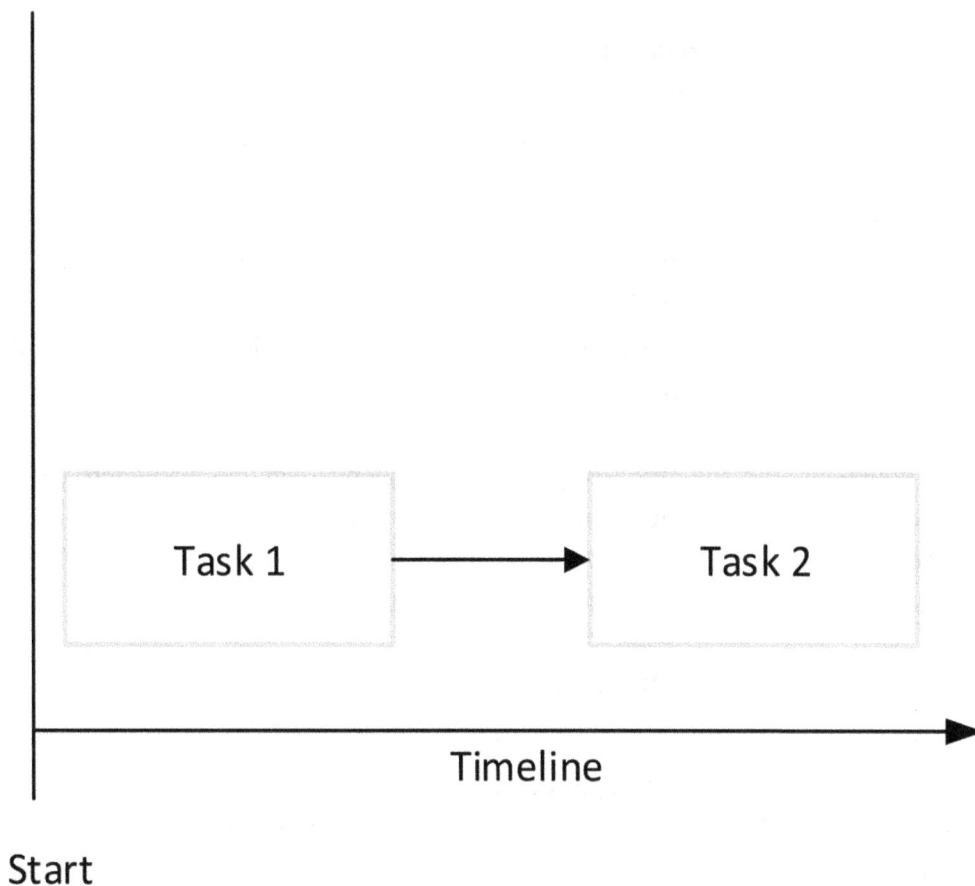

Figure 6-7. Illustration of dependent tasks.

6.5.5   Implementing Independent and Dependent Tasks in MS Project

Step-by-Step Guide:
1. **Enter Tasks**:
   - List all tasks in the "Task Name" column.
   - Assign appropriate durations to each task in the "Duration" column.
2. **Set Dependencies**:
   - In the "Predecessors" column, enter the task numbers of preceding tasks for dependent tasks.
   - For independent tasks, leave the "Predecessors" column blank.
3. **Add Lag or Lead Time**:
   - To add lag time, enter the predecessor task number followed by "+" and the number of days (e.g., "1+3d" for a 3-day lag).
   - To add lead time, enter the predecessor task number followed by "-" and the number of days (e.g., "1-2d" for a 2-day lead).
4. **Visualize Task Relationships**:
   - Use the Gantt chart view to visualize the relationships between tasks. Dependent tasks will be connected with arrows, while independent tasks will appear as parallel bars.
5. **Adjust and Optimize**:
   - Regularly review and adjust task dependencies and durations to reflect any changes or new information.
   - Use MS Project's features to optimize the schedule and ensure efficient resource utilization.

By understanding and effectively managing independent and dependent tasks, capstone design teams can create realistic schedules, optimize resource allocation, and ensure timely project completion. Proper use of MS Project enhances these capabilities, providing a visual and interactive platform for project management.

6.6   Assigning and Accepting Responsibility

In professional engineering organizations, projects are typically managed by experienced engineers who serve as project managers. These managers oversee a group of engineers and have the authority to allocate human and financial resources to ensure the project's successful completion. However, this hierarchical model does not directly translate to capstone design projects, where a team of peers with equal responsibility and authority collaborates on the same project. This section explores the dynamics of assigning and accepting responsibility within capstone design teams, emphasizing the importance of individual accountability and teamwork.

6.6.1   The Capstone Team Model

In capstone design projects, each team member plays a crucial role in the project's success. Unlike in the industry, where a project manager leads the team, capstone projects rely on a flat structure where all members share equal responsibility. This model requires every team member to be an active and productive participant, with the success or failure of the project resting on everyone's shoulders.

6.6.2   Responsibilities and Expectations:

- **Active Participation**: Every team member must contribute meaningfully to the project.
- **Self-Assignment and Team Assignment**: Team members may volunteer for tasks or be assigned responsibilities by the group.
- **Acceptance of Responsibility**: Once tasks are assigned, team members must accept and commit to their roles and responsibilities.

6.6.3   Key Responsibilities of Team Members

As a vested team member, accepting responsibility involves several vital commitments:

1. **Goal-Oriented Conduct**:
   o   Stay motivated by the project's goals and strive for successful completion.
   o   Be willing to listen and compromise with team members, sponsors, and mentors.
2. **Understanding Project Constraints**:
   o   Understand and agree with the project's purpose.
   o   Work within realistic constraints, including schedule, resources, costs, and capabilities.
3. **Consensus and Collaboration**:
   o   Reach consensus with the team on methods, tools, facilities, task sequencing, scheduling, locations, division of tasks, and technical approaches.
   o   Work professionally, collaboratively, and with integrity to achieve reasonable compromises, mitigate conflicts, and resolve problems.
4. **Motivation and Attitude**:
   o   Stay motivated to achieve the best design and outcomes for the project.
   o   Maintain a positive attitude towards the project and fellow team members.

6.6.4   Attitude and Motivation

Attitude and motivation are critical attributes of successful capstone design teams. A positive attitude and high motivation levels are reflected in the quality of work and interactions within the team. Here are some key points to consider:

1. **Positive Attitude**:
   - o A positive attitude is contagious and energizing for everyone involved.
   - o It is noticed and appreciated by team members, sponsors, mentors, and professors.
2. **Motivation**:
   - o Motivation can be derived from successes, feedback, and progress in project tasks.
   - o For capstone design, good grades and evaluations are often the result of excellent technique, rigor, and the quality of work and results achieved.

6.6.5   Tips for Success in Engineering Design

Here are some additional tips for success in your capstone design project:

1. **Be Inspirational**:
   - o Communicate your passion for the project and its potential impact.
   - o Inspire others by sharing why the project is essential and how it can make a difference.
2. **Identify and Overcome Obstacles**:
   - o Be proactive in identifying obstacles early.
   - o Develop strategies to remove or work around these obstacles.
3. **Enjoy the Process**:
   - o Enjoy the experience of working on a meaningful project with your team.
   - o Appreciate the unique opportunity to collaborate in a supportive environment.
4. **Exhibit Good Work Habits**:
   - o Always be polite, rational, and calm.
   - o Listen to your team members, mentors, sponsors, and professors, and learn from their insights.
   - o Respect the opinions and wisdom of others.
   - o Accept responsibility or blame when appropriate.
5. **Effective Communication**:
   - o Communicate important information clearly during meetings and in written reports.
   - o Ensure that interactions with sponsors and mentors are professional and informative.
6. **Act with Integrity and Ethics**:
   - o Always uphold high ethical standards and integrity in all your actions and decisions.
7. **Respect and Cultivate Diversity**:
   - o Embrace diversity within your team to enhance creativity and innovation.

     ○ Leverage diverse perspectives to achieve better results for your project.

In capstone design projects, assigning and accepting responsibility is a shared effort among team members. Each member acts as a project manager for their assigned tasks, ensuring they are accountable and committed to the project's success. By fostering a positive attitude, staying motivated, and embracing collaboration and integrity, capstone design teams can achieve outstanding results and gain invaluable experience that prepares them for professional engineering practice.

6.7 Working with the Project Plan

Working with a project plan is an iterative and dynamic process. A project plan is not static; it requires continuous updates and refinements as the project progresses. This section covers how to effectively work with your project plan using Microsoft Project, including updating tasks, managing calendars, and utilizing different project views for analysis.

6.7.1 Iterative Nature of Project Planning

Project planning involves constant iteration. As the project unfolds, new information emerges, tasks evolve, and schedules may need adjustments. Here are critical activities involved in maintaining and refining a project plan:

1. **Updating Task Details**:
   - ○ Regularly update task descriptions, durations, and dependencies as more information becomes available.
   - ○ Adjust start and finish dates based on the progress and any changes in the project scope.
2. **Progress Tracking**:
   - ○ Continuously monitor the progress of each task.
   - ○ Use the "% Complete" field in MS Project to update the status of tasks.
   - ○ Identify any delays or issues early and make necessary adjustments.
3. **Adding Additional Details**:
   - ○ As tasks become clearer, add more detailed descriptions and requirements.
   - ○ Include any additional resources or constraints that were not initially identified.
4. **Financial Analysis and Cost Estimation**:
   - ○ Use the project plan to perform financial analysis and estimate costs.
   - ○ Regularly review and update the budget and resource allocations.

6.7.2 Utilizing Project Views in MS Project

MS Project provides various views to help analyze the project comprehensively:
1. **Gantt Chart View**:

o   This default view displays tasks as bars along a timeline.
o   Use it to see the overall project schedule, dependencies, and critical path.

2. **Task Usage View**:
    o   This view shows task details along with resource assignments.
    o   Useful for analyzing resource allocation and identifying potential overloads.

3. **Resource Usage View**:
    o   Focuses on resources rather than tasks.
    o   Helps in managing individual team members' workloads and availability.

4. **Calendar View**:
    o   Displays tasks on a calendar layout.
    o   Ideal for visualizing task schedules and deadlines.

5. **Timeline View**:
    o   Provides a high-level overview of the project timeline.
    o   Useful for communicating key milestones and deadlines to stakeholders.

6.7.3   Changing the Project Start Date

When you create a new project in MS Project, the start date is set to the current date by default. You may need to change this to reflect the actual project start date:

1. **Accessing Project Information**:
    o   Click on the "Project" tab.
    o   Select "Project Information" from the properties group, as shown in Figure 6-8..

2. **Setting the Start Date**:
    o   In the "Project Information" dialog box, change the "Start date" field to the desired start date of the project.
    o   Adjust the "Schedule from" option if necessary (usually set to "Project Start Date").

3. **Adjusting the Work Calendar**:
    o   Choose the appropriate work calendar for the project and team members.

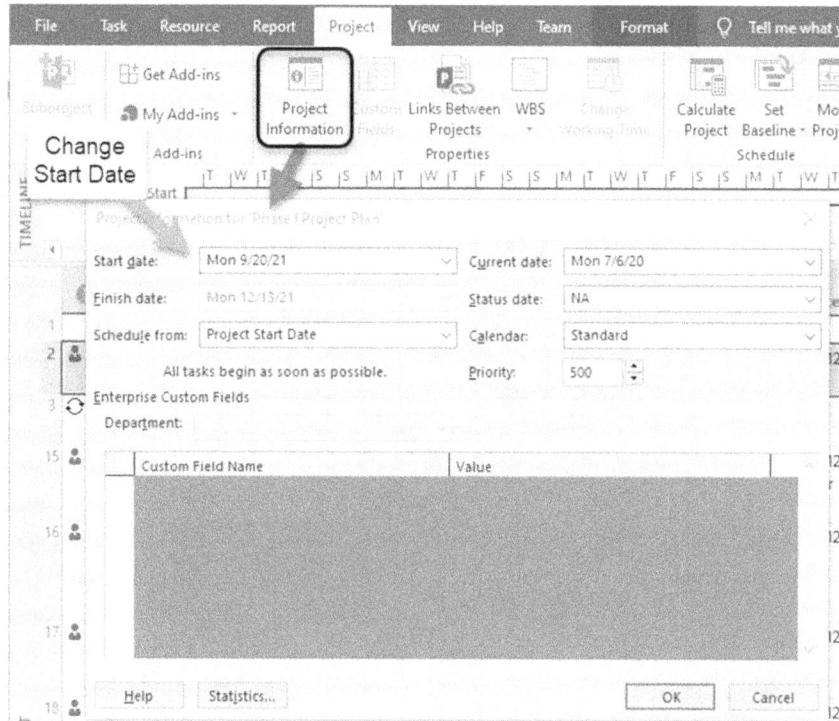

Figure 6-8. Changing project start date in Project Information.

6.7.4   Customizing the Project Calendar

As a student team, your availability might differ from a typical professional work calendar. MS Project allows customization of the calendar to fit your specific needs:

1. **Changing Working Time**:
    - Go to the "Project" tab and click on "Change Working Time."
    - This opens the Calendar and Project Options, as shown in Figure 6-9.
2. **Creating a Customized Calendar**:
    - Create a new calendar that reflects your team's availability.
    - Define working days, non-working days, and specific working hours.
3. **Setting Exceptions**:
    - Use the "Exceptions" tab in the "Change Working Time" dialog box to add exceptions (e.g., holidays, conferences).
    - Each team member can indicate personal exceptions, such as travel or other commitments.

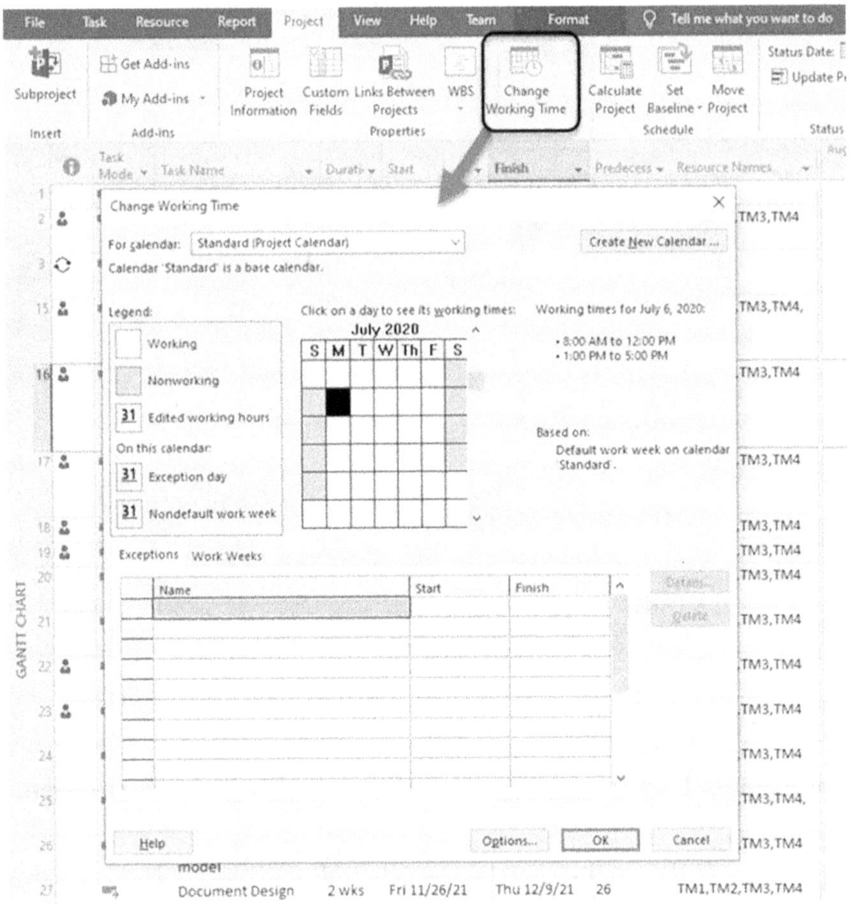

Figure 6-9. Changing working time for a resource.

6.7.5 Practical Steps for Updating the Project Plan

1. **Regular Updates**:
   - Schedule regular meetings to review and update the project plan.
   - Ensure all team members provide updates on their tasks and progress.
2. **Incorporating Changes**:
   - Adjust the project plan to incorporate any changes in scope, requirements, or timelines.
   - Use change management processes to document and approve significant changes.
3. **Communicating Updates**:
   - Share updated project plans with all stakeholders, including team members, sponsors, and professors.
   - Use project views and reports in MS Project to communicate updates effectively.

4. **Reviewing Project Performance**:
   - Regularly review the project's performance against the plan.
   - Identify any variances and take corrective actions to keep the project on track.

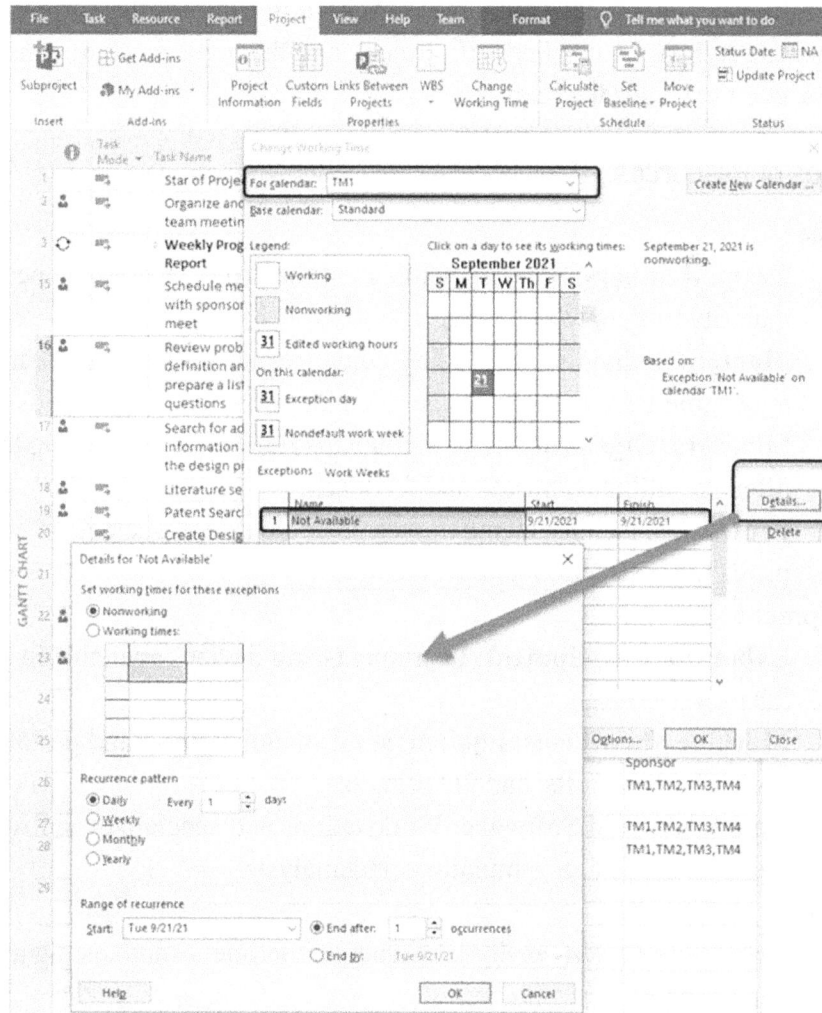

Figure 6-10. Team member customized calendar.

Working with a project plan is an ongoing process that requires attention to detail, regular updates, and effective communication. By utilizing the features and views in MS Project, capstone design teams can manage their projects efficiently, ensuring that tasks are completed on time and within budget. Customizing calendars and incorporating detailed updates help accommodate the unique schedules and requirements of student teams, leading to successful project outcomes.

6.8 Resources

Project resources encompass people, equipment, and facilities. Proper management of these resources is crucial for the successful completion of any project. In a capstone design project, understanding and effectively utilizing these resources can significantly influence the project's outcome. This section explores the various types of resources, how to set them up in Microsoft Project, and the importance of considering resource costs, even in an academic setting.

6.8.1   Types of Resources

1. **People**:
   - **Team Members**: Student members of the project team who perform the majority of the project tasks.
   - **Mentors**: Individuals providing guidance and advice to the team. They may come from the industry, academia, or other relevant fields.
   - **Sponsors**: Organizations or individuals funding or supporting the project, often providing resources and feedback.
   - **Professors**: Faculty members overseeing the capstone design course, providing instruction and evaluation.
2. **Equipment**:
   - **Laboratory Equipment**: Instruments and devices available in university labs.
   - **Machine Shop Tools**: Equipment for manufacturing and prototyping, such as lathes, mills, and 3D printers.
   - **Computers and Software**: Workstations and specialized software required for design, simulation, and analysis.
3. **Facilities**:
   - **Laboratories**: University labs where experiments and testing are conducted.
   - **Workshops**: Spaces where physical prototypes are built and tested.
   - **Classrooms and Meeting Rooms**: Areas used for team meetings, presentations, and collaboration.
4. **Supplies and Materials**:
   - **Consumables**: Items like raw materials, components, and office supplies required for the project.

6.8.2   Setting Up Resources in MS Project

Microsoft Project provides robust functionality for setting up and managing project resources. Here's how to set up resources in MS Project:
1. **Accessing the Resource Tab**:

o Click on the "Resource" tab to access resource-related features, as shown in Figure 6-11.

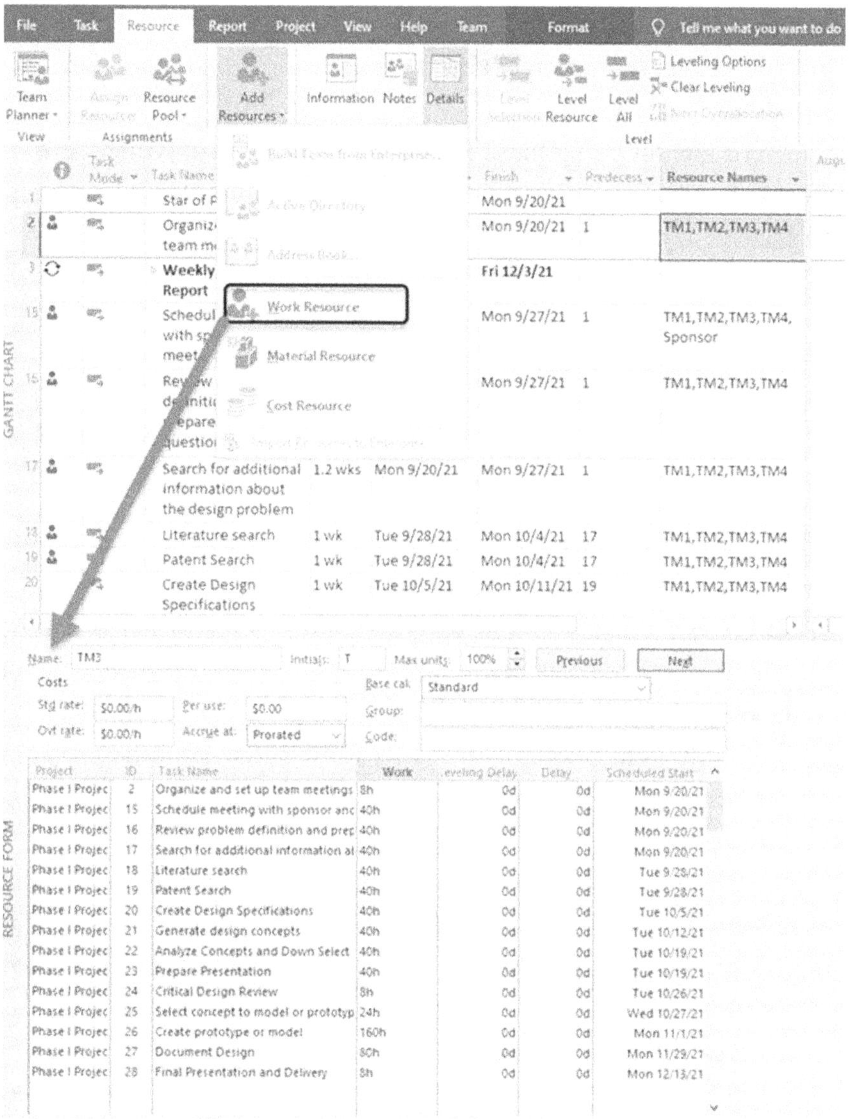

Figure 6-11. Setting up project resources.

2. **Adding People Resources**:
   o Click on "Resource Sheet" in the Resource tab to open the resource sheet view.
   o Enter the names of all team members, mentors, sponsors, and professors.
   o Set the availability of each person's resource by specifying their work schedules and any exceptions (e.g., holidays, travel).

3. **Specifying Resource Costs**:
   o Although team members are not typically paid, entering a nominal pay rate can help in financial analysis.

- o For mentors, sponsors, and professors, you may enter an estimated hourly rate based on industry standards.

4. **Adding Equipment and Facilities**:
   - o List all equipment and facilities used in the project.
   - o Include details such as availability, usage rates, and maintenance schedules.
   - o Specify costs for using these resources, even if the university provides them at no charge. This provides a more accurate financial analysis of the project.

5. **Including Supplies and Materials**:
   - o Enter all consumable items required for the project.
   - o Specify quantities, costs, and availability dates to ensure timely procurement and usage.

6.8.3   Importance of Resource Cost in Financial Analysis

Incorporating resource costs in your project plan is essential, even in a university setting where direct costs might not be incurred. Here's why:

1. **Financial Analysis**:
   - o Including resource costs helps in performing a thorough financial analysis of the project.
   - o It allows you to understand the actual value and cost of the project, which can be beneficial for sponsors and stakeholders.

2. **Resource Allocation**:
   - o Understanding resource costs can help in better allocation and utilization of resources.
   - o It ensures that high-cost resources are used efficiently and justifiably.

3. **Project Valuation**:
   - o Assigning costs to resources provides a clear picture of the project's value.
   - o It aids in evaluating the project's return on investment (ROI) and overall feasibility.

4. **Budgeting and Funding**:
   - o Accurate resource costing can help in preparing project budgets and seeking additional funding if necessary.
   - o It enables the team to justify their resource needs to sponsors and funding bodies.

6.8.4   Practical Steps in MS Project

1. **Resource Sheet Setup**:
   - o Open the Resource Sheet view from the Resource tab.
   - o Enter all relevant information for each resource, including name, type

(work, material, cost), and availability.

2. **Assigning Costs**:
   - For each resource, enter cost rates under the "Standard Rate" and "Overtime Rate" columns.
   - For material resources, specify the unit cost.
3. **Managing Resource Calendars**:
   - Customize working time for each resource by clicking "Change Working Time" in the Project tab.
   - Define standard working hours, non-working days, and exceptions for each resource.
4. **Resource Allocation**:
   - Assign resources to tasks by selecting the task and then using the "Assign Resources" button in the Task tab.
   - Ensure that resource allocation matches the availability and does not overburden any single resource.
5. **Review and Adjust**:
   - Regularly review resource allocation and costs to ensure optimal usage.
   - Adjust assignments and schedules as necessary to address any conflicts or overallocations.

Effective resource management is crucial for the success of a capstone design project. By setting up and managing people, equipment, facilities, and materials in MS Project, teams can ensure that all necessary resources are accounted for and utilized efficiently. Including resource costs in the project plan enhances financial analysis and provides a realistic view of the project's value. Through careful planning and continuous monitoring, teams can optimize resource usage, maintain project schedules, and achieve their project goals successfully.

6.9   Gantt Chart

A Gantt chart is a widely used project management tool that provides a visual representation of a project schedule. It displays tasks, their durations, dependencies, and assigned resources in a graphical bar chart format. Named after Henry Gantt, who developed it between 1910 and 1915, the Gantt chart is a powerful tool for planning, scheduling, and tracking project progress. In Microsoft Project, the Gantt chart is the default view, offering an intuitive interface for managing and sharing project information.

6.9.1   Key Components of a Gantt Chart

1. **Time Scale**:
   - The horizontal axis of a Gantt chart represents time, which can be scaled to various resolutions (e.g., days, weeks, months).

- o The time scale can be modified by double-clicking on the timeline bar above the chart, allowing customization to fit different needs, such as daily or monthly views.

2. **Tasks and Task Bars**:
   - o Tasks are listed on the left side of the chart, with corresponding bars on the right side representing the duration of each task.
   - o Taskbars are color-coded to enhance visual communication and differentiation between tasks.

3. **Dependencies**:
   - o Arrows between task bars indicate dependencies, showing the order in which tasks must be completed.
   - o Dependencies are crucial for understanding the sequence of tasks and identifying the critical path.

4. **Milestones**:
   - o Milestones are significant points in the project timeline, often representing the completion of critical phases or deliverables.
   - o Milestones are displayed as diamonds on the Gantt chart.

5. **Resources**:
   - o Resources assigned to tasks are listed alongside task names, indicating who is responsible for each task.
   - o This helps in resource management and ensures that tasks are appropriately allocated.

6.9.2   Sample Gantt Chart for Phase I of Capstone Design:

Figure 6-12. illustrates a sample Gantt chart for Phase I of a capstone design project. The chart shows tasks, durations, dependencies, and resources in a clear and organized manner.

6.9.3   Practical Use of Gantt Charts

1. **Project Planning**:
   - o During the planning phase, tasks are entered into MS Project, and their durations and dependencies are defined.
   - o The Gantt chart helps visualize the entire project timeline, making it easier to identify potential bottlenecks and optimize task sequences.

2. **Progress Tracking**:
   - o As the project progresses, the Gantt chart is updated with the actual start and finish dates, as well as task completion percentages.
   - o This allows the team to monitor progress against the original plan and make necessary adjustments.

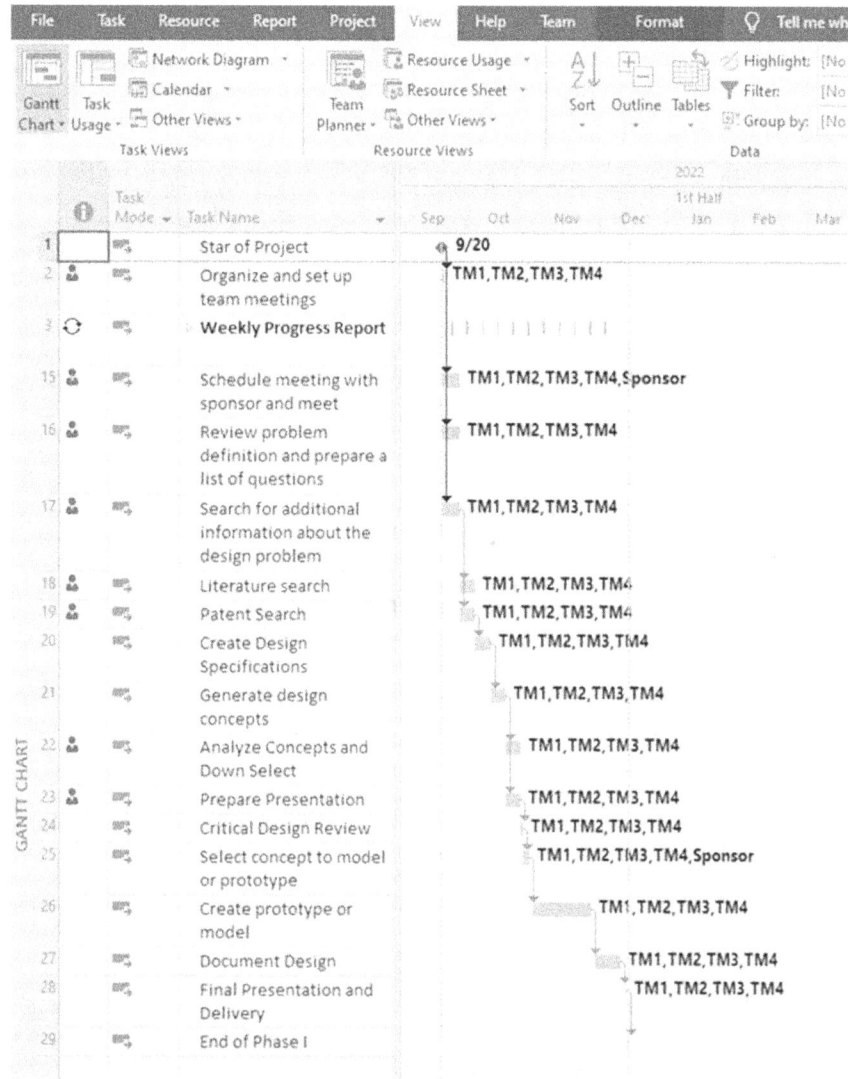

Figure 6-12. Sample capstone design Phase I Gantt chart.

3. **Communication**:
   - Gantt charts are helpful for communicating project status to stakeholders, including team members, sponsors, and professors.
   - They can be printed on large format paper for display in meeting rooms or shared electronically.

4. **Customizing the Gantt Chart**:
   - Customize the Gantt chart by changing the time scale, color-coding tasks, and adding notes to task bars.
   - Use the "Format" tab in MS Project to apply these customizations.

6.9.4   Network Diagram and Optimizing the Project Plan

While Gantt charts provide a clear visual representation of the project schedule, network diagrams offer an alternative view focusing on task dependencies and the flow of activities. Network diagrams, also known as PERT charts, help identify critical paths and optimize project execution.

6.9.4.1   Network Diagrams

1. **Task Dependencies**:
   - Network diagrams emphasize the dependencies between tasks, showing the flow from one task to another.
   - This helps in understanding the logical sequence of activities and identifying potential delays.

2. **Critical Path Analysis**:
   - The critical path is the most extended sequence of tasks that determines the minimum project duration.
   - Tasks on the critical path must be completed on time to avoid delaying the entire project.

3. **Slack Time**:
   - Slack time, or float, is the amount of time a task can be delayed without affecting the project's end date.
   - Identifying slack helps in reallocating resources and optimizing the schedule.

6.9.4.2   Creating a Network Diagram in MS Project:

1. **Accessing the Network Diagram**:
   - Click on the "View" tab and select "Network Diagram" to switch from the Gantt chart view to the network diagram view.
   - The diagram displays tasks as nodes connected by arrows representing dependencies.

2. **Analyzing the Critical Path**:
   - Highlight the critical path by selecting "Critical Tasks" under the "Format" tab.
   - Tasks on the critical path are displayed in a different color, making them easily identifiable.

3. **Optimizing Task Dependencies**:
   - Examine task dependencies and look for opportunities to perform tasks in parallel or adjust lag times.
   - Use the "Task Information" dialog box to change dependencies and fine-tune the schedule.

Sample Network Diagram:
Figure 6-13. shows a network diagram for a sample Phase I project plan. It highlights the critical path and illustrates task dependencies.

6.9.5 Practical Steps for Using Gantt Charts and Network Diagrams

1. **Initial Setup**:
    - o Enter all tasks, durations, and dependencies in MS Project.
    - o Use the Gantt chart view to visualize the project timeline and adjust as needed.
2. **Regular Updates**:
    - o Update task statuses regularly to reflect actual progress.
    - o Adjust the schedule to accommodate changes and unforeseen delays.
3. **Communication and Reporting**:
    - o Use Gantt charts and network diagrams to communicate project status to stakeholders.
    - o Print large format charts for display or share digital versions for remote collaboration.
4. **Continuous Optimization**:
    - o Regularly analyze the critical path and slack time to identify opportunities for optimization.
    - o Make adjustments to task sequences and resource allocations to keep the project on track.

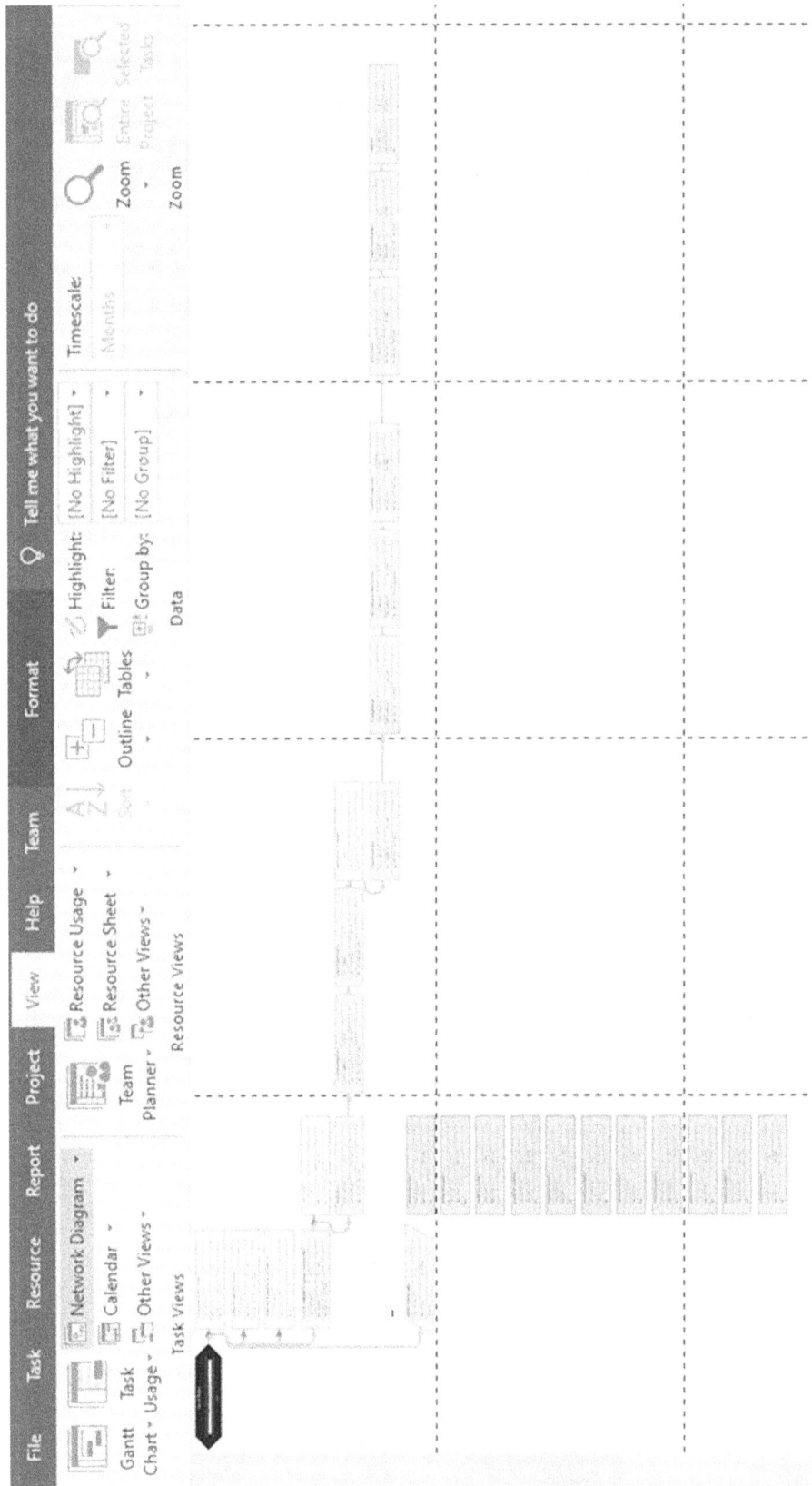

Figure 6-13. Sample capstone design Phase I network diagram chart.

6.10 Resource Leveling

Resource leveling is a crucial process in project management that ensures no resource is over-allocated or underutilized throughout the project. When assigning resources in Microsoft Project, the default assumption is a full-time commitment to the task. For instance, assigning Team Member 1 (TM1) to a patent search task assumes an eight-hour workday dedicated solely to that task. This assumption is often unrealistic, especially in a capstone design project where students juggle multiple responsibilities. Resource leveling helps distribute the workload evenly across all available resources, optimizing their use and ensuring the project stays on track.

6.10.1   Understanding Resource Leveling

1. **Adjusting Time Commitment**:
   - Initially, MS Project assigns a full-time commitment (8 hours/day) to tasks. This default setting must be adjusted to reflect actual availability.
   - Team members can be assigned to multiple tasks simultaneously, requiring adjustments to ensure no one exceeds their available working hours.
2. **Reassigning Tasks**:
   - If a team member is over-allocated, tasks must be reassigned to other members to balance the workload.
   - This iterative process of balancing and equalizing the workload across all resources is known as resource leveling.

6.10.2   Practical Steps for Resource Leveling in MS Project

1. **Initial Data Entry**:
   - Enter initial data for team member participation, which may be estimated or left as default.
   - As the team reviews and starts working with the project plan, these estimates need correction and fine-tuning.
2. **Task Usage View**:
   - The Task Usage view (Figure 6-14.) is useful for viewing and correcting workload estimates for team members.
   - It displays tasks assigned to each resource and allows adjustments to ensure no one exceeds their total available time.

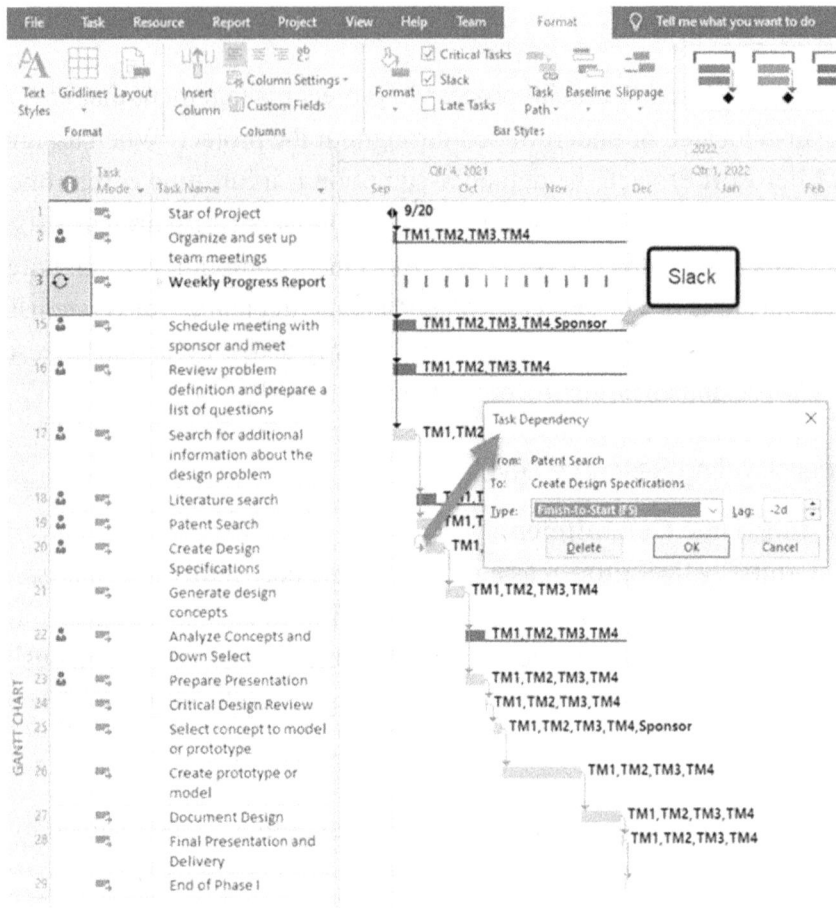

Figure 6-14. Critical path and tasks with slack on the Gantt chart.

3. **Team Planner View**:
   o The Team Planner view (Figure 6-15.) helps team members see their task assignments on the project timeline.
   o It allows them to compare their schedules with other commitments and adjust their engagement with project tasks accordingly.

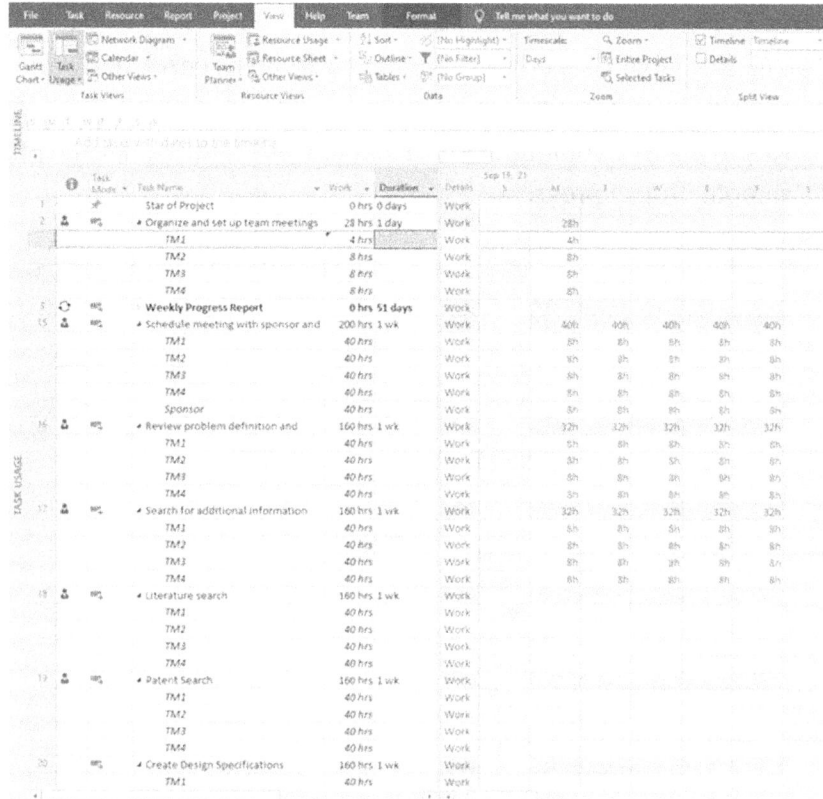

Figure 6-15. Task usage view.

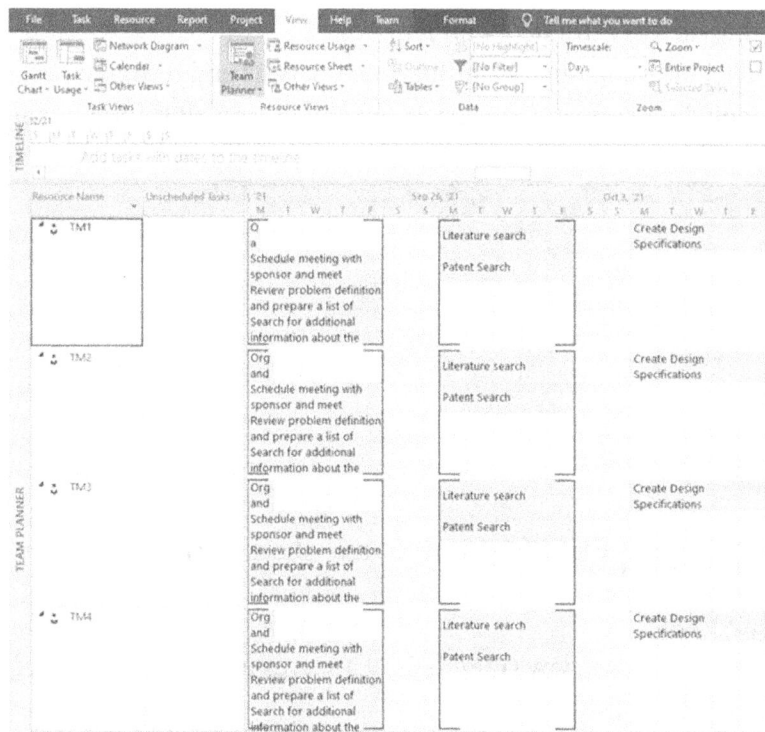

Figure 6-16. Team planner view.

4. **Resource Usage View**:
   - The Resource Usage view (Figure 6-17.) helps check daily allocations and identify overcommitments.
   - Resolving overcommitments quickly avoids bottlenecks and ensures smooth project progress.

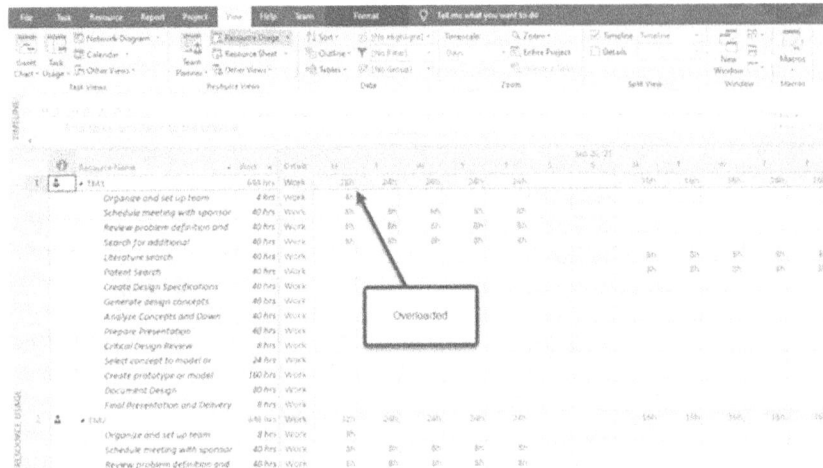

Figure 6-17. Resource usage view.

5. **Resource Sheet View**:
   - Use the Resource Sheet view (Figure 6-18.) to enter hourly rates for individuals affiliated with the project.
   - This information aids in financial analysis and resource cost accounting, including software, materials, machines, travel, shipping, and services.

Figure 6-18. Resource sheet view.

1. **Calendar View**:
   - The Calendar view (Figure 6-19.) serves as a reminder for current and upcoming tasks.
   - It is useful for progress tracking and reporting, ensuring team members are aware of their responsibilities and deadlines.

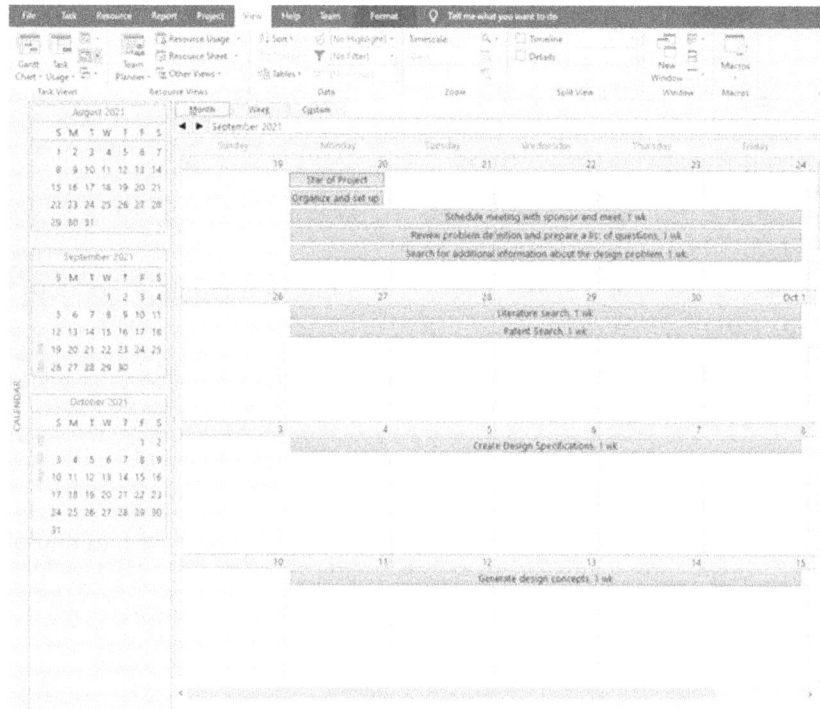

Figure 6-19. Calendar view.

6.10.3 Implementing Resource Leveling

1. **Adjusting Work Hours**:
   - Change the default 40-hour workweek to a more realistic estimate for student workload, such as 12 hours per week.
   - Use the Task Usage view to adjust individual task allocations based on this new estimate.
2. **Balancing Workloads**:
   - Regularly review the Resource Usage view to identify and resolve over-allocations.
   - Reassign tasks or adjust durations to ensure an even distribution of work.
3. **Using MS Project Features**:
   - Utilize the "Level Resources" tool in MS Project to adjust assignments and resolve conflicts automatically.
   - Access this tool under the "Resource" tab and configure settings to suit your project needs.
4. **Continuous Monitoring**:
   - Continuously monitor resource allocations using the various views in MS Project.
   - Make adjustments as necessary to accommodate changes in project scope,

availability, or resource performance.

6.10.4   Benefits of Resource Leveling

1. **Optimized Resource Utilization**:
   o Ensures all resources are used efficiently without overburdening any individual.
   o Balances workloads to prevent burnout and maintain productivity.
2. **Improved Project Planning**:
   o Provides a realistic view of resource availability and workload distribution.
   o Enhances the accuracy of project schedules and timelines.
3. **Financial Accuracy**:
   o Accurate resource costing and allocation improve financial analysis and budgeting.
   o Helps in preparing detailed reports for stakeholders, sponsors, and professors.
4. **Enhanced Collaboration**:
   o Clear visibility of task assignments and schedules promotes better collaboration among team members.
   o Encourages accountability and responsibility within the team.

Resource leveling is an essential process in project management, ensuring that no resource is over-allocated or underutilized. By effectively managing and balancing workloads, capstone design teams can optimize their resources, maintain project schedules, and achieve their project goals. Utilizing the various views and tools in MS Project, such as Task Usage, Team Planner, and Resource Usage, teams can continuously monitor and adjust their resource allocations to ensure successful project completion.

6.11 Project Reports

A comprehensive project plan guides the entire execution of the project. This plan, known as the base project plan, is created and refined by the team to track progress on each task. Regular updates to the project plan are crucial to reflect actual performance, identify deviations, and make necessary adjustments. Project reports generated from this updated plan provide valuable insights into the project's status, resource utilization, and financial health. This section explores the importance of project reports, the types of reports available in Microsoft Project, and how to use these reports effectively.

6.11.1   Importance of Project Reports

1. **Tracking Progress**:
   o Project reports track the progress of tasks against the base project plan.

o Regular updates, ideally on a weekly basis, ensure that the project stays on schedule and that any deviations are promptly addressed.

2. **Identifying Variances**:
   o Reports help identify variances between the actual project performance and the planned performance.
   o This allows the team to analyze the reasons for deviations and take corrective actions.

3. **Resource Management**:
   o Reports provide insights into resource utilization, helping to manage people, equipment, and other resources effectively.
   o They highlight any changes in resource availability and their impact on the project.

4. **Financial Analysis**:
   o Cost and cash flow reports offer a detailed view of the project's financial status.
   o They help in budgeting, financial planning, and ensuring the project remains within its financial constraints.

6.11.2   Guiding Questions for Analyzing the Project Plan

To ensure the project is on track and to identify areas that need attention, the team should regularly ask the following questions:

1. **Project Status**:
   o Is the project on track?
   o Will the project be completed before the deadline?
   o Have the duration and cost of the project changed? If so, what has changed?

2. **Resource Availability**:
   o Have there been any changes in the availability of resources (people, funding, equipment, software)?

3. **Milestones and Requirements**:
   o Are the capstone course milestones and requirements achievable with the current project plan?

4. **Plan Optimization**:
   o What changes are needed to the project plan?
   o Can the plan be further optimized?

Answering these questions helps the team measure progress and identify necessary changes or corrections to the project plan.

6.11.3   Types of Project Reports in Microsoft Project

Microsoft Project offers several report views that are incredibly informative and useful for progress reports, design presentations, and design reports. Below are critical types of project reports and their uses:

1. **Project Overview Report**:
   o   Provides a high-level summary of the project's status, including critical metrics such as task completion, milestones, and deadlines (Figure 6-20.).
   o   Useful for presenting the overall progress to stakeholders and identifying critical issues that need attention.

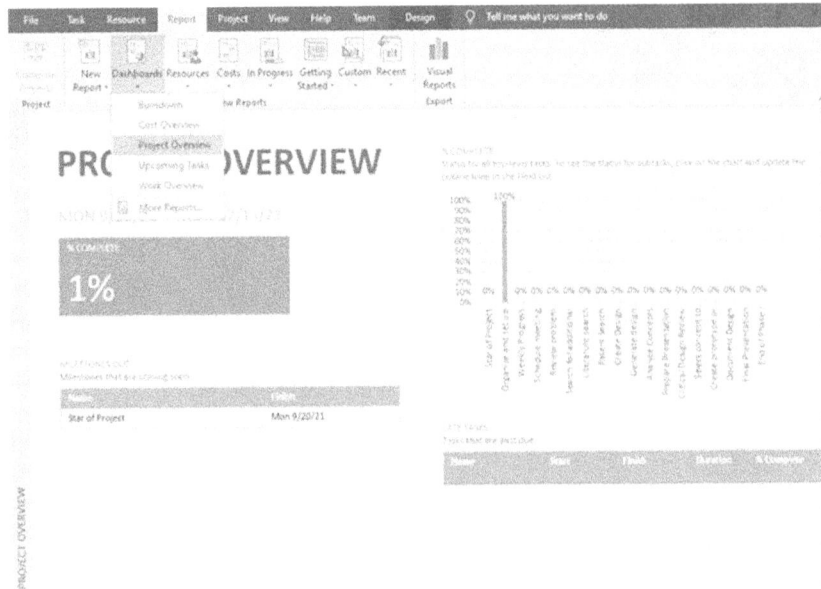

Figure 6-20. Project overview report.

2. **Cost Overview Report**:
   o   Summarizes the project's costs, including planned versus actual costs and variances (Figure 6-21.).
   o   Helps track the financial health of the project and ensure that it stays within budget.
3. **Cash Flow Report**:
   o   Provides a detailed view of the project's cash flow, showing the inflow and outflow of funds over time (Figure 6-22.).
   o   Essential for financial planning and ensuring that the project has sufficient funds at each stage.

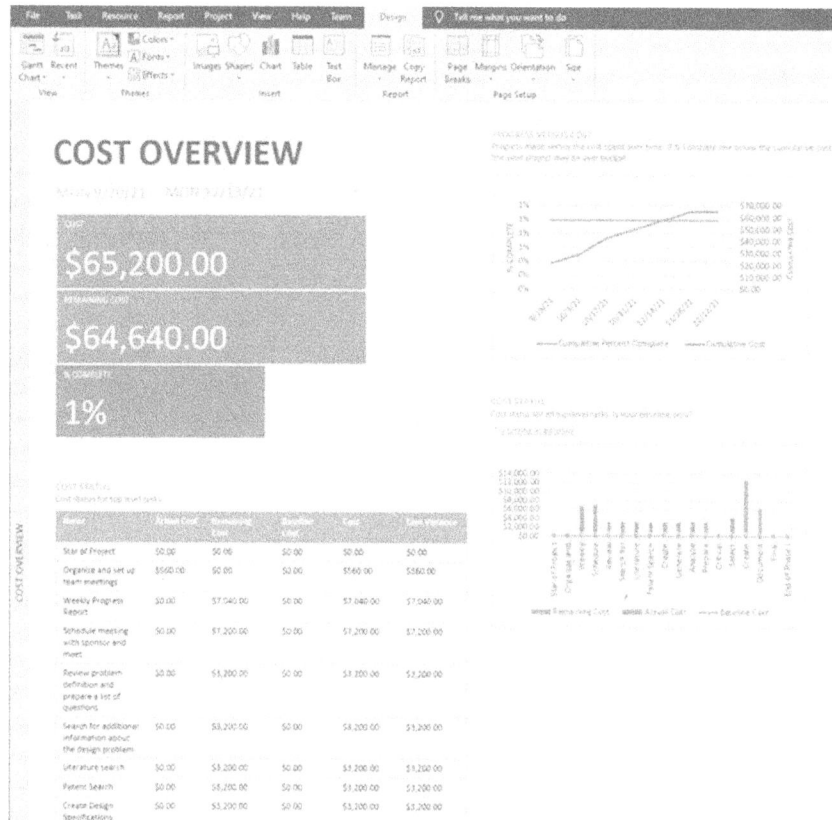

Figure 6-21. Project cost overview report.

6.11.4   Using Reports for Project Management

1. **Regular Updates**:
   - Update the project plan regularly, preferably weekly, to reflect actual progress and any changes in tasks or resources.
   - Ensure that the updated project plan is attached to weekly progress reports submitted to sponsors and professors.

2. **Analyzing Reports**:
   - Use the project overview report to get a snapshot of the project's status.
   - Analyze the cost overview report to track financial performance and identify any budget issues.
   - Review the cash flow report to ensure the project has sufficient funds and to plan for future financial needs.

3. **Communicating with Stakeholders**:
   - Share relevant reports with stakeholders to keep them informed about the project's progress and any issues that need their attention.

    o Use visual aids such as charts and graphs to enhance the clarity and impact of the reports.

4. **Making Informed Decisions**:
    o Use the insights gained from the reports to make informed decisions about resource allocation, task prioritization, and schedule adjustments.
    o Continuously refine the project plan based on the feedback and data from the reports.

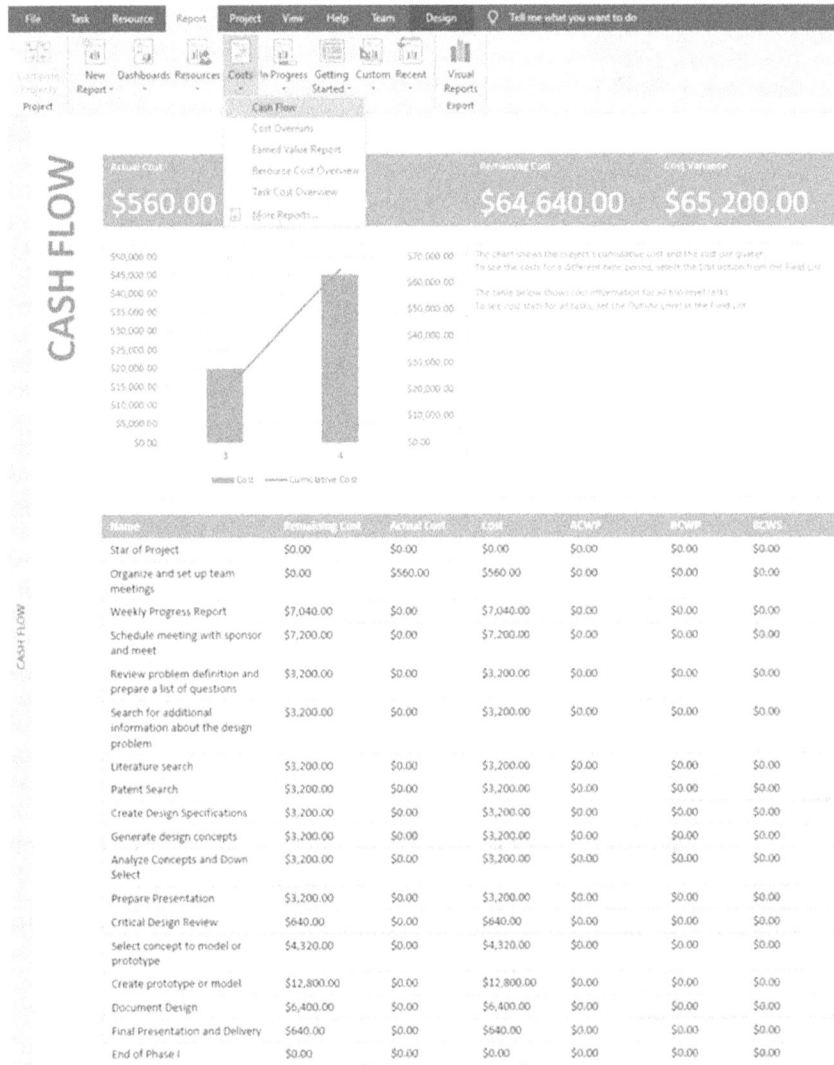

Figure 6-22. Project cash flow report.

6.11.5   Practical Steps for Generating Reports in MS Project

1. **Accessing the Reports Tab**:
   - Click on the "Report" tab to access various reporting options.
   - Select the type of report you need from the available categories, such as "Overview," "Cost," and "Cash Flow."

2. **Customizing Reports**:
   - Customize the reports to include specific data and visualizations that are relevant to your project.
   - Use the "Design" tab to modify the layout, add charts, and adjust the formatting.

3. **Exporting and Sharing Reports**:
   - Export reports to PDF or other formats for easy sharing with stakeholders.
   - Print reports for meetings or presentations to provide a clear and professional summary of the project status.

4. **Continuous Improvement**:
   - Regularly review and improve your reporting process to ensure that the reports provide accurate and actionable insights.
   - Seek feedback from stakeholders to enhance the usefulness of the reports.

Project reports are essential tools for tracking progress, managing resources, and ensuring financial control in capstone design projects. By regularly updating the project plan and generating detailed reports, teams can stay informed about the project's status, identify issues early, and make data-driven decisions. Microsoft Project provides a range of reporting options that are easy to customize and share, making it an invaluable tool for effective project management. Through diligent use of project reports, capstone design teams can enhance their planning, execution, and communication, leading to successful project outcomes.

6.12 Assignments

Assignment 6-1: Create an Initial Project Management Plan for Your Capstone Design Project

Objective: Develop a comprehensive initial project management plan for your capstone design project. The plan should outline the project timeline, key milestones, tasks, dependencies, resource allocation, and team responsibilities. By creating a well-structured project management plan, you will establish a roadmap for successful project execution and ensure effective collaboration among team members.

Instructions:

1. Review Project Requirements
   - Carefully review the project description, objectives, and any provided guidelines.
   - Identify the key deliverables and milestones specific to your project.
2. Define Project Phases and Milestones
   - Break down the project into distinct phases, such as:
     - Project Initiation
     - Research and Problem Definition
     - Concept Generation and Selection
     - Design and Development
     - Testing and Refinement
     - Final Presentation and Report
   - Identify the significant milestones within each phase, including:
     - Project Qualification Application
     - Team Assignment
     - Literature Search and Patent Search
     - Problem Definition
     - Customer Requirements and Engineering Requirements
     - Design Specification
     - Concept Generation and Pugh Analysis
     - QFD Analysis
     - Critical Design Review Presentation
     - Proof of Concept Design and Presentation
     - Preliminary Design Report
     - Peer Evaluation
3. Create a Project Timeline
   - Determine the start and end dates for each project phase and milestone.
   - Establish a timeline that reflects the project's overall duration and critical dates.
   - Consider the project work calendar, including any holidays, breaks, or other constraints.

4. Identify Tasks and Dependencies
    - Break down each project phase and milestone into specific tasks.
    - Identify the dependencies between tasks, indicating which tasks must be completed before others can begin.
    - Estimate the duration of each task based on its complexity and the resources required.

5. Assign Resources and Responsibilities
    - Identify the team members responsible for each task.
    - Allocate resources, such as equipment, materials, or software, needed for each task.
    - Ensure a balanced distribution of workload among team members based on their skills and availability.

6. Develop a Resource Plan
    - Create a detailed resource plan that includes:
        - Human resources: Team members' roles, responsibilities, and time allocation.
        - Financial resources: Budget for materials, equipment, software, or other expenses.
        - Physical resources: Workspace, tools, or facilities required for the project.

7. Establish a Communication Plan
    - Define how the team will communicate and share information throughout the project.
    - Set up regular team meetings, progress updates, and status reports.
    - Determine the tools and platforms to be used for communication and file sharing (e.g., email, collaboration software, shared drives).

8. Create a Risk Management Plan
    - Identify potential risks and challenges that may impact the project.
    - Assess the likelihood and impact of each risk.
    - Develop contingency plans and mitigation strategies to address identified risks.

9. Prepare the Project Management Plan Document
    - Use a suitable template or format for your project management plan document.
    - Include the following sections:
        - Project Overview and Objectives
        - Project Phases and Milestones
        - Project Timeline
        - Task Breakdown and Dependencies
        - Resource Allocation and Responsibilities

- Communication Plan
- Risk Management Plan
  - Ensure that the document is well-organized, clear, and concise.

10. Review and Refine the Plan
    - Review the project management plan with your team members.
    - Gather feedback and make necessary revisions to ensure everyone is aligned and agrees with the plan.
    - Ensure that the plan is realistic, achievable, and aligned with the project requirements and constraints.

11. Submit and Update the Plan
    - Submit the initial project management plan as an attachment with your first weekly progress report on Brightspace.
    - Regularly update the plan throughout the project to reflect progress, changes, or adjustments.
    - Include the updated project management plan with each subsequent weekly progress report.

Remember, the project management plan is a living document that should be regularly reviewed and updated as the project progresses. It serves as a guide to keep your team organized, on track, and accountable. By proactively managing tasks, dependencies, resources, and risks, you will increase the likelihood of completing your capstone design project.

Assignment 6-2: Weekly Progress Report

Objective: Create a comprehensive weekly progress report for your capstone design project using the provided template on Brightspace. The report should provide a detailed account of the team's efforts, accomplishments, challenges, and plans for the upcoming week. By regularly submitting progress reports, you will keep your project sponsor and instructors informed about your project's status and ensure that your team stays on track.

Instructions:

1. Access the Progress Report Template
   - Navigate to the designated assignment on Brightspace.
   - Download the provided progress report template.
   - Familiarize yourself with the sections and structure of the template.

2. Individual Team Member Updates
   - Each team member should fill out their section in the template, providing a detailed account of their efforts and contributions during the reporting period.
   - Include the following information for each team member:
     - Name and Role
     - Tasks Completed
     - Time Spent on Each Task
     - Challenges Encountered (if any)
     - Plans for the Upcoming Week
   - Encourage team members to be specific and provide enough detail to convey their progress and contributions clearly.

3. Team Accomplishments
   - In the designated section of the template, summarize the key accomplishments and progress made by the team as a whole during the reporting period.
   - Highlight any significant milestones reached or deliverables completed.
   - Provide a brief description of how these accomplishments align with the project's overall goals and timeline.

4. Challenges and Issues
   - Use the provided section in the template to identify any challenges, problems, or delays encountered during the reporting period.
   - Describe the impact of these challenges on the project's progress and timeline.
   - Be specific and provide enough context for the project sponsor and instructors to understand the nature and severity of the issues.

5. Proposed Solutions and Mitigation Plans
   o In the corresponding section of the template, propose potential solutions or mitigation strategies for each challenge or issue identified.
   o Explain how these solutions will address the problems and get the project back on track.
   o Assign responsibilities and timelines for implementing the proposed solutions.
6. Plans for the Upcoming Week
   o Fill out the section in the template outlining the team's plans and objectives for the upcoming week.
   o Break down the planned tasks and activities for each team member.
   o Set specific goals and milestones to be achieved during the next reporting period.
   o Ensure that the plans align with the project's overall timeline and goals.
7. Attachments
   o If applicable, include any relevant attachments, such as:
     ▪ Updated project timeline or Gantt chart
     ▪ Technical drawings, diagrams, or schematics
     ▪ Meeting minutes or notes
     ▪ Supporting documents or research materials
   o Ensure that the attachments are clearly labeled and referenced within the appropriate section of the progress report template.
8. Review and Finalize
   o As a team, review the completed progress report to ensure accuracy, clarity, and completeness.
   o Make any necessary revisions or additions based on team feedback.
   o Proofread the report for spelling, grammar, and formatting errors.
9. Submit the Progress Report
   o Upload the final progress report, using the provided template, to the designated assignment on Brightspace.
   o Send a copy of the progress report to your project sponsor via email.
   o Ensure that the report is submitted by the specified deadline each week.
10. Continuously Improve
    o Reflect on the feedback received from your project sponsor and instructors regarding your progress reports.
    o Identify areas for improvement and make necessary adjustments to enhance the quality and effectiveness of future reports.
    o Continuously refine your reporting process and content based on lessons learned and best practices.

By consistently submitting detailed and informative weekly progress reports using the provided template on Brightspace, you will maintain clear communication with your project sponsor and instructors, identify and address challenges in a timely manner, and ensure that your capstone design project stays on track. Regular reporting also promotes accountability and helps your team stay focused on achieving your project goals. Adhering to the given template ensures consistency and clarity in your progress reports throughout the project.

7 Defining the Design Problem

Capstone design problem ideas originate from various sources, often involving sponsors or motivators who identify a need or desire to create or improve something. The motivation behind these problems is to solve an issue, fix a perceived defect, or enhance a system or device. The initial problem statement is usually vague, representing an open-ended challenge with multiple potential solutions. This vagueness is intentional, allowing for creativity and innovation in finding the best solution within the given (or sometimes undefined) constraints. The first task for the design team is to understand the problem domain thoroughly and resolve any unknown aspects through critical thinking, research, and iteration.

7.1   The Importance of Problem Definition

Defining the problem is arguably the most critical aspect of any project. Engineering students are often used to well-defined problems with a single correct solution. However, capstone design projects present a different challenge, where the problem definition itself is incomplete, requiring the team to identify what is missing and gather additional information to understand the design challenge fully. This process involves:

- Learning the problem domain
- Identifying missing pieces of information
- Engaging with sponsors and experts
- Conducting research and iterative refinement

7.2   Methods for Understanding Open-Ended Problems

There are several approaches to understanding open-ended problems. Here are some effective techniques to gather more information about the design problem:

7.2.1   Brainstorming

Brainstorming is a well-known method for group problem-solving, introduced by Alex Osborne in his 1953 book "Applied Imagination: Principles and Procedures of Creative Thinking." Osborne described brainstorming as a method where an organized group can generate more and better ideas than individuals working alone. The fundamental principles of effective brainstorming include:

- **Focus on Quantity**: Encourage the generation of a large number of ideas without worrying about their quality initially.
- **Withhold Criticism**: Avoid any form of criticism during the idea generation phase to encourage free thinking and the sharing of all ideas.
- **Welcome Unusual Ideas**: Encourage creative and far-fetched ideas that may lead to innovative solutions.
- **Combine and Improve Ideas**: Build upon the ideas presented by others to develop them further.
- **Example: Designing a Five-Axis 3D Printer**

A team of five students was tasked with designing a five-axis 3D printer for the Naval Undersea Warfare Center in Newport, Rhode Island. The motivation was to create a printer capable of printing complex geometries without relying on a fixed z-direction. The brainstorming session led to several innovative ideas that eventually shaped the final design.

7.3   Attributes of a Good Problem Definition

A well-defined design problem should include:

- **Clarity of Needs**: The problem should clearly state the needs and requirements that the design will address.
- **Achievability**: The design should be feasible and within the capabilities of the team to develop and manufacture.
- **Maintenance**: The design should require minimal maintenance, calibration, and upkeep.
- **User-Friendly**: The design should be easy to use, and minimal training is required.
- **Cost-Effective**: The design should save materials and money for the sponsor.
- **Error-Proof**: The design should minimize the possibility of user errors.

7.4   Common Pitfalls in Problem Definition

Poorly defined problems can lead to various issues, including:

- **Over-Design**: Wasting materials and resources on unnecessary features.
- **Excessive Precision**: Imposing unnecessary tight tolerances that complicate manufacturing.

- **Untested Solutions**: Implementing methods beyond the team's expertise.
- **Limited Functionality**: Failing to meet the full scope of user needs.
- **Behavioral Changes**: Requiring users to change their behavior significantly to use the design.
- **High Costs**: Resulting in a design that is too expensive to be practical.
- **Limited Use**: Creating a design that is not widely applicable or available.

7.4.1   Research and Iteration

To define the problem accurately, the team must engage in thorough research and iterative refinement. This process includes:

- **Literature Review**: Conducting a comprehensive review of existing literature, including academic papers, articles, and patents related to the problem domain.
- **Expert Interviews**: Engaging with experts in the field to gain insights and gather additional information.
- **User Surveys**: Conducting surveys with potential users to understand their needs and preferences.
- **Competitive Analysis**: Analyzing similar products or solutions to identify strengths and weaknesses.
- **Prototyping and Testing**: Developing prototypes and testing them to validate assumptions and refine the problem definition.

7.5   Engaging with Sponsors and Stakeholders

Engaging with sponsors and stakeholders is crucial for gathering additional information and refining the definition of the problem. This process involves:

- **Initial Meetings**: Holding initial meetings with sponsors to discuss the problem, gather their perspectives, and understand their expectations.
- **Regular Updates**: Providing regular updates to sponsors and stakeholders to keep them informed about the progress and gather feedback.
- **Collaborative Workshops**: Organizing workshops and brainstorming sessions with sponsors and stakeholders to generate ideas and refine the problem definition.

7.6   Tools and Techniques for Problem Definition

Several tools and techniques can assist in defining the design problem, including:

- **Mind Mapping**: Creating mind maps to visually organize information and identify relationships between different aspects of the problem.
- **SWOT Analysis**: Conducting a SWOT (Strengths, Weaknesses, Opportunities, Threats) analysis to evaluate the problem from different perspectives.
- **Root Cause Analysis**: Using techniques like the "5 Whys" to identify the root

causes of the problem.
- **Fishbone Diagrams**: Developing fishbone diagrams (also known as Ishikawa diagrams) to explore the potential causes of the problem.

7.7 Example Case Study: Defining a Capstone Design Problem

Consider a capstone design project aimed at developing a sustainable water filtration system for a rural community. The initial problem statement might be:

"The community faces challenges with access to clean drinking water. Develop a sustainable water filtration system that can be easily maintained and operated by the community members."

To define this problem more clearly, the team would:

1. **Conduct Research**: Gather information about water quality issues, existing filtration technologies, and the community's specific needs.
2. **Engage with Stakeholders**: Meet with community leaders, potential users, and experts in water filtration to gather their insights and perspectives.
3. **Brainstorm Solutions**: Generate a wide range of ideas through brainstorming sessions, focusing on innovative and feasible solutions.
4. **Develop Prototypes**: Create initial prototypes and conduct testing to validate the assumptions and refine the design.
5. **Iterate and Refine**: Continuously iterate on the problem definition and design based on feedback and testing results.

Defining the design problem is a critical step in the capstone design process. It involves understanding the problem domain, engaging with stakeholders, conducting thorough research, and using various tools and techniques to refine the problem definition. A well-defined problem sets the stage for successful project execution, ensuring that the team can develop a feasible, user-friendly, and cost-effective solution. By following a structured approach to problem definition, capstone design teams can enhance their creativity, collaboration, and problem-solving skills, leading to innovative and impactful design solutions.

7.8 Assignments

Assignment 7-1: Problem Definition Assignment (Team Assignment)
Objective:

      The goal of this assignment is to develop a clear and comprehensive problem definition based on the initial information provided by the sponsor. You will critically evaluate the sponsor's problem statement, identify areas that need further clarification, and document the process of gathering the necessary information to refine the problem definition. This will involve asking critical questions, conducting research, and using various methods to fill in any gaps.

Instructions:

  1. **Initial Review of Sponsor's Problem Definition:**
   - Carefully read the problem statement provided by the sponsor.
   - Highlight key points and any initial questions or uncertainties you have about the problem.

    Response:
   - Key Points from Sponsor's Problem Definition:

   - Initial Questions and Uncertainties:

  2. **Critical Questions to Evaluate the Problem Definition:**
   - Use the following questions to critically evaluate the problem definition

and identify areas that need further information or clarification:

- What is the main goal or objective of the design project?
- Who are the primary users or beneficiaries of the solution?
- What specific needs or problems does the solution aim to address?
- What are the constraints (e.g., budget, time, materials) and requirements specified by the sponsor?
- Are there any assumptions made in the problem definition that need to be verified?
- What are the expected deliverables or outcomes of the project?

Response:

o   Main Goal or Objective:

_____

_____

_____

o   Primary Users or Beneficiaries:

_____

_____

_____

o   Specific Needs or Problems:

_____

_____

_____

o   Constraints and Requirements:

_____

_____

_____

o   Assumptions to be Verified:

_____

_____

_____

o   Expected Deliverables or Outcomes:

_____

_____

3. **Identify Areas Lacking Specificity:**

o   Document the areas where the problem definition lacks specificity or enough information.

o Categorize these areas into those that need assumptions, research, interviews, surveys, or other methods to gather more information.

Response:

o Areas Lacking Specificity:

_____

_____

_____

_____

_____

o Categorized by Information Needed:

- **Assumptions**:

_____

_____

_____

- **Research**:

_____

_____

_____

- **Interviews**:

_____

_____

_____

- **Surveys**:

_____

_____

_____

- **Other Methods**:

_____

_____

_____

4. **Developing a Plan to Gather Missing Information:**

o Create a plan outlining how you will gather the missing information. This plan should include the methods you will use (e.g., research, interviews, surveys), the people you will contact, and the resources you will need.

Response:
- o   Plan for Gathering Missing Information:

Method:

_____

_____

_____

Target Contacts:

_____

_____

_____

Resources Needed:

_____

_____

_____

Timeline: _____

_____

_____

_____

5. **Conducting Research and Information Gathering:**
   - o   Implement your plan to gather the missing information. Document the process and findings from each method you use.
   - o   Ensure that you record interviews, survey responses, and any other relevant data collected.

Response:
- o   Documentation of Research and Information Gathering:
     Research Findings:

     _____

     _____

     _____

     _____

     _____

     Interview Summaries:

     _____

_____

_____

_____

_____

Survey Results:

_____

_____

_____

_____

Other Methods:

_____

_____

_____

_____

6. **Refining the Problem Definition:**
   o Use the information gathered to refine the initial problem definition. Ensure that your refined problem statement is clear, specific, and comprehensive.
   o Address any previously identified gaps and ensure that all critical aspects of the problem are covered.

   Response:
   o Refined Problem Definition:

7. **Documentation of the Entire Process:**
   o Compile all the documentation from the previous steps into a

comprehensive report. This report should include:
- Initial problem definition and initial questions/uncertainties.
- Critical questions and responses.
- Areas lacking specificity and categorized information needs.
- Plan for gathering missing information.
- Process and findings from research and information gathering.
- Refined problem definition.
- Reflection on the process and lessons learned.

Response:
- Comprehensive Report:

_____
_____
_____
_____
_____
_____
_____
_____
_____

8. **Reflection on the Process:**
   - Reflect on the process of defining the problem. Discuss the challenges you faced, how you overcame them, and what you learned from the experience.

Response:
- Reflection on the Process:

_____
_____
_____
_____
_____
_____
_____
_____
_____
_____

Submission:
Submit your completed assignment, including all responses and the

comprehensive report, by the due date. Ensure that your responses are clear, concise, and well-organized.

Grading Criteria:
- Completeness and clarity of the initial review and critical questions.
- Thoroughness in identifying areas lacking specificity and categorizing information needs.
- Feasibility and detail of the plan for gathering missing information.
- Quality and relevance of the gathered information.
- Clarity, specificity, and comprehensiveness of the refined problem definition.
- Quality of the documentation and the comprehensive report.
- Depth of reflection on the process and lessons learned.

By completing this assignment, you will develop critical skills in problem definition, information gathering, and project planning. These skills are essential for successfully tackling complex design projects and will be valuable in your future professional endeavors.

8 . Analyzing Market Conditions for a Product or Process

8.1 Introduction

In the realm of capstone engineering design, the creation of innovative products or processes is only half the battle. The other crucial component, often overlooked by engineering students, is understanding and analyzing the market conditions in which these innovations will exist. This analysis is fundamental to ensuring that the engineered solution not only solves a technical problem but also meets a real market need and has the potential for commercial success.

8.1.1 The Importance of Market Analysis in Engineering Design

Engineering design does not occur in a vacuum. Every product or process developed is intended to serve a purpose, fulfill a need, or solve a problem for a specific group of users or customers. Therefore, a thorough understanding of the market landscape is essential for several reasons:

1. Validation of Need: Market analysis helps verify that there is a genuine need or demand for the proposed solution. It prevents the development of products that, while technically impressive, may lack practical application or market appeal.
2. User-Centered Design: Understanding the market allows engineers to design with the end-user in mind, ensuring that the product or process meets actual user needs and preferences.
3. Competitive Positioning: Analysis of market conditions provides insight into existing solutions, allowing engineers to position their designs to offer unique

value or competitive advantages.

4. Economic Viability: Market analysis helps estimate the innovation's potential economic value, which is crucial for securing funding, resources, and stakeholder buy-in.

5. Regulatory and Standards Compliance: Understanding the market includes awareness of relevant regulations and standards that the design must adhere to for successful commercialization.

8.1.2   Key Components of Market Analysis

When analyzing market conditions for a capstone engineering design project, several critical components should be considered:

1. Market Size and Growth Potential Understanding the current size of the market and its projected growth is crucial. This involves researching:
   o Total addressable market (TAM)
   o Serviceable available market (SAM)
   o Serviceable obtainable market (SOM)

These metrics help in gauging the potential scale of the opportunity and inform decisions about resource allocation and design scalability.

2. Customer Segmentation Identifying and understanding different customer segments within the market is essential. This includes:
   o Demographic analysis
   o Psychographic profiling
   o Needs assessment for each segment

This information guides design decisions to ensure the product meets the specific needs of target users.

3. Competitive Landscape A thorough analysis of existing competitors and alternative solutions is crucial. This involves:
   o Identifying direct and indirect competitors
   o Analyzing their strengths and weaknesses
   o Understanding their market positioning and strategies

This analysis helps identify gaps in the market and opportunities for differentiation.

4. Market Trends and Dynamics Understanding current trends and future projections in the market is vital. This includes:
   o Technological trends
   o Socio-economic factors
   o Regulatory changes
   o Shifts in consumer behavior

Awareness of these trends ensures that the design remains relevant and adaptable to future market conditions.

5. Economic Factors Analyzing economic aspects of the market is crucial for the viability of the design. This includes:
   o Pricing strategies of existing solutions
   o Cost structures in the industry
   o Economic barriers to entry
   o Potential revenue models for the proposed solution
6. Distribution Channels Understanding how products or services reach end-users in the market is important. This involves analyzing:
   o Existing distribution networks
   o Potential partnerships or collaborations
   o Direct-to-consumer opportunities
   o Digital vs. physical distribution considerations

8.1.3   Integrating Market Analysis into the Design Process

The insights gained from market analysis should inform various stages of the engineering design process:

1. Problem Definition: Market analysis helps in refining the problem statement to address real market needs.
2. Conceptualization: Understanding market conditions guides the generation of ideas that are not only technically feasible but also marketable.
3. Design Development: Market insights inform design decisions, ensuring that features and specifications align with user needs and market demands.
4. Prototyping and Testing: Knowledge of user preferences and competitive offerings helps in creating more relevant prototypes and designing more effective user testing protocols.
5. Final Design and Implementation: Market analysis informs decisions about manufacturing, pricing, and go-to-market strategies.

Analyzing market conditions is an integral part of the capstone engineering design process. It bridges the gap between technical innovation and practical, marketable solutions. By incorporating market analysis, engineering students not only enhance the relevance and potential success of their designs but also develop a more holistic understanding of the role of engineering in society and the economy.

As future engineers, developing the skills to analyze and interpret market conditions alongside technical problem-solving abilities is crucial. This dual focus ensures that the innovative solutions developed in capstone projects have the best chance of making a real-world impact, whether through commercialization or addressing significant societal needs. In an increasingly competitive and rapidly evolving global market, the ability to align technical innovation with market realities is what distinguishes genuinely impactful engineering solutions.

8.2   Market Analysis

In mechanical engineering capstone design projects, the fusion of technical innovation with market realities is crucial for developing successful products. While engineering students are often well-versed in technical aspects, the importance of market analysis cannot be overstated. This essay explores various market analysis methods and provides practical guidelines for mechanical engineering students to effectively integrate market considerations into their capstone design projects.

8.2.1   Importance of Market Analysis in Mechanical Engineering Design

Market analysis in mechanical engineering design serves several critical purposes:
- Ensures that the product meets real-world needs and demands
- Guides design decisions to align with user preferences and market trends
- Helps in identifying potential commercialization opportunities
- Informs cost considerations and pricing strategies
- Assists in understanding regulatory requirements and industry standards

For mechanical engineering students, incorporating market analysis into their capstone projects not only enhances the relevance of their designs but also prepares them for the realities of professional engineering practice.

8.2.2   Key Market Analysis Methods

8.2.2.1   Porter's Five Forces Analysis

This framework helps in understanding the competitive forces within an industry:
a) Threat of New Entrants: Assess barriers to entry in the market for your mechanical design. b) Bargaining Power of Suppliers: Evaluate how supplier dynamics might affect your design choices and costs. c) Bargaining Power of Buyers: Understand how customer demands influence design specifications. d) Threat of Substitute Products: Identify alternative solutions that could compete with your design. e) Rivalry Among Existing Competitors: Analyze current market players and their offerings.

Practical Application: Use this analysis to identify market gaps and opportunities for your mechanical design to offer unique value.

8.2.2.2   SWOT Analysis

Analyze Strengths, Weaknesses, Opportunities, and Threats:
a) Strengths: Identify unique features or capabilities of your mechanical design. b) Weaknesses: Recognize limitations or areas for improvement in your design. c) Opportunities: Explore market trends or unmet needs your design could address. d) Threats: Consider external factors that could hinder the success of your design.

Practical Application: Use SWOT analysis to refine your design concept and develop strategies to maximize strengths and opportunities while mitigating weaknesses and threats.

8.2.2.3   Market Segmentation

Divide the market into distinct groups of consumers with similar needs or characteristics:

a) Demographic Segmentation: Age, gender, income, occupation b) Geographic Segmentation: Location, climate, urban/rural c) Psychographic Segmentation: Lifestyle, values, attitudes d) Behavioral Segmentation: Usage patterns, brand loyalty, benefits sought

Practical Application: Tailor your mechanical design to meet the specific needs of identified market segments.

8.2.2.4   Customer Journey Mapping

Map out the stages a customer goes through in interacting with your product:

a) Awareness: How customers become aware of the need for your product b) Consideration: How they evaluate options, including your design c) Decision: Factors influencing the purchase decision d) Use: Actual experience of using the product e) Loyalty/Advocacy: Post-use experience and potential for repeat purchase or recommendation

Practical Application: Use this method to identify key touchpoints and optimize your design for better user experience throughout the product lifecycle.

8.2.2.5   Competitive Analysis

Analyze existing products or solutions in the market:

a) Feature Comparison: Compare specifications and capabilities b) Price Point Analysis: Understand pricing strategies in the market c) Performance Benchmarking: Evaluate performance metrics d) Market Positioning: Analyze how competitors position their products

Practical Application: Identify areas where your mechanical design can outperform existing solutions or fill gaps in the market.

8.2.3   Data Collection Methods for Market Analysis

8.2.3.1   Primary Research

a) Surveys: Design and distribute surveys to potential users or customers. b) Interviews: Conduct in-depth interviews with industry experts or potential users. c) Focus Groups: Organize small group discussions to gather qualitative insights. d) Observational

Studies: Observe users interacting with similar products.

8.2.3.2   Secondary Research

a) Industry Reports: Analyze reports from reputable sources like McKinsey, Deloitte, or industry-specific publications. b) Academic Journals: Review relevant engineering and market research journals. c) Government Data: Utilize data from sources like the U.S. Census Bureau or Bureau of Labor Statistics. d) Trade Associations: Explore resources provided by relevant trade associations.

8.2.3.3   Online Tools and Platforms

a) Google Trends: Analyze search trends related to your product category. b) Social Media Listening: Use tools to monitor social media discussions about similar products or needs. c) E-commerce Platforms: Analyze customer reviews and ratings of similar products.

8.2.4   Practical Guidelines for Market Analysis in Mechanical Engineering Capstone Design

8.2.4.1   4.1 Start Early

Integrate market analysis from the project's inception. This ensures that market considerations inform every stage of the design process.

8.2.4.2   Define Your Target Market

Clearly identify who your product is for. Consider factors like industry sector, company size, or specific user demographics.

8.2.4.3   Understand User Needs

Conduct thorough user research to understand pain points, preferences, and unmet needs in the market.

8.2.4.4   Analyze the Competitive Landscape

Identify and analyze direct and indirect competitors. Understand their strengths and weaknesses to position your design effectively.

8.2.4.5   Consider Regulatory Environment

Research relevant industry standards, safety regulations, and certification requirements that may affect your design.

8.2.4.6   Evaluate Manufacturing and Supply Chain

Assess the feasibility of manufacturing your design, considering costs, materials availability, and supply chain logistics.

8.2.4.7   Pricing Strategy

Develop a preliminary pricing strategy based on manufacturing costs, competitor pricing, and perceived value to the customer.

8.2.4.8   Forecast Market Demand

Use available data to estimate potential market demand for your product. This helps in scaling decisions and assessing economic viability.

8.2.4.9   Identify Potential Partnerships

Explore possibilities for partnerships or collaborations that could enhance the value or reach of your design.

8.2.4.10   Consider Sustainability and Environmental Factors

Analyze market trends and regulations related to sustainability, as this is increasingly important in mechanical engineering.

8.2.5   Integrating Market Analysis into the Design Process

8.2.5.1   Concept Development Phase

Use market insights to generate and refine design concepts that align with user needs and market gaps.

8.2.5.2   Design Specification Phase

Incorporate market requirements into your technical specifications, ensuring that your design meets both technical and market needs.

8.2.5.3   Prototyping and Testing

Involve potential users in prototype testing to gather feedback and validate market assumptions.

8.2.5.4   Final Design and Documentation

Include a market analysis section in your final design report, demonstrating how your design addresses market needs and stands out from competitors.

8.2.6   Challenges and Considerations

8.2.6.1   Balancing Technical and Market Considerations

Strive to find a balance between pushing technical boundaries and meeting market demands.

8.2.6.2   Handling Conflicting Data

Be prepared to encounter conflicting information from different sources. Develop critical thinking skills to evaluate and reconcile disparate data.

8.2.6.3   Staying Current

Markets can change rapidly, especially in technology-driven fields. Continuously update your market analysis throughout the project.

8.2.6.4   Ethical Considerations

Be mindful of ethical considerations in market research, such as data privacy and unbiased representation of diverse user groups.

Effective market analysis is a crucial component of successful mechanical engineering capstone design projects. By employing a variety of market analysis methods and following practical guidelines, students can develop designs that are not only technically sound but also commercially viable and user-centric.

The skills developed through conducting market analysis – such as critical thinking, data analysis, and understanding user needs – are invaluable for mechanical engineers entering the professional world. These skills enable engineers to create products that not only solve technical problems but also address real market needs and have the potential for successful commercialization.

As the field of mechanical engineering continues to evolve, the ability to integrate technical expertise with market understanding will become increasingly important. By embracing market analysis as an integral part of the design process, mechanical engineering students can position themselves as well-rounded professionals capable of developing innovative solutions that make a meaningful impact on the market and society at large.

8.3 Market Analysis Guidelines for Capstone Design Projects

This guide provides a comprehensive approach to conducting market analysis for mechanical engineering capstone design projects. Following these steps will help you understand your design's market potential and align your project with real-world needs.

8.3.1 Step-by-Step Market Analysis Process

8.3.1.1 Step 1: Define Your Product/Service

a) Clearly articulate what your mechanical engineering design project aims to achieve. b) Identify the primary features and benefits of your product/service. c) Determine the problem your design solves or the need it fulfills.

8.3.1.2 Step 2: Identify Your Target Market

a) Define potential user groups or industries that could benefit from your design. b) Consider both direct users and other stakeholders who might influence adoption. c) Create initial customer profiles based on your understanding of the problem.

8.3.1.3 Step 3: Conduct Secondary Research

a) Industry Reports and Market Studies:
- Search databases like IBISWorld, Statista, or MarketResearch.com
- Review reports from industry associations relevant to your product

b) Government Data:
- Explore data from sources like the U.S. Census Bureau, Bureau of Labor Statistics, or equivalent agencies in other countries
- Look for industry-specific data from government departments (e.g., Department of Energy for energy-related projects)

c) Academic and Technical Publications:
- Search engineering databases like IEEE Xplore or ScienceDirect
- Review recent conference proceedings in your field

d) News and Trade Publications:
- Subscribe to relevant industry newsletters
- Review trade magazines and websites specific to your field

8.3.1.4 Step 4: Perform Primary Research

a) Surveys:
- Design a survey to gather quantitative data from potential users
- Use tools like Google Forms, SurveyMonkey, or Qualtrics
- Aim for at least 50-100 responses for meaningful data

b) Interviews:
- Conduct 5-10 in-depth interviews with industry experts or potential users
- Prepare a semi-structured interview guide
- Record and transcribe interviews for analysis

c) Observations:
- If possible, observe potential users interacting with similar products or in

relevant environments
- Take detailed notes on user behaviors and pain points

8.3.1.5   Step 5: Analyze the Competitive Landscape

a) Identify Competitors:
- List direct competitors (similar solutions)
- Identify indirect competitors (alternative solutions to the same problem)

b) Competitive Analysis:
- Create a comparison matrix of key features and specifications
- Analyze pricing strategies of competitors
- Assess the market positioning and unique selling propositions of each competitor

c) SWOT Analysis:
- Conduct a SWOT (Strengths, Weaknesses, Opportunities, Threats) analysis for your design and key competitors

8.3.1.6   Step 6: Assess Market Size and Growth Potential

a) Total Addressable Market (TAM):
- Estimate the total market size for your type of product/service
- Use industry reports and government data to support your estimates

b) Serviceable Available Market (SAM):
- Narrow down to the portion of the market you can realistically target
- Consider geographical limitations, specific industries, or customer segments

c) Serviceable Obtainable Market (SOM):
- Estimate the portion of the market you can realistically capture
- Consider factors like competition, resources, and go-to-market strategy

d) Growth Projections:
- Research industry growth forecasts
- Identify factors that could drive or hinder market growth

8.3.1.7   Step 7: Analyze Market Trends and Dynamics

a) Technological Trends:
- Identify emerging technologies in your field
- Assess how these trends might impact your design

b) Regulatory Environment:
- Research current regulations affecting your product/industry
- Investigate potential future regulatory changes

c) Economic Factors:
- Consider how economic conditions might affect demand for your product

- Analyze pricing trends in the industry

d) Social and Environmental Trends:
- Identify relevant social or environmental movements (e.g., sustainability)
- Assess how these trends might influence product adoption

8.3.1.8 Step 8: Customer Segmentation and Profiling

a) Segment your market based on:
- Demographics (age, income, education)
- Psychographics (values, interests, lifestyles)
- Behavioral factors (usage patterns, brand loyalty)
- Firmographics (for B2B products: company size, industry)

b) Create detailed customer personas for each key segment

8.3.1.9 Step 9: Analyze the Value Chain

a) Map out the entire value chain for your product:
- Raw material suppliers
- Manufacturing processes
- Distribution channels
- End-users

b) Identify critical players at each stage of the value chain c) Assess potential partnerships or collaborations within the value chain

8.3.1.10 Step 10: Develop Pricing Strategy

a) Analyze costs:
- Estimate manufacturing costs
- Consider overhead and distribution costs

b) Research pricing strategies of competitors c) Assess price sensitivity in your target market d) Develop initial pricing models (cost-plus, value-based, competitive)

8.3.1.11 Step 11: Identify Potential Risks and Challenges

a) Technical risks:
- Feasibility of mass production
- Potential for technical obsolescence

b) Market risks:
- Changes in customer preferences
- New competitive entries

c) Regulatory risks:
- Potential for new regulations affecting your product
- Compliance challenges

d) Economic risks:

- Market volatility
- Supply chain disruptions

8.3.1.12   Step 12: Synthesize Findings and Draw Conclusions

a) Summarize key findings from your research b) Assess the overall market viability of your design c) Identify critical success factors for market entry d) Develop recommendations for design refinements based on market insights

8.3.2   Guidelines for Effective Market Analysis

1. Be Objective: Avoid bias towards your design. Be willing to accept and act on negative findings.
2. Use Diverse Sources: Don't rely on a single source of information. Cross-verify data from multiple sources.
3. Quantify Where Possible: Use numerical data to support your analysis whenever available.
4. Consider Global Perspectives: Even if targeting a local market, be aware of global trends and potential international opportunities or threats.
5. Stay Current: Markets can change rapidly. Ensure you're using the most up-to-date information available.
6. Engage with Potential Users: Direct interaction with your target market provides invaluable insights.
7. Iterate: Market analysis is not a one-time activity. Be prepared to revisit and update your analysis as your project evolves.
8. Document Your Process: Keep detailed records of your sources, methods, and reasoning.
9. Consider Ethical Implications: Ensure your market research methods are ethical and respect privacy concerns.
10. Collaborate: Work with teammates and seek input from advisors or industry contacts to broaden your perspective.

8.3.3   Example Market Analysis: Smart Home Energy Management System

Project Description: A smart home energy management system that uses IoT sensors and AI to optimize energy usage in residential buildings.

8.3.3.1   Step 1: Define Your Product/Service

- Smart home system using IoT sensors to monitor energy usage
- AI-powered algorithms to optimize energy consumption
- User interface (mobile app and web portal) for monitoring and control
- Integration with smart home devices (thermostats, appliances, etc.)

8.3.3.2 Step 2: Identify Your Target Market

- Primary: Homeowners in urban and suburban areas
- Secondary: Property developers, energy companies, smart home installers

8.3.3.3 Step 3-4: Conduct Secondary and Primary Research (Summary of key findings)

- Global smart home market expected to reach $135.3 billion by 2025 (MarketsandMarkets report)
- 63% of homeowners interested in smart energy management solutions (survey of 150 homeowners)
- Average energy savings of 15-20% reported in similar systems (DOE study)
- Key customer pain points: high energy bills, complexity of existing systems, lack of actionable insights (interviews with eight homeowners and three energy experts)

8.3.3.4 Step 5: Analyze the Competitive Landscape Competitors:

1. Nest (Google)
2. Ecobee
3. Sense
4. Neurio

SWOT Analysis (for our product): Strengths:

- More comprehensive integration with various home systems
- Advanced AI algorithms for predictive optimization Weaknesses:
- New entrant in a market with established players
- Limited brand recognition Opportunities:
- Growing awareness of energy efficiency
- Increasing smart home adoption Threats:
- Potential for tech giants to enter the market
- Privacy concerns regarding data collection

8.3.3.5 Step 6: Assess Market Size and Growth Potential

- TAM: $135.3 billion (global smart home market by 2025)
- SAM: $28.6 billion (energy management segment of the smart home market)
- SOM: Targeting 1% market share in first three years = $286 million

8.3.3.6 Step 7: Analyze Market Trends and Dynamics

- Increasing focus on renewable energy integration
- Growing concerns about climate change driving energy efficiency

initiatives
- Regulatory push for smart grid technologies

8.3.3.7   Step 8: Customer Segmentation/Profiling Primary Persona: "Eco-conscious Emma"

- Age: 35-50
- Income: $100,000+
- Homeowner in suburban area
- Environmentally conscious, tech-savvy
- Motivated by reducing carbon footprint and saving money

8.3.3.8   Step 9: Analyze the Value Chain

- Sensor Manufacturers: STMicroelectronics, Texas Instruments
- Software Development: In-house
- Installation: Partnerships with local electricians and smart home installers
- Distribution: Direct to consumers and through home improvement retailers

8.3.3.9   Step 10: Develop Pricing Strategy

- Hardware cost: $200 (based on component costs and manufacturing estimates)
- Software subscription: $10/month
- Target retail price: $499 for hardware + subscription model
- Competitive with Sense ($299) but justified by additional features

8.3.3.10   Step 11: Identify Potential Risks and Challenges

- Technical: Ensuring reliability and accuracy of energy predictions
- Market: Competition from established brands, potential market saturation
- Regulatory: Data privacy regulations, varying energy policies across regions

8.3.3.11   Step 12: Synthesize Findings and Draw Conclusions

- Strong market potential driven by increasing energy consciousness and smart home adoption
- Differentiation through comprehensive integration and advanced AI capabilities is key
- Focus on a user-friendly interface and verifiable energy savings to drive adoption
- Consider partnerships with energy companies or property developers for faster market penetration

8.3.3.12   Recommendations:

1. Refine AI algorithms to achieve or exceed the 20% energy savings benchmark
2. Develop strong data privacy and security measures to address consumer concerns
3. Create a straightforward, intuitive user interface that provides actionable energy-saving recommendations
4. Explore partnerships with renewable energy providers for enhanced functionality

This example demonstrates how to apply the market analysis process to a specific mechanical engineering capstone project. Students should adapt this framework to their unique projects, focusing on the aspects most relevant to their design and target market.

8.5 Assignments

Assignment 8-1: Market Analysis for Capstone Design Project

Objective: Conduct a comprehensive market analysis to evaluate the commercial viability and potential market opportunities for your capstone design project. This analysis will help inform your design decisions and business strategy.

Instructions: Follow these steps to complete your market analysis:

1. Define your product/service: a. Write a clear, concise description (200-300 words) of your design project. b. List 5-7 key features and their corresponding benefits to users. c. Create a problem statement that articulates the specific issue your product solves. d. Develop a unique selling proposition (USP) in one sentence, highlighting what sets your product apart. e. Create a visual representation (e.g., infographic or diagram) of your product's key features and benefits.
2. Identify your target market: a. Develop 3-5 detailed customer personas, including:
   - Demographic information (age, gender, income, education, occupation)
   - Psychographic details (values, interests, lifestyle, personality)
   - Behavioral traits (buying habits, brand preferences, technology usage) b. Online tools like Facebook Audience Insights or Google Analytics can be used to gather demographic data. c. Conduct at least ten interviews with potential users to refine your personas. d. Estimate your total addressable market (TAM), serviceable addressable market (SAM), and serviceable obtainable market (SOM) using available industry data and your assumptions.
3. Analyze the industry and market trends: a. Identify and summarize findings from at least three recent industry reports (e.g., from IBISWorld, Gartner, or Forrester). b. Create a timeline of significant industry developments over the past five years. c. Identify and explain 5-7 key trends affecting your product category. d. Use tools like Google Trends to analyze search interest over time for relevant keywords. e. Create graphs or charts to visualize market growth projections for the next 3-5 years.
4. Conduct competitor analysis: a. Create a spreadsheet listing at least ten direct and indirect competitors. b. For each competitor, analyze:
   - Product features and benefits
   - Pricing strategy
   - Target audience
   - Marketing channels and messaging
   - Estimated market share (if available) c. Perform a detailed SWOT analysis for your top 3-5 competitors. d. Create a positioning map to visualize how

your product compares to competitors in terms of key attributes. e. Identify potential barriers to entry and how you plan to overcome them.

5. Assess market demand: a. Design a customer survey with 10-15 questions to gauge interest in your product concept. b. Aim for at least 100 responses from your target demographic. c. Conduct 5-10 in-depth interviews with potential customers or industry experts. d. Use tools like Google Keyword Planner to analyze search volumes for relevant keywords. e. Monitor social media platforms and online forums to identify discussions related to your product category. f. If applicable, analyze app store or product review data for similar products.

6. Analyze pricing and profitability: a. Create a detailed cost breakdown for your product, including:
   o Material costs
   o Labor costs
   o Overhead expenses
   o Marketing and distribution costs b. Research pricing strategies of at least five competitors or similar products. c. Develop three potential pricing models (e.g., cost-plus, value-based, competitive) and calculate potential profit margins for each. d. Conduct a break-even analysis to determine the minimum sales volume needed for profitability. e. Create a pricing sensitivity analysis to understand how different price points might affect demand.

7. Identify distribution channels: a. List and describe at least five potential distribution channels for your product. b. Create a pros and cons table for each channel, considering factors like:
   o Reach and accessibility to the target market
   o Cost and complexity of implementation
   o Control over customer experience
   o Potential for scalability c. Research and summarize case studies of successful distribution strategies in your industry. d. Develop a multi-channel distribution strategy, outlining primary and secondary channels. e. Create a flowchart illustrating the proposed distribution process from production to end-user.

8. Assess regulatory and legal considerations: a. Identify and summarize vital industry-specific regulations or standards (e.g., safety standards, data privacy laws). b. Consult with a legal professional or use online resources to understand intellectual property protection options. c. Conduct a preliminary patent search to identify any potential conflicts or opportunities. d. Create a checklist of legal requirements for bringing your product to market (e.g., certifications, licenses). e. Estimate costs and timeline for addressing legal and regulatory requirements.

9. Conduct a PESTEL analysis: a. For each PESTEL factor, identify and explain 3-5

relevant trends or issues:
- o Political: Government policies, political stability, trade regulations
- o Economic: Economic growth, inflation rates, consumer spending trends
- o Social: Demographic shifts, cultural trends, lifestyle changes
- o Technological: Emerging technologies, R&D activity, automation
- o Environmental: Sustainability concerns, environmental regulations, climate change impacts
- o Legal: Changes in legislation, consumer protection laws, industry-specific regulations b. Create a matrix to evaluate the potential impact (high, medium, low) and likelihood of each factor. c. Identify specific opportunities and threats based on your PESTEL analysis.

10. Synthesize findings and conclude: a. Create an executive summary (1-2 pages) highlighting key findings from each analysis section. b. Develop a SWOT analysis for your product based on all the research conducted. c. Identify 3-5 key market opportunities and explain how your product addresses them. d. List potential challenges or risks and propose mitigation strategies. e. Provide specific recommendations for:
- o Product development or refinement
- o Target market focus
- o Pricing strategy
- o Distribution approach
- o Marketing and positioning f. Create a high-level action plan with the following steps and timelines.

Additional guidance:
- • Use data visualization tools (e.g., Tableau, PowerBI) to create compelling charts and graphs.
- • Regularly meet as a team to discuss findings and ensure consistency across different analysis sections.
- • Consider creating a shared document or project management tool to track progress and share resources.
- • Seek feedback from your project advisor at key milestones to ensure you're on the right track.

By following these detailed instructions, your team will conduct a thorough market analysis that provides valuable insights for your capstone design project. Remember to document your sources and assumptions throughout the process, and be prepared to adjust your approach as you uncover new information.

Deliverables:
1. A comprehensive written report (5-10 pages) detailing your market analysis findings and recommendations.
2. A concise executive summary (1-2 pages) highlighting key insights and conclusions.
3. Include in your preliminary design report.

Timeline:
- Week 1-2: Steps 1-3
- Week 3-4: Steps 4-6
- Week 5-6: Steps 7-9
- Week 7-8: Step 10 and preparation of deliverables

Tips for success:
- Use a mix of primary (e.g., surveys, interviews) and secondary (e.g., industry reports, academic papers) research sources.
- Be objective in your analysis and support your conclusions with data.
- Consider seeking input from industry professionals or mentors to validate your findings.
- Regularly communicate with your team members and divide tasks efficiently.
- Keep your project advisor updated on your progress and seek guidance when needed.

This comprehensive market analysis will provide valuable insights to guide your capstone design project and demonstrate its potential for real-world application. Good luck!

9 Define the Customer Requirements

Defining customer requirements is a foundational step in the engineering design process. It sets the stage for a successful project by ensuring that the final product meets the needs and expectations of the end-users. This phase involves understanding who the customers are and what their needs and desires are, as well as translating these into precise, actionable requirements that will guide the design and development of the product.

The definition of customer requirements is critical because it bridges the gap between a project's conceptual phase and its practical implementation. Without a thorough understanding of what customers need, there is a significant risk of developing a technically sound product that fails to address the core problems or enhance user satisfaction. This section will delve into the importance of customer requirements, methods for gathering and defining these requirements, and the tools and techniques used to ensure they are accurately captured and effectively utilized throughout the project.

9.1 The Importance of Customer Requirements

1. **Alignment with Customer Needs**:
   o Ensuring the project aligns with customer needs and expectations is vital for its success. A product that meets or exceeds customer requirements is more likely to be adopted and valued by the end-users.
2. **Risk Mitigation**:
   o Defining customer requirements early in the project helps to identify potential risks and issues that could arise. Addressing these early can save time and resources and prevent costly changes later in the project.
3. **Enhanced Communication**:
   o Clear customer requirements improve communication among stakeholders, including project team members, customers, and sponsors. This shared understanding helps to align the project goals and ensures

everyone is working towards the same objectives.

4. **Foundation for Design and Development**:
   o Customer requirements provide a concrete foundation for the design and development phases of the project. They guide decision-making and prioritization, ensuring that the design efforts are focused on features and functionalities that matter most to the customers.

5. **Performance Measurement**:
   o Well-defined customer requirements allow for measurable performance criteria to be established. This makes it easier to assess whether the final product meets the defined goals and delivers the expected value to the customers.

9.2   Methods for Gathering Customer Requirements

To accurately define customer requirements, it is essential to gather comprehensive and reliable information from the customers. Several methods can be employed:

1. **Interviews**:
   o Conducting one-on-one interviews with customers can provide deep insights into their needs, preferences, and pain points. Open-ended questions and active listening are key to extracting valuable information.

2. **Surveys and Questionnaires**:
   o Surveys and questionnaires can reach a larger audience and collect quantitative data. They should be carefully designed to capture a wide range of customer opinions and preferences.

3. **Focus Groups**:
   o Bringing together a diverse group of customers in a focus group setting allows for dynamic discussions and the generation of ideas. This method can uncover needs and requirements that individual interviews might miss.

4. **Observation**:
   o Observing customers in their natural environment can reveal unspoken needs and behaviors. This method is particularly useful for understanding how customers interact with current products and identifying areas for improvement.

5. **Workshops**:
   o Interactive workshops involving customers and stakeholders can facilitate collaborative requirement gathering. Techniques such as brainstorming, mind mapping, and role-playing can be employed to explore different perspectives and generate comprehensive requirements.

9.3 Translating Customer Needs into Requirements

Once customer needs are gathered, the next step is to translate these needs into precise, actionable requirements. This involves:

1. **Categorization**:
   - Organizing customer needs into categories such as functional, non-functional, and user experience requirements. This helps in systematically addressing each aspect of the project.

2. **Prioritization**:
   - Not all customer needs are equally important. Prioritizing requirements based on factors such as customer value, feasibility, and project constraints ensures that the most critical needs are addressed first.

3. **Specification**:
   - Writing detailed and unambiguous requirement statements. Each requirement should be specific, measurable, achievable, relevant, and time-bound (SMART).

4. **Validation**:
   - Validating the requirements with customers and stakeholders to ensure accuracy and completeness. This step often involves reviewing the requirements and seeking feedback to confirm that they correctly represent customer needs.

9.4 Tools and Techniques

Several tools and techniques can aid in defining and managing customer requirements:

1. **Requirements Management Software**:
   - Tools like JIRA, Trello, and Microsoft Project can help track and manage requirements throughout the project lifecycle. These tools facilitate collaboration, documentation, and tracking of changes.

2. **Use Cases and User Stories**:
   - Use cases and user stories describe how customers will interact with the product. They provide a narrative that helps in understanding the context and flow of user interactions, making it easier to derive specific requirements.

3. **Quality Function Deployment (QFD)**:
   - QFD is a method used to transform customer needs into engineering characteristics. The House of Quality, a part of QFD, is a matrix that helps visualize the relationships between customer requirements and the technical features of the product.

4. **Prototyping**:
   - Developing prototypes based on initial requirements allows for early

testing and feedback. Prototypes help refine requirements and ensure they align with customer expectations.

5. **SWOT Analysis**:
   - o Conducting a SWOT analysis (Strengths, Weaknesses, Opportunities, Threats) on the gathered requirements helps in identifying potential challenges and opportunities for improvement.

Defining customer requirements is a crucial step that lays the groundwork for the entire project. It ensures that the final product aligns with customer needs, mitigates risks, enhances communication among stakeholders, and provides a solid foundation for design and development. By employing various methods to gather, translate, and manage requirements, project teams can create a clear roadmap that guides the project to successful completion. Through diligent attention to customer requirements, teams can deliver products that not only meet but exceed customer expectations, leading to higher satisfaction and success.

9.5   Gathering Information

Before an applicable definition of the design problem can be achieved, the team must thoroughly understand the problem. A crucial step in this process is identifying the motivation behind the design problem—why it was posed in the first place. Understanding this motivation is essential for developing a comprehensive set of design specifications that align with user requirements. Several methods can be employed to gather the necessary information, each offering unique insights and benefits.

9.5.1   Customer, User, and Sponsor Interviews

Customer/User/Sponsor Interviews are among the most effective methods for gathering detailed information about the design problem and user requirements. These interviews provide direct insights from the people who will be using or benefiting from the design. Here are key steps and considerations for conducting successful interviews:

1. **Planning the Interview**:
   - o **Develop a Plan**: Before meeting with the sponsor or stakeholders, develop a detailed plan outlining the objectives of the interview, key questions to ask, and the information needed.
   - o **Prepare Questions**: Prepare a list of questions to guide the interview. Focus on understanding the users' needs, preferences, and pain points. Example questions include:
     - ▪ What benefit will the new design bring to your work environment?
     - ▪ What is currently deficient, or what could be improved?
     - ▪ What features are most important to you?
     - ▪ What are the safety considerations that need to be included?

o **Arrange the Meeting**: Coordinate with the sponsor to arrange the initial meeting, often at their company, lab, or another relevant location. This visit allows the team to observe the environment and gather context-specific information.

2. **Conducting the Interview**:
   o **Engage Stakeholders**: During the meeting, engage with potential users or customers to gather their insights and feedback. Understand whether the design will be used internally or if it will become a product for sale.
   o **Active Listening**: Practice active listening to capture detailed information and understand the nuances of user feedback.
   o **Document Insights**: Take detailed notes during the interview. With permission, recording the session (audio or video) can be valuable for later reference.

3. **Post-Interview Analysis**:
   o **Review Notes and Recordings**: After the interview, review the notes and recordings to extract key information and insights.
   o **Identify Patterns and Themes**: Look for common themes, recurring issues, and critical requirements that emerge from the discussions.

9.5.2 Group Interviews

Group Interviews involve interviewing multiple users or stakeholders simultaneously. This method can be highly effective due to the synergy created when people build on each other's ideas and concerns. Here are the steps for conducting group interviews:

1. **Organize the Group**:
   o **Select Participants**: Choose a diverse group of users or stakeholders to ensure a wide range of perspectives.
   o **Set a Moderator**: Assign a team member to act as the moderator. The moderator's role is to guide the discussion, keep it on track, and ensure that all participants have an opportunity to contribute.

2. **Facilitate the Discussion**:
   o **Encourage Interaction**: Foster an environment where participants feel comfortable sharing their ideas and building on others' contributions.
   o **Inject Questions**: The moderator should have a list of prepared questions and be ready to inject new questions when the discussion stalls or needs redirection.

3. **Document the Session**:
   o **Take Notes**: Have one or more team members take detailed notes. If permitted, record the session (audio or video) for more comprehensive documentation.

4. **Analyze Group Dynamics**:
   - o **Review Notes and Recordings**: Analyze the recorded session to capture all insights and ideas generated during the discussion.
   - o **Synthesize Information**: Identify key themes, unique insights, and any consensus or disagreements among participants.

9.5.3   User/Operator/Customer Surveys

Surveys are an excellent tool for collecting information from a large number of people. They can be distributed widely, allowing the team to gather a broad range of data. Here's how to effectively use surveys for gathering information:

1. **Designing the Survey**:
   - o **Create Clear Questions**: Design the survey questions carefully to avoid bias and ensure clarity. Avoid multi-barrel questions that may confuse respondents or lead to ambiguous answers.
   - o **Types of Questions**: Use a mix of question types, such as multiple-choice, Likert scale, and open-ended questions, to gather both quantitative and qualitative data.
2. **Selecting Recipients**:
   - o **Assemble a Target List**: Identify a representative sample of users, operators, or customers to ensure the survey results are valid and reliable.
   - o **Ensure Diversity**: Include a diverse group of respondents to capture a wide range of perspectives and experiences.
3. **Distributing the Survey**:
   - o **Use Online Tools**: Utilize online survey tools like SurveyMonkey or Qualtrics to create and distribute the survey efficiently.
   - o **Promote Participation**: Encourage participation by explaining the survey's purpose and the importance of respondents' input.
4. **Analyzing Survey Results**:
   - o **Collect and Compile Data**: Gather the survey responses and compile the data for analysis.
   - o **Identify Trends and Patterns**: Look for common trends, significant findings, and any outliers that may require further investigation.
   - o **Extract Actionable Insights**: Translate the survey results into actionable insights that inform the design specifications and problem definition.

9.5.4   Integrating Information

Once the information is gathered through interviews, group discussions, and surveys, it must be integrated and synthesized to form a comprehensive understanding of the design problem. This involves:

1. **Creating a Knowledge Base**:
   - o **Centralize Information**: Compile all gathered data into a centralized repository or knowledge base.
   - o **Organize by Themes**: Organize the information by themes or categories for easier analysis.
2. **Developing User Personas**:
   - o **Create Personas**: Develop detailed user personas that represent different segments of the target audience. These personas help in visualizing the users' needs, preferences, and behaviors.
3. **Mapping Customer Journeys**:
   - o **Journey Mapping**: Create customer journey maps to illustrate the steps users take when interacting with the product. This helps identify pain points and opportunities for improvement.
4. **Refining Requirements**:
   - o **Iterative Refinement**: Continuously refine the requirements based on new information and feedback. Ensure that the requirements remain aligned with user needs and project goals.

Gathering information is a critical step in defining the design problem for a capstone project. By employing various methods such as interviews, group discussions, and surveys, the team can gather comprehensive insights into the users' needs, preferences, and pain points. This information is then synthesized to form a clear and actionable problem definition that guides the design process. Through careful planning, execution, and analysis, the team can ensure that they have a deep understanding of the design problem, leading to the development of solutions that truly meet the users' needs and expectations.

9.5.5   Surveys

When the design project involves a product or software application that many people will use, conducting a survey can be an invaluable method for gathering information about customer and user needs and desires. Surveys are particularly useful when data collection from a broad audience or field experts is required. This section explores the purpose and design of surveys, the tools available for conducting them, and how to analyze and utilize the collected data.

9.5.5.1   Purpose of Surveys

Surveys are designed to collect specific information that can help define design specifications or requirements. Before creating a survey, it is crucial to identify the information you need from stakeholders, users, or customers. This could include their needs, preferences, pain points, and any other variables that will influence the design.

Surveys can help answer questions such as:
- What features do users find most important in the product?
- What are the common pain points or problems users face with existing solutions?
- What are the demographic characteristics of the target user group?
- What improvements or new functionalities do users desire?

By gathering this information, the design team can ensure that the final product meets user needs and expectations.

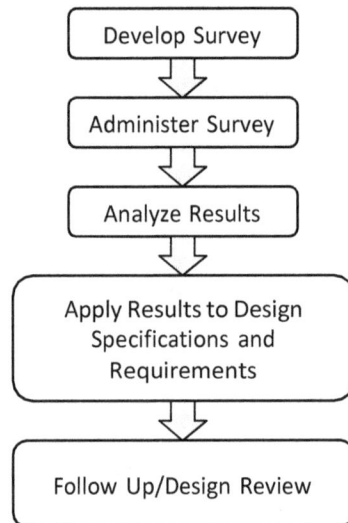

```
┌─────────────────────┐
│   Develop Survey    │
└─────────────────────┘
          ⇩
┌─────────────────────┐
│  Administer Survey  │
└─────────────────────┘
          ⇩
┌─────────────────────┐
│   Analyze Results   │
└─────────────────────┘
          ⇩
┌─────────────────────┐
│ Apply Results to Design │
│   Specifications and    │
│      Requirements       │
└─────────────────────┘
          ⇩
┌─────────────────────┐
│ Follow Up/Design Review │
└─────────────────────┘
```

Figure 9-1. Survey process flow.

9.5.5.2   Designing the Survey

The survey was designed as a collaborative process that involved group thinking. Sponsors of the project can also contribute by identifying relevant questions and variables. The steps involved in designing an effective survey are as follows:

1. **Identify Variables**:
   o   Start by making a list of variables that need to be measured in the survey. These could include user preferences, frequency of use, desired features, and any other relevant factors.

2. **Develop Questions**:
   o   Develop a set of questions based on the identified variables. Questions should be clear, concise, and free from bias. Include a mix of question types such as multiple-choice, Likert scale, and open-ended questions to gather both quantitative and qualitative data.

3. **Select Survey Method**:
   o   Choose a survey method based on the expected number of participants.

For smaller groups, interviews might be more effective, while larger groups may require an electronic survey tool.

4. **Create a Trial Survey**:
   o Before sending out the survey to actual participants, create a trial version and have it reviewed by team members, friends, family, fellow students, sponsors, and mentors. Gather constructive feedback and make necessary corrections.

5. **Finalize the Survey**:
   o After incorporating feedback from the trial survey, finalize the survey for distribution. Ensure that the survey includes a preamble explaining the motivation behind the survey and how it will benefit the design project and participants.

9.5.5.3 Example: Automated Pill Dispenser Survey

Consider a design project aimed at creating an automated pill dispenser for elderly patients who have difficulty managing their medications. The survey could target patients, medical doctors, pharmacists, and caretakers to gather comprehensive insights.

Variables to Measure:
- Frequency and types of medications taken
- Common challenges in medication management
- Desired features in an automated pill dispenser
- User preferences for reminders and logs
- Safety and reliability concerns

Sample Survey Questions:
1. How many different medications do you take daily?
   o 1-2
   o 3-4
   o 5-6
   o More than 6
2. What is the most challenging aspect of managing your medications?
   o Remembering to take them
   o Keeping track of dosages
   o Refilling prescriptions
   o Other (please specify)
3. How would you prefer to be reminded to take your medication?
   o Audible alarm
   o Mobile app notification
   o Visual indicator on the device
   o Other (please specify)
4. What features would you like to see in an automated pill dispenser? (Open-

ended question)

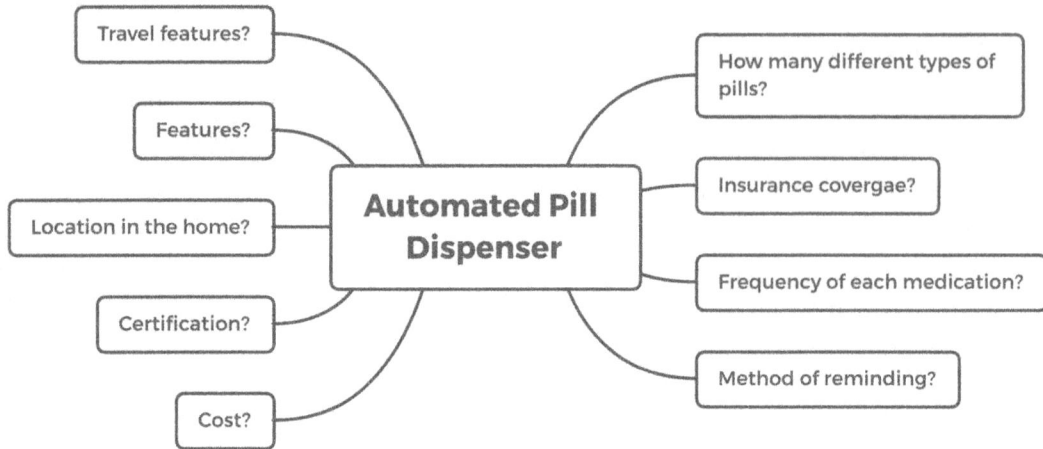

Figure 9-2. Automated pill dispenser survey variables/questions relationship diagram.

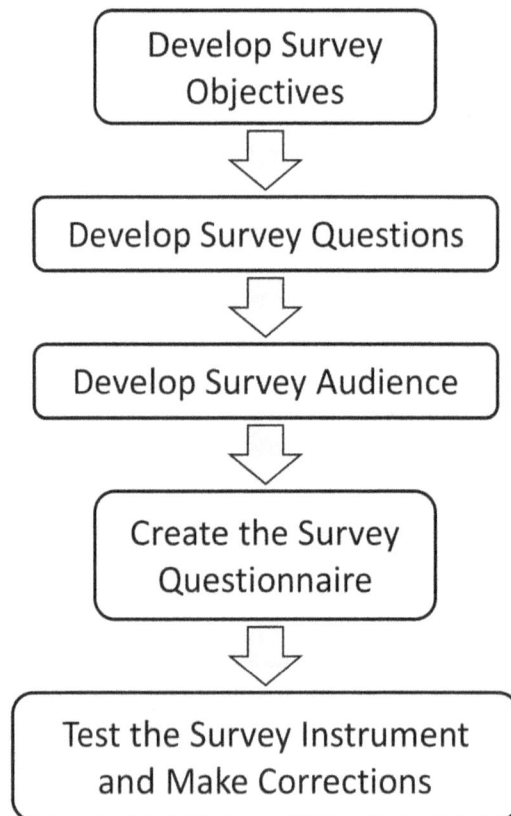

Figure 9-3. Develop survey.

9.5.5.4   Survey Tools

Several tools are available for creating and administering surveys, each with its own set of features and benefits:

1. **Google Forms**:
   - Google Forms is a free tool available through Google's G-Suite, commonly used in academic settings. It allows for easy creation and distribution of surveys and provides basic analytics.
2. **SurveyMonkey**:
   - SurveyMonkey offers more advanced features for survey creation and analysis, including templates, customizable themes, and detailed analytics. It is suitable for more complex surveys and larger participant groups.
3. **Qualtrics**:
   - Qualtrics is a powerful survey tool with extensive features for survey design, distribution, and analysis. It is particularly useful for academic research and professional surveys requiring advanced statistical analysis.

9.5.5.5   Administering the Survey

The steps involved in administering a survey include:

1. **Select Participants**:
   - Create a list of potential survey participants, ensuring a diverse sample in terms of age, gender, race, and other demographics.
2. **Distribute the Survey**:
   - Use the selected survey tool to distribute the survey. Email invitations can be sent to participants for electronic surveys. For in-person surveys, schedule interviews or distribute paper forms.
3. **Collect Responses**:
   - Monitor the survey responses and ensure a good response rate by sending reminders if necessary.

9.5.5.6   Analyzing Survey Data

Once the survey is completed, the data must be analyzed to extract meaningful insights:

1. **Compile Data**:
   - Gather all responses and compile the data for analysis. Most survey tools provide built-in analytics to help with this process.
2. **Statistical Analysis**:
   - Perform statistical analysis to identify trends, patterns, and significant findings. Tools like Excel, SPSS, or the analytics of survey software can be used for this purpose.

3. **Report Findings**:
   - ○ Create a comprehensive report summarizing the survey results. Highlight key insights, common themes, and any outliers. Include charts and graphs to visualize the data.
4. **Inform Design Specifications**:
   - ○ Use the findings from the survey to develop detailed design specifications. These specifications should be directly informed by the needs and preferences identified through the survey.

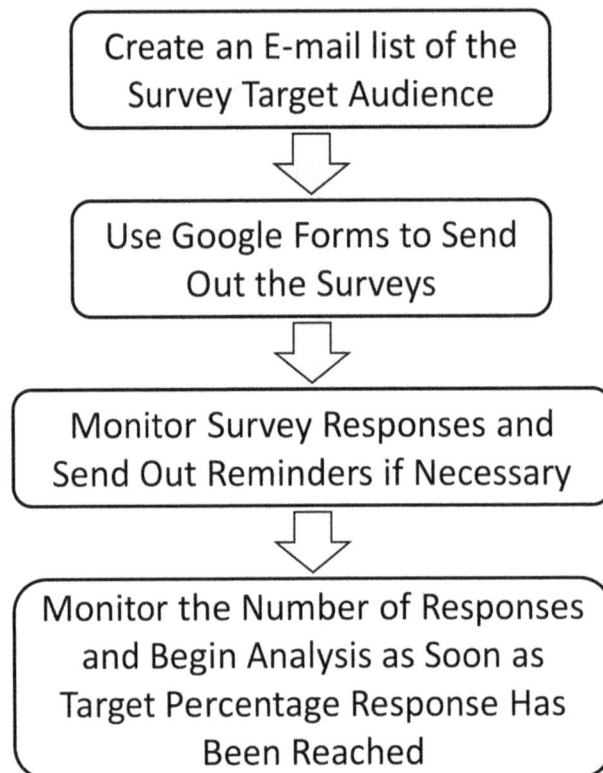

```
┌─────────────────────────────┐
│   Create an E-mail list of the  │
│     Survey Target Audience      │
└─────────────────────────────┘
              ⬇
┌─────────────────────────────┐
│   Use Google Forms to Send      │
│        Out the Surveys          │
└─────────────────────────────┘
              ⬇
┌─────────────────────────────┐
│  Monitor Survey Responses and   │
│  Send Out Reminders if Necessary│
└─────────────────────────────┘
              ⬇
┌─────────────────────────────┐
│  Monitor the Number of Responses│
│   and Begin Analysis as Soon as │
│  Target Percentage Response Has │
│          Been Reached           │
└─────────────────────────────┘
```

Figure 9-4. Administer survey.

Surveys are powerful tools for gathering detailed information about user needs and preferences. By carefully designing and administering surveys and then analyzing the collected data, design teams can ensure that their projects are aligned with user requirements. This process not only enhances the likelihood of project success but also ensures that the final product effectively addresses the real-world problems it is intended to solve. Through thoughtful application of survey research, capstone design teams can develop innovative, user-centered solutions that meet and exceed stakeholder expectations.

9.6   Sources of Information

Design project sources can vary significantly depending on the approach and philosophy taken within the engineering program, department, and college. Most engineering programs strive to maintain close relationships with the employers of their students, seeking out design project opportunities with these companies. As a student, you may have specific projects in mind, particularly those involving companies where your friends or family members work. Once you become aware of the design projects through your professor or company presentations, it is crucial to conduct thorough background research on the problems that interest you. This background research not only aids in your understanding but also enhances your qualifications for the project.

9.6.1   Importance of Background Research

Conducting background research is essential for several reasons:
1. **Understanding the Problem**: It provides a deep understanding of the design problem, which is critical for developing effective solutions.
2. **Demonstrating Initiative**: It shows your initiative and willingness to go beyond basic requirements, a trait highly valued in professional settings.
3. **Enhancing Team Contribution**: It enables you to contribute more effectively to the project team, leveraging your knowledge and insights.
4. **Improving Qualifications**: A well-researched background can be a significant asset when writing a statement of qualifications, helping to secure your position on a preferred project.

9.6.2   Steps for Conducting Background Research

1. **Consulting with Your Professor**:
   o **Initial Discussion**: Begin by discussing the project with your professor. Professors often have valuable insights and background knowledge about the project.
   o **Taking Notes**: Take detailed notes during your conversation to ensure you capture all the critical points. These notes will be invaluable for further research.
2. **Engaging with the Project Sponsor**:
   o **Direct Contact**: Reach out to the project sponsor to ask detailed questions about the project. The sponsor is likely the most knowledgeable person about the project's specifics.
   o **Additional Contacts**: If the sponsor directs you to other knowledgeable individuals within their company, follow up with them to gather more information.
3. **Exploring Patent Databases**:

- o **Patent Research**: If the project involves a product, search through patent databases for similar or related inventions. This research can provide insights into existing solutions and potential areas for innovation.
- o **Company Patents**: Look for patents filed by the sponsoring company, as these can reveal necessary information about their technological focus and previous work.
4. **Reviewing Scholarly Works**:
   - o **Academic Sources**: Use resources like Google Scholar, ResearchGate, and Academia.edu to find scholarly articles and publications related to the design project. Your university library's database is another excellent resource for accessing academic papers.
   - o **Open Source Projects**: For software projects, look for open-source repositories and communities that might have relevant information or similar projects.
5. **Analyzing Competitor Products and Processes**:
   - o **Competitor Research**: Identify and study competitor products or processes. This can provide a benchmark for your project and highlight areas for improvement or differentiation.
   - o **Sponsor's Input**: Ask the sponsor about their competitors to gain a better understanding of the market landscape and the specific challenges they face.

9.6.3   Flexibility and Broad Interest

While focusing on specific projects of interest is essential, maintaining flexibility in your project preferences can be beneficial. Being open to a broad range of technical areas can serve you well in your professional career, as it exposes you to diverse challenges and solutions. This flexibility can also increase your chances of being assigned to a project that matches your skills and interests.

9.6.3.1   Final Thoughts

Conducting thorough background research is a highly worthwhile effort that significantly influences your likelihood of being considered for a desired project assignment. It demonstrates your dedication, enhances your understanding, and positions you as a valuable team member. Engaging in this research prepares you for the project and sets the foundation for a successful professional engineering career.

9.6.4   Literature Search

A literature search is a foundational step in any scientific or engineering research and development project. It is also crucial for design projects that rely on previous research, discoveries, studies, or other design projects. Conducting a literature search involves collecting and compiling information from books, articles, papers, presentations, reports, theses, or dissertations related to the design project's topic. This process helps provide the background, methodologies, comparisons, context, and identification of knowledge gaps necessary for the design team to work with credibility and rigor.

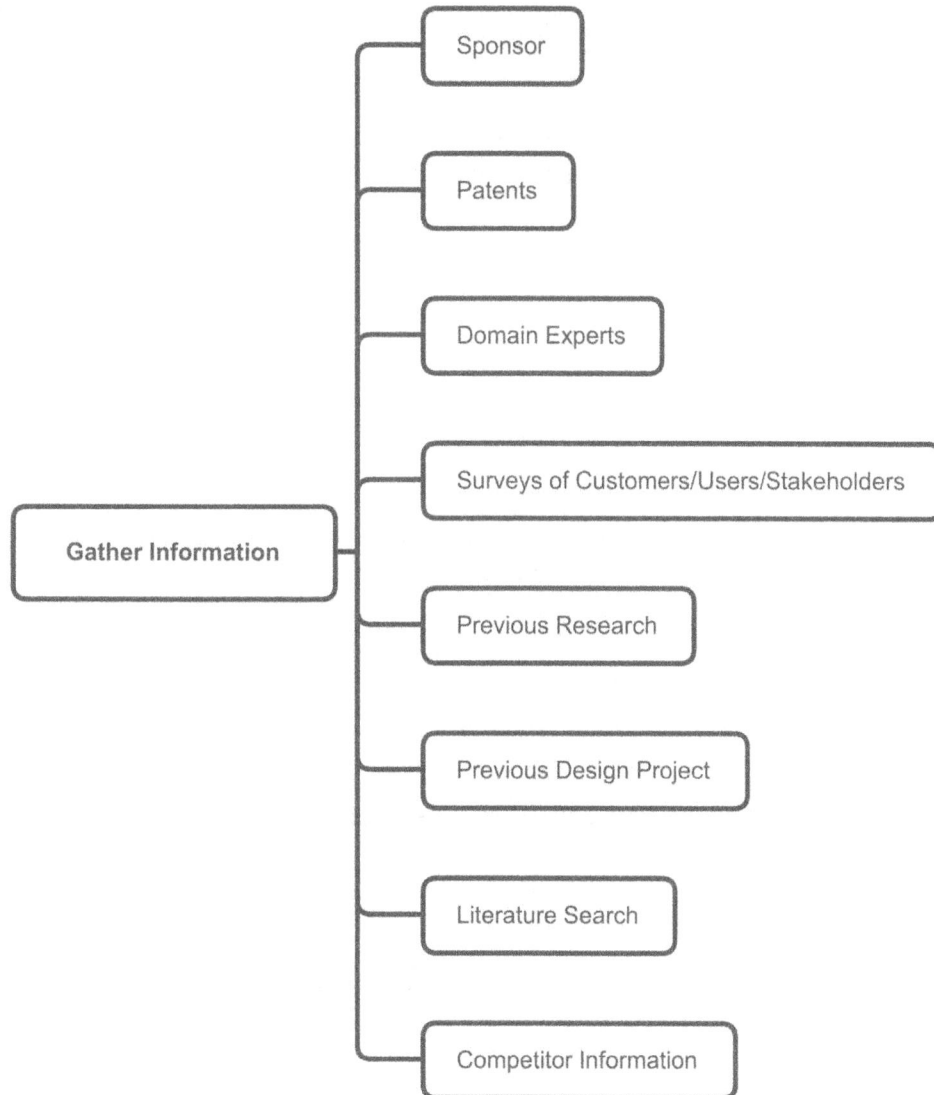

Figure 9-5. Sources of information for design projects.

9.6.4.1   The Importance of Literature Search

1. **Background and Context**:
   - Provides a historical and contextual understanding of the problem.
   - Helps understand the evolution of the topic and current trends.
2. **Methodologies**:

    o   Identifies methodologies used in previous research that can be adopted or adapted for the current project.

    o   Helps avoid past mistakes and leverage successful strategies.

3. **Comparative Analysis**:

    o   Allows comparison of various approaches and solutions to similar problems.

    o   Helps identify the strengths and weaknesses of different methods.

4. **Identifying Knowledge Gaps**:

    o   Highlights areas that have not been explored or need further research.

    o   Guides the focus of the new design project to contribute new knowledge or solutions.

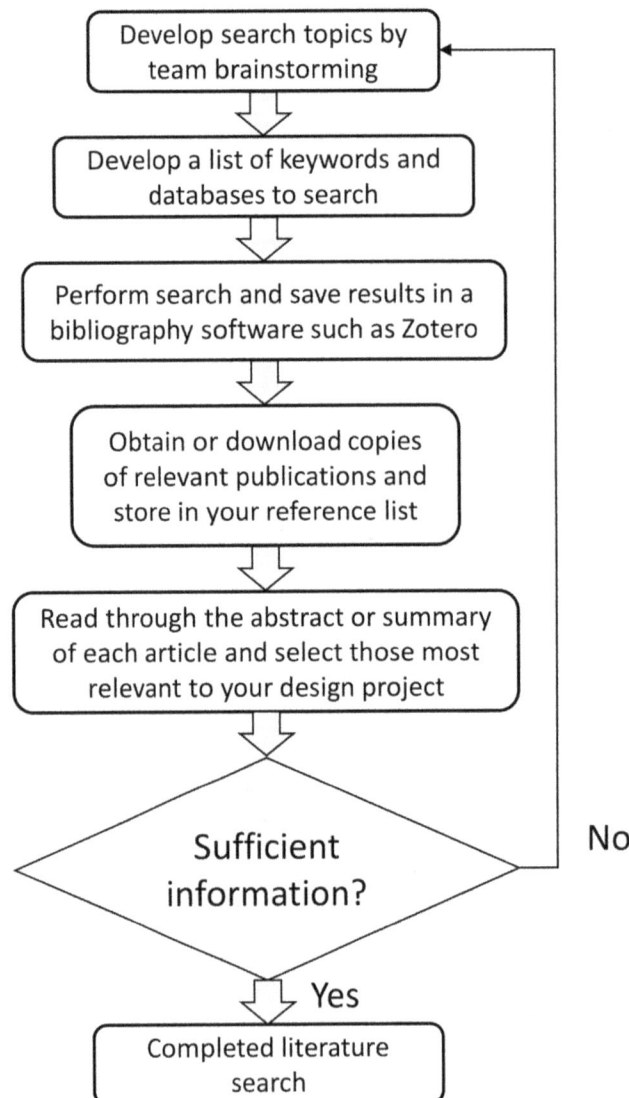

    o

Figure 9-6. Literature search process.

9.6.4.2   Process Flow for a Literature Search

Figure 9-6. illustrates the literature search process. The process begins with brainstorming topics and keywords related to your project. If a report or article has been provided as a starting point, use it to develop your list of topics and keywords. References in the provided report or article can also be an excellent source of topics and keywords. Your sponsor may assist in providing additional topics and keywords for your literature search.

1. **Brainstorming Topics and Keywords**:
    o   Identify the main themes and sub-themes of your design project.
    o   Develop a comprehensive list of keywords and phrases relevant to your project.
2. **Finding Sources of Literature**:
    o   **University Libraries**: Utilize the resources and tutorials provided by your university libraries. They offer access to databases and the ability to download articles. Libraries may also provide interlibrary loan agreements with other institutions.
    o   **Online Databases**: Use free online databases such as Google Scholar, Academia.edu, and ResearchGate. These platforms offer extensive repositories of academic publications and research articles.
    o   **Creating Accounts**: Some online services may require you to create an account. While basic search capabilities are often free, some additional features may require a subscription.
3. **Conducting the Search**:
    o   Enter your keywords and topics into the search engines of your chosen databases.
    o   Review the titles and abstracts of the results to determine their relevance to your project.
4. **Expanding the Search**:
    o   Read relevant articles or book sections thoroughly.
    o   Use the reference lists in these sources to find additional related articles.
5. **Evaluating Sources**:
    o   Focus on peer-reviewed journals and books, as they provide validated and credible information.
    o   Be cautious of non-peer-reviewed internet documents or blogs, as they may contain misinformation.
6. **Compiling and Integrating Information**:
    o   Collect and organize the information gathered from various sources.
    o   Summarize key findings, methodologies, and insights that are pertinent to your design project.

7. **Developing a Comprehensive Understanding**:
   o Integrate the information to build a solid foundation for your project.
   o Identify gaps that your project can address and formulate your research questions or design specifications accordingly.

9.6.4.3   Tools and Techniques for Literature Search

1. **Google Scholar**:
   o A free search engine for scholarly literature across various disciplines. It indexes articles, theses, books, conference papers, and patents.
2. **University Library Databases**:
   o Most universities provide access to databases such as IEEE Xplore, ScienceDirect, JSTOR, and PubMed. These databases are excellent sources for high-quality, peer-reviewed research.
3. **ResearchGate and Academia.edu**:
   o Online platforms where researchers share their publications. They often provide access to full-text articles and allow interaction with authors.
4. **Library Services**:
   o Utilize the services offered by your university library, including access to specialized databases, interlibrary loans, and librarian assistance.
5. **Citation Management Tools**:
   o Use tools like EndNote, Zotero, or Mendeley to organize your references and manage citations. These tools help in keeping track of your sources and generating bibliographies.

9.6.4.4   Example: Automated Pill Dispenser

Consider a design project aimed at creating an automated pill dispenser for elderly patients. The literature search for this project might involve the following steps:
1. **Brainstorming Topics and Keywords**:
   o Keywords might include "automated pill dispenser," "medication adherence," "elderly medication management," "smart health devices," "pill reminder systems," and "medication tracking."
2. **Finding Sources of Literature**:
   o Search databases like PubMed for articles on medication adherence and smart health devices.
   o Use Google Scholar to find papers on existing automated pill dispensers and related technologies.
   o Consult university library databases for books and theses on medication management.
3. **Conducting the Search**:
   o Use combinations of keywords to narrow down the search results.

o Review abstracts to determine the relevance of the articles.

4. **Expanding the Search**:
   o Read relevant articles and explore their reference lists for additional sources.
   o Identify seminal papers and authors in the field to further expand the search.

5. **Evaluating Sources**:
   o Prioritize peer-reviewed journals and books for credible information.
   o Be wary of non-peer-reviewed sources unless they provide unique and valuable insights.

6. **Compiling and Integrating Information**:
   o Summarize key findings from the literature.
   o Identify common themes, methodologies, and gaps in the existing research.

7. **Developing a Comprehensive Understanding**:
   o Use the information to define the problem statement and design specifications.
   o Identify areas where your project can contribute new knowledge or solutions.

Conducting a literature search is an essential skill for any design team. It provides the necessary background, context, and rigor for the design project. By systematically collecting, reading, and integrating information from a variety of sources, the design team can build a solid foundation for their project. This process not only ensures that the team is well-informed but also helps in identifying gaps and opportunities for innovation. A well-executed literature search enhances the credibility and success of the design project, preparing the team for the challenges ahead.

9.6.5 Patents and Patent Search

A patent is a legal right granted by the government to an inventor, giving them the exclusive right to make, use, sell, or import an invention for a certain period, typically 20 years from the filing date, provided they pay maintenance fees. The definition from the United States Patent and Trademark Office (USPTO) states, "A patent for an invention is the grant of a property right to the inventor, issued by the United States Patent and Trademark Office. Generally, the term of a new patent is 20 years from the date on which the application for the patent was filed in the United States or, in special cases, from the date an earlier related application was filed, subject to the payment of maintenance fees."

A patent grants the right to exclude others from making, using, offering for sale, or selling the invention within the United States, or importing the invention into the United States. It does not grant the right to make, use, or sell the invention; instead, it provides the means to prevent others from doing so without permission. The inventor can license or sell the rights to others, potentially profiting from their invention.

US Patents Issued Per Year

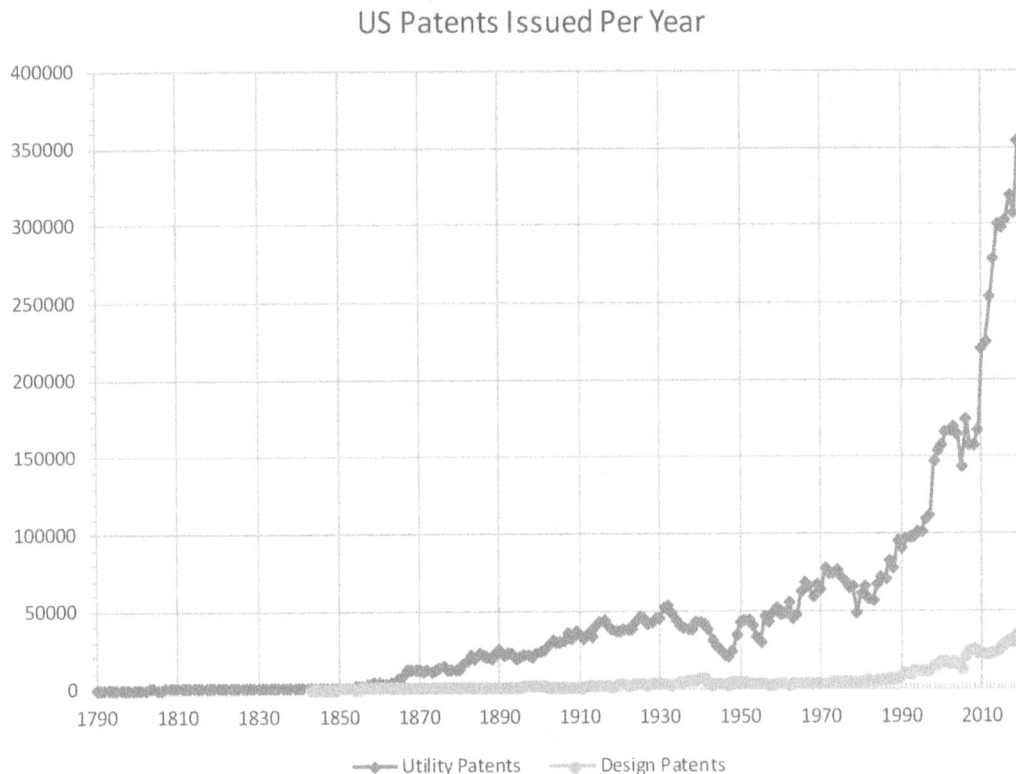

Figure 9-7. **The number of US Utility and Design patents issued per year since 1790 (source: U.S. Patent and Trademark Office.)**

9.6.5.1   Types of Patents

U.S. laws provide for three types of patents:

1. **Utility Patents**: Granted for new and useful processes, machines, articles of manufacture, or compositions of matter, or any new and useful improvements thereof.
2. **Design Patents**: Granted for new, original, and ornamental designs for articles of manufacture.
3. **Plant Patents**: Granted for distinct and new varieties of plants reproduced asexually.

In capstone design projects, utility and design patents are most relevant. Utility patents cover the functional aspects of an invention, while design patents cover the aesthetic aspects.

9.6.5.2   Importance of Patents in Capstone Design

Many capstone design projects have the potential to generate new inventions. If the project sponsor is a corporate entity or an individual entrepreneur, they will likely want full rights to the intellectual property generated from the project. Negotiating intellectual property rights should be done at the project's start to avoid conflicts later.

Past inventions, documented in patents, provide valuable information and insights that can guide the design team. Reviewing existing patents can help identify prior art, avoid infringement, and inspire new ideas.

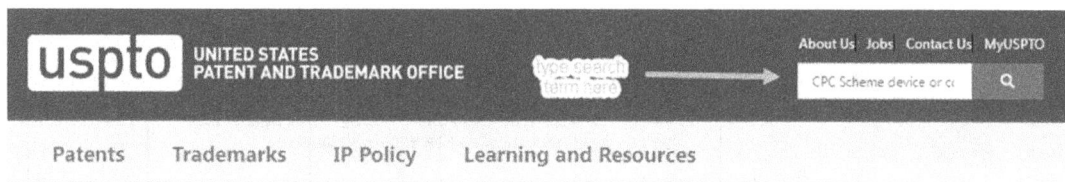

Figure 9-8. The first step in patent search is classification search.

9.6.5.3   The Patent Search Process

Searching the patent database is different from conducting an internet search and requires understanding the database's structure and how to explore it effectively. The U.S. Patent Office recommends a seven-step strategy for patent searches:

1. **Brainstorm Keywords**:
   - Identify keywords describing the purpose, composition, and use of your design. Avoid using general search terms typically used in internet searches.

   Example: For an automated pill dispenser, possible keywords could be "device," "container," "dispensing pills," and "tablets."

2. **Classification Search**:
   - Use the keywords to perform a classification search at www.uspto.gov. Type "CPC Scheme" followed by your keywords in the site search box. CPC stands for Cooperative Patent Classification, based on the International Patent Classification (IPC).

   Example: Typing "CPC Scheme device or container for dispensing pills and tablets" might yield relevant classifications.

3. **Find Classification Numbers**:
   - Review the search results to find applicable CPC or IPC classification numbers. This step requires persistence and may involve multiple iterations of keyword adjustments.

Example: For the pill dispenser, you might find class A61J: CONTAINERS SPECIALLY ADAPTED FOR MEDICAL OR PHARMACEUTICAL PURPOSES.

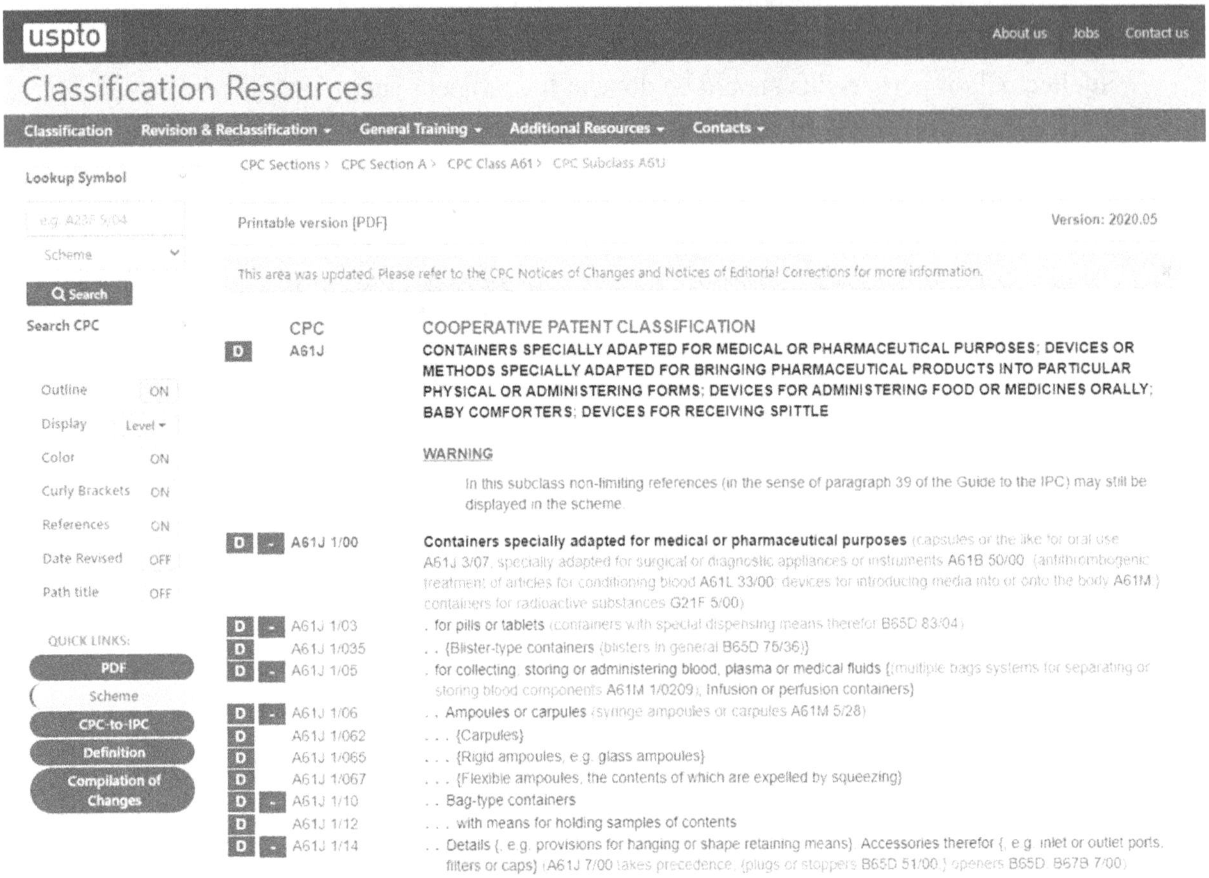

Figure 9-9. Results from classification search for pill dispenser.

4. **Application Full-Text and Image Database Search**:
   o Use the CPC classification number to search the Application Full-Text and Image Database (AppFT) at http://appft.uspto.gov. Utilize the "Quick Search" or "Advanced Search" options to locate relevant patents.

Example: Using the classification number A61J, you can perform a search to find patents related to medical containers and dispensing devices.

5. **Analyze Search Results**:
   o Examine the search results to identify relevant patents. If initial results are not useful, refine your keyword list and search again.

Example: If no relevant patents are found, consider modifying or adding keywords to improve search accuracy.

6. **Review Patents**:

- o Review the full text and images of identified patents to understand their scope and claims. Pay attention to the similarities and differences with your design project.

Example: Reviewing patents related to pill dispensers can provide insights into existing solutions and potential improvements.

7. **Refine Search Using Advanced Options**:
   - o If too many results are found, use the "Advanced Search" option to narrow down the search. This can include Boolean operators and specific search fields to target more precise results.

9.6.5.4 Tools and Databases for Patent Search

Several tools and databases are available for conducting patent searches:

1. **USPTO Database**:
   - o The official database of the United States Patent and Trademark Office provides access to full-text patents and applications.

2. **Google Patents**:
   - o Google has indexed many patents and patent applications, making it accessible via scholar.google.com. It allows for searching using familiar internet search techniques.

3. **World Intellectual Property Organization (WIPO)**:
   - o WIPO provides an international patent classification system and search tools at
     https://www.wipo.int/classifications/ipc/ipcpub/?notion=catchword.

9.6.5.5 Example: Automated Pill Dispenser Patent Search

1. **Brainstorm Keywords**:
   - o Device, container, dispensing pills, tablets, medication management.

2. **Classification Search**:
   - o Use keywords to find CPC classification: "CPC Scheme device or container for dispensing pills and tablets."

3. **Find Classification Numbers**:
   - o CPC class A61J: CONTAINERS SPECIALLY ADAPTED FOR MEDICAL OR PHARMACEUTICAL PURPOSES.

4. **Search Application Database**:
   - o Use classification A61J to search the AppFT database.

5. **Analyze Results**:
   - o Review search results and refine keywords if necessary.

6. **Review Patents**:
   - o Examine relevant patents to understand existing solutions and identify areas for innovation.

7. **Advanced Search**:
   o   Use advanced search options to narrow down results and find specific
       patents related to your design.

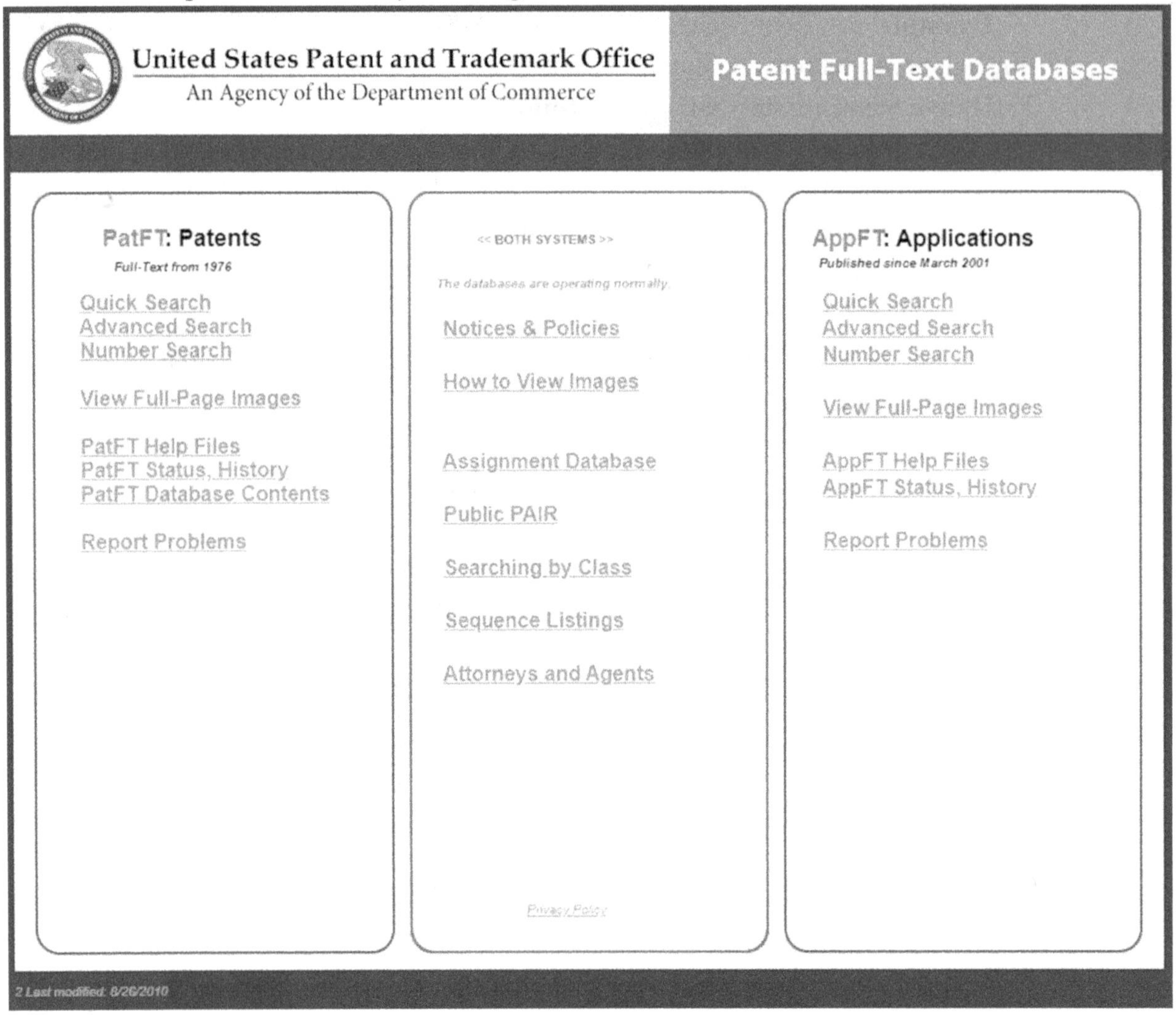

Figure 9-10.   USPTO Patent Full-Text Database

Conducting a patent search is a critical step in the design process. It helps ensure
that your design is innovative and does not infringe on existing patents. It provides a
wealth of information on past and current inventions, guiding the design team toward a
successful and original solution. By following the structured approach outlined above,
design teams can effectively navigate the patent landscape and leverage existing
knowledge to enhance their projects.

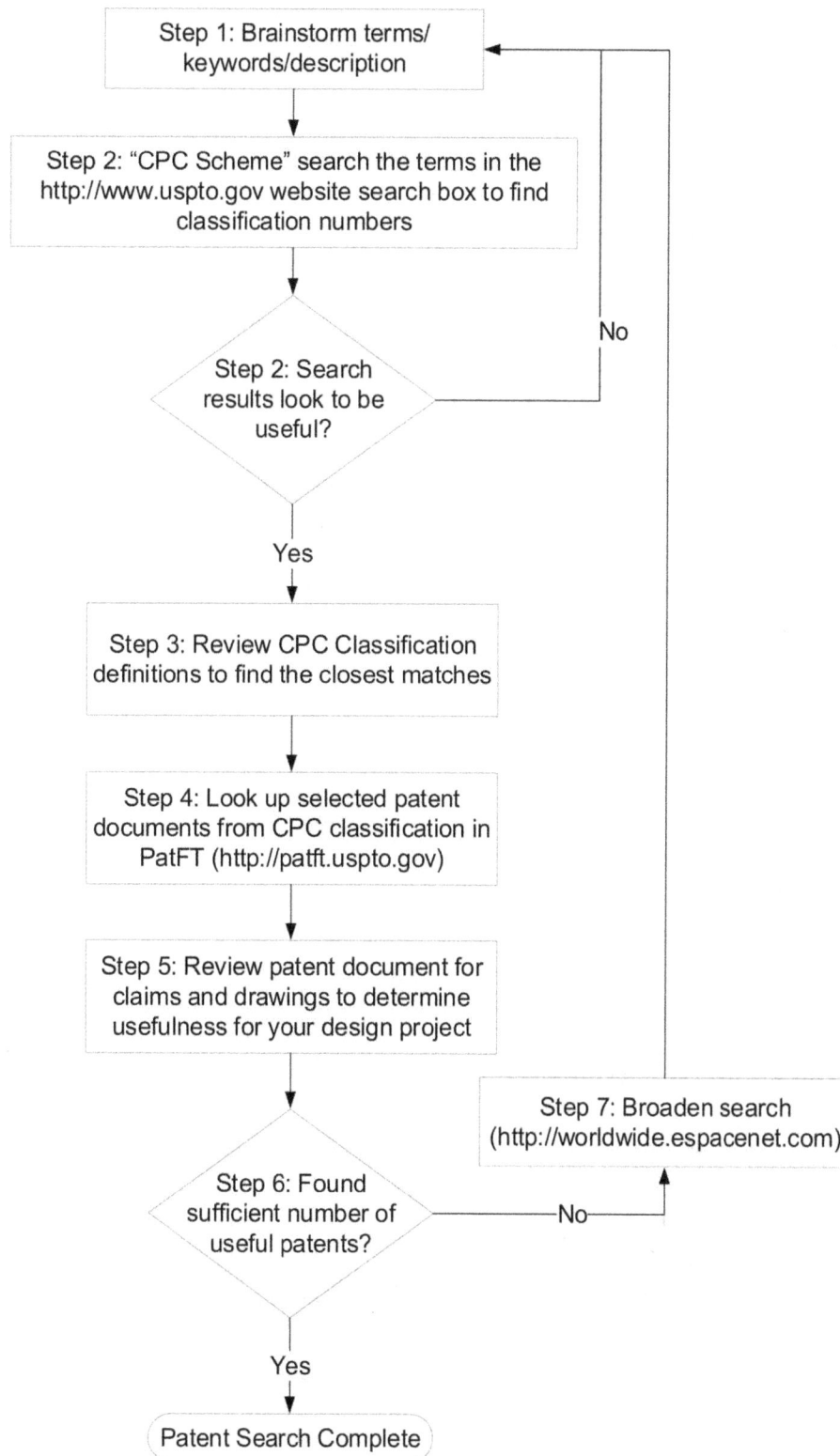

Figure 9-11. Patent search flowchart

USPTO PATENT FULL-TEXT AND IMAGE DATABASE

| Home | Quick | Advanced | Pat Num | Help |

| View Cart |

Data current through August 4, 2020.

Query [Help]

Term 1: [A61J]   in Field 1: [Current CPC Classification Class ▾]

[AND ▾]

Term 2: [   ]   in Field 2: [All Fields ▾]

Select years [Help]
[1976 to present [full-text] ▾]   [Search] [Reset]

Patents from 1790 through 1975 are searchable only by Issue Date, Patent Number, and Current US Classification.
When searching for specific numbers in the Patent Number field, utility patent numbers are entered as one to eight numbers in length, excluding commas (which are optional, as are leading zeroes)

Figure 9-12. Classification search using Quick Search.

Results of Search in US Patent Collection db for:
CPCL/A61J: 11392 patents.
Hits 1 through 50 out of 11392

[Next 50 Hits]

[Jump To] [   ]

[Refine Search] [CPCL/A61J   ]

PAT. NO.   Title
1 RE48,136   Pill dispensing method and apparatus
2 10,733,397   Method for managing at least one container and associated methods and devices
3 10,732,083   Thawing biological substances
4 10,730,687   Intelligent medicine dispenser
5 10,730,682   Connecting and container system
6 10,730,207   Process for manufacturing a resulting pharmaceutical film
7 10,730,058   Apparatus, systems and methods for storing, treating and/or processing blood and blood components
8 10,729,859   Mask for administration of inhaled medication
9 10,729,854   Housing for mounting a container on an injection pen, assembly forming an injectable product reservoir for an injection pen and injection pen equipped with such an assembly
10 10,729,842   Medical vial and injector assemblies and methods of use
11 10,729,828   Closed disposable multiple sterile blood bag system for fractionating blood with the corresponding method
12 10,729,722   Dialysis solution, formulated and stored in two parts, comprising phosphate
13 10,729,721   Dialysis solution, formulated and stored in two parts, comprising phosphate
14 10,729,673   Taxane particles and their use
15 10,729,666   Use of GABAA receptor reinforcing agent in preparation of sedative and anesthetic medicament
16 10,729,655   Orally disintegrating tablets
17 10,729,621   Acoustic reflectometry device in catheters
18 10,729,620   Baby bottle apparatus
19 10,729,619   Capsule sealing composition and its sealing method thereof
20 10,729,618   Liquid medicine filling device and liquid medicine filling method

Figure 9-13. CPC classification search results for class A61J.

9.7 Design for X

Design is an iterative process. The initial idea or rendition of a design solution is rarely the final design. Quality in design is one attribute that necessitates iteration. Each iteration requires establishing specific objectives to test against, allowing the team to check for convergence and determine when to stop the iteration. It's crucial to develop important objective attributes or functions before starting the design solution activity and integrate those into the process.

Design for X (DFX) is a methodology that leverages the knowledge gained by designers in various fields and applies it to the design activity. The "X" in Design for X is a variable that can represent many different values, such as cost, manufacture, assembly, maintainability, operability, quality, reliability, safety, sustainability, ergonomics, installability, user-friendliness, portability (software), modularity, etc. Considering a set of X values during the design process may lengthen the time needed to achieve a design solution. Still, ito.gov will result in a better and more comprehensive design.

9.7.1 Choosing Relevant X Values

You must decide which X values apply to your design and incorporate those into your design process. For a typical capstone design project, it is advisable to limit the number of X values to three or four, which are most important for your design objectives. It is also desirable to pick X values that demonstrate how you incorporated those methods into your design activity. You need to show, explain, or demonstrate how your design achieves the X value.

9.7.2 Examples of Design for X

1. **Design for Manufacture (DFM)**:
   o **Objective**: Ensure that the design can be easily and cost-effectively manufactured.
   o **Application**: When designing a new mechanical component, consider the manufacturing processes available, such as injection molding, CNC machining, or 3D printing.
   o **Demonstration**: If the design includes complex geometries, simplify the shapes to reduce manufacturing time and costs. For instance, converting a part with multiple undercuts into a design that can be made with a single mold will significantly reduce manufacturing complexity and expense.
2. **Design for Assembly (DFA)**:
   o **Objective**: Simplify the assembly process to reduce time and errors.
   o **Application**: When designing an electronic device, ensure that components can be easily aligned and fastened together.
   o **Demonstration**: Reduce the number of fasteners and make parts self-

locating, which minimizes assembly steps. For example, using snap-fit features instead of screws can expedite assembly and reduce the need for tools.

3. **Design for Safety (DFS)**:
    o **Objective**: Ensure that the design is safe for users and operators.
    o **Application**: When designing a consumer product, consider all potential hazards and incorporate safety features.
    o **Demonstration**: Add guards to moving parts, use non-toxic materials, and ensure that electrical components are properly insulated. For example, a power tool should have safety switches and blade guards to prevent accidental injuries.

4. **Design for Cost (DFC)**:
    o **Objective**: Minimize the cost of production and operation.
    o **Application**: When designing a new product, select materials and processes that are cost-effective without compromising quality.
    o **Demonstration**: Choose less expensive materials that meet the necessary performance requirements and design components that can be produced using less expensive manufacturing techniques. For instance, using plastic instead of metal for certain non-structural parts can reduce material and production costs.

5. **Design for Sustainability (DFS)**:
    o **Objective**: Minimize environmental impact throughout the product's lifecycle.
    o **Application**: When designing a product, consider its entire lifecycle, from material selection to end-of-life disposal.
    o **Demonstration**: Use recyclable materials, design for disassembly, and minimize energy consumption during manufacturing and operation. For example, designing a product with easily separable components can facilitate recycling at the end of its life.

9.7.3   Design for X and Engineering Requirements

Engineering requirements are specifications that a design must meet to be successful. These requirements encompass various aspects, including functionality, performance, reliability, and regulatory compliance. Integrating DFX principles helps ensure that the design meets these engineering requirements efficiently and effectively. Let's explore how DFX principles relate to engineering requirements:

1. **Functional Requirements**:
    o **DFM and DFA** ensure that the design is not only manufacturable but also easy to assemble, which directly impacts the functionality and reliability of the final product. If a product is designed with manufacturability and

assembly in mind, it is more likely to function correctly because the chances of manufacturing or assembly errors are minimized.

- o **Example**: A car engine designed with DFM and DFA principles will have components that fit together precisely and can be assembled without mistakes, leading to a reliable and functional engine.

2. **Performance Requirements**:
   - o **DFS** helps ensure that the product performs safely under all expected conditions. By integrating safety considerations into the design, the product will meet performance requirements related to user safety and operational reliability.
   - o **Example**: A medical device designed with DFS will include features like fail-safes and error detection to ensure it performs reliably in critical situations.

3. **Reliability Requirements**:
   - o **Design for Reliability (DFR)** focuses on ensuring that the product operates as intended over its expected lifespan. By designing for reliability, the team ensures that the product meets stringent reliability requirements.
   - o **Example**: An aerospace component designed with DFR principles will undergo rigorous testing and validation to ensure it meets the reliability standards required for space missions.

4. **Regulatory Requirements**:
   - o **Design for Compliance (DFC)** involves ensuring that the product meets all relevant regulatory and industry standards. This includes considerations for safety, environmental impact, and quality.
   - o **Example**: An electronic device designed with DFC principles will comply with standards like CE marking or FCC certification, ensuring it meets regulatory requirements for safety and electromagnetic compatibility.

5. **Sustainability Requirements**:
   - o **Design for Sustainability (DFS)** ensures that the product minimizes its environmental impact throughout its lifecycle, meeting sustainability requirements.
   - o **Example**: A consumer product designed with DFS principles will use recyclable materials and energy-efficient manufacturing processes, aligning with sustainability goals.

9.7.4 Integrating DFX into the Design Process

To effectively integrate DFX into your design process, follow these steps:

1. **Identify Relevant X Values**:
   - o Based on the nature of your project and the objectives, choose three or

four most critical X values.

Example: The relevant X values for an automated pill dispenser might be safety, manufacture, user-friendliness, and cost.

2. **Conduct a Literature Search**:
   o Research contemporary publications, articles, and patents relevant to your chosen X values. This will provide insights into best practices and current trends.

Example: Search for articles on safe medical device design, efficient manufacturing techniques for plastic components, and user interface design for elderly users.

3. **Develop Design Guidelines**:
   o Create specific guidelines based on your research to address each X value in your design process.

Example: For safety, include guidelines on using non-toxic materials and incorporating safety interlocks. For user-friendliness, ensure the interface is simple and intuitive.

4. **Apply DFX Principles in Iterations**:
   o Implement the guidelines in your design iterations and continually test against these objectives to ensure they are met.

Example: In each design iteration of the pill dispenser, test the ease of use by having potential users interact with prototypes and provide feedback.

5. **Document the Process**:
   o Keep detailed records of how you applied each DFX principle, the changes made to the design, and the results of each iteration.

Example: Document the modifications made to the pill dispenser to enhance manufacturability, such as simplifying the design to reduce the number of parts and using standard components.

6. **Evaluate and Refine**:
   o Continuously evaluate the design against the DFX objectives and refine it as necessary.

Example: After each iteration, assess whether the design improvements have reduced manufacturing costs or increased user safety and make further adjustments based on these evaluations.

Design for X is a powerful methodology that ensures the design process considers multiple critical aspects, leading to a more robust, efficient, and effective final product. By carefully selecting and integrating relevant X values, design teams can create solutions that meet high standards of quality, safety, cost-efficiency, and user satisfaction. This iterative process not only improves the design but also equips the team with valuable insights and experience in applying industry best practices. Integrating DFX principles with engineering requirements ensures that the design is not only innovative but also meets the practical and regulatory standards necessary for successful implementation.

9.8 Design Specifications

Design specifications, or engineering design specifications, are a set of constraints or criteria that the design must satisfy. These specifications must be quantitative and measurable with a range or specific target value and must have units.

Design parameters are specifications, requirements, or customer desires explicitly expressed in a quantitative form. For example, for the automated pill dispenser, we may state that the device can hold five different pills, where the maximum dimension of any tablet is less than 1 inch.

Designers should avoid vague design specifications which the customer may directly state. For example, the customer may indicate that they want the software to be user-friendly. How can we quantify user-friendliness? At the end of the design process, we should measure our design's performance against the design specification. If the specification is not quantitative (numeric), it won't be easy to measure and compare.

The numerical values associated with design specifications may be expressed as a range when there is insufficient information or during an early design phase. For example, we may require that the pill dispenser be priced between $300 and $500 when it is introduced into the market. The price range can be learned from surveys of potential customers for the product or by competitive analysis if similar products are sold in the market.

As the design process proceeds, the design team will develop a deeper understanding of the design problem. Therefore, updating the design specifications is a necessary part of the design process. For example, suppose you encounter a design specification that cannot be met without disturbing the balance of the design. In that case, you should consider modifying the design spec in consultation with your project sponsor. Be prepared to explain why you are asking to change a design specification.

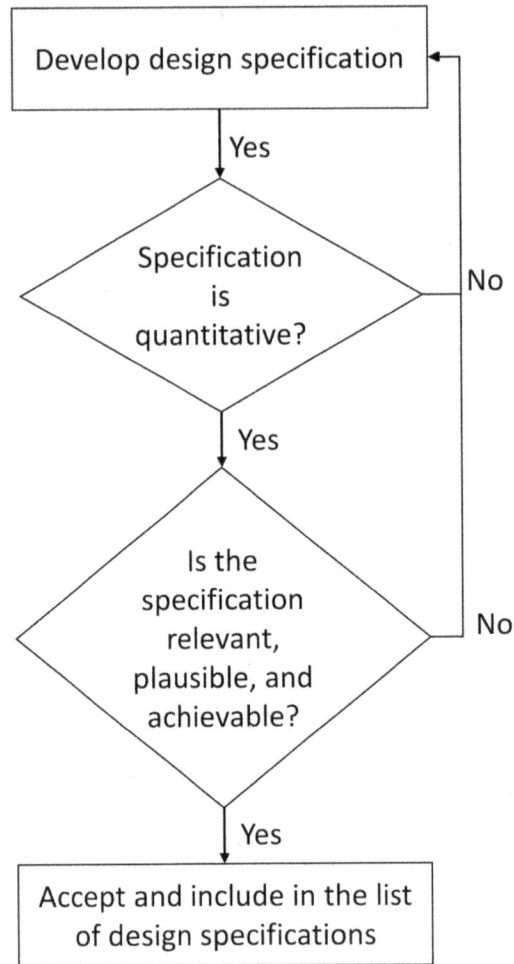

Figure 9-14. Check design specifications for correctness and relevance.

9.8.1   Customer Needs and Requirements

Capstone design projects generally have sponsors who are the customers for the design or represent a grouping of customers for the desired solution. The design problems are expressed in an open-ended form such that the design team must investigate and discover the list of needs or requirements. Finding the customer's needs and requirements can be challenging for the design team. The design team must translate the statements or definitions provided by the customer into a form that will be useful to the design process. Engineering designers should focus on stating the customer needs or requirements in the form of features or functions of the product and avoid including specifications on how the design performs that function or delivers that feature.

Customer needs or requirements should be expressed as clearly and accurately as possible. The team should share and solicit feedback from the sponsor on the statements. The statements should also be stated as positive attributes of the final design product or process. For example, instead of saying that the product weight should not be too heavy, you can say that the product should be light and within some numerical range. Also, express the customer needs and requirements with some priority importance level instead of a hard constraint. You can work with the importance level to achieve more flexibility in meeting a plausible design.

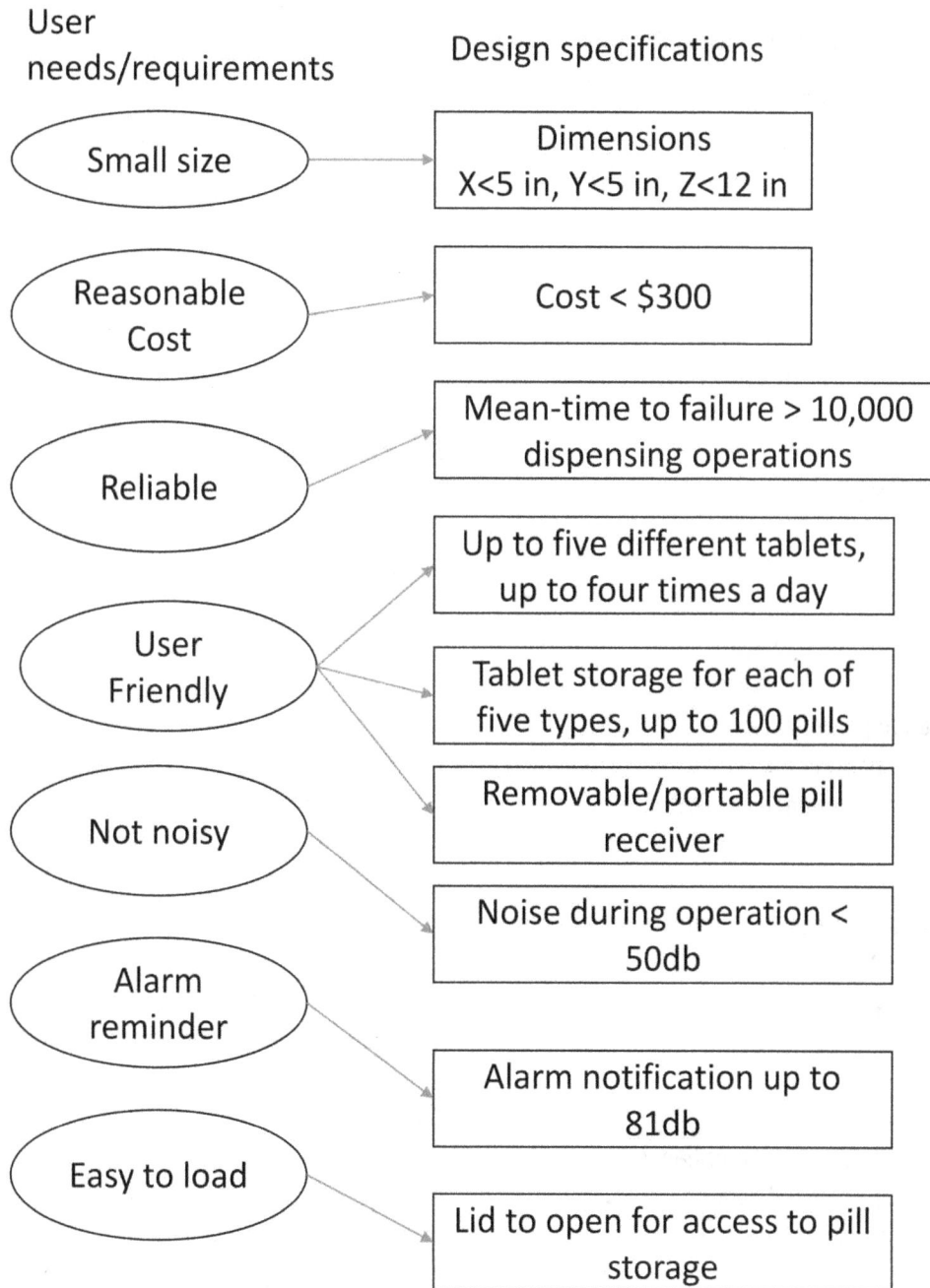

Figure 9-15. Translation of customer needs/requirements to design specifications.

If the customer has many requirements, it will be necessary to group those requirements into logical groupings. Assign a relative importance value (say between 0 and 100) to each need and sort the list such that the most critical items appear first. By scanning the list, you should be able to eliminate redundant requirements. If you can group the requirements list into different objectives, you may need to move some items to balance and prune the hierarchy.

9.8.2   Design Specifications Leading Questions

In this section, we present a template to help the design team develop the specifications for a capstone design project. The questions presented, and the answers to them can be stated as design specifications in a tabular form. This template will be helpful to both product and process-type design projects. Software design projects may need additional changes to this template.

Table 9-1. Working template for product or process capstone design projects.

Parameters	Specific Values or Ranges
Name of project	Project name
Costs	Values and units
Functions	List each one if it can be quantified, including values and units.
Performance parameters	E.g., acceleration, efficiency, power, reliability; values and units
Installation parameters	Values and units
Operation parameters	Values and units
Maintenance parameters	Values and units
Safety parameters	Values and units
Dimensions	Values and units
Weight, load, pressure, stress, temperatures, humidity, etc.	Values and units
Special requirements	E.g., clean environment, values, and units
Environmental impact	Sustainability, durability, reliability, etc.; values and units
User parameters	Values and units
Standards and codes compliance	Numbers and references
Manufacturing specifications	Numbers and references
Materials specifications	References, values, and units

9.8.3 Detailed Steps for Creating Design Specifications

9.8.3.1 Identify Customer Needs and Requirements

- **Interviews**: Conduct interviews with customers, sponsors, and stakeholders to gather needs and requirements.
- **Surveys**: Distribute surveys to potential users to collect broader input on their needs and expectations.
- **Observations**: Observe users interacting with similar products or systems to identify unmet needs or areas for improvement.
- **Feedback Analysis**: Review feedback from previous products or iterations to understand common issues or desires.

9.8.3.2 Translate Needs into Specifications

- **Functional Specifications**: Define what the product or system must do. For instance, "The pill dispenser must dispense five different types of pills."
- **Performance Specifications**: Specify how well the product must perform its functions. For example, "The pill dispenser must dispense each pill within ±2 seconds of the programmed time."
- **Operational Specifications**: Detail the conditions under which the product will operate. For example, "The pill dispenser must operate at temperatures between 0°C and 40°C."
- **Maintenance Specifications**: Outline requirements for maintaining the product. For instance, "The pill dispenser must require maintenance no more frequently than once per year."

9.8.3.3 Quantify the Specifications

- Assign measurable values to each specification to ensure they are clear and testable.
- **Example**: Instead of stating, "The pill dispenser must be lightweight," specify, "The pill dispenser must weigh no more than 500 grams."

9.8.3.4 Validate Specifications with Stakeholders

- Review the specifications with stakeholders to ensure they accurately reflect their needs and expectations.
- Make necessary adjustments based on stakeholder feedback.

9.8.3.5 Document and Organize Specifications

- Create a comprehensive document that lists all design specifications in a clear, organized manner.

- Use tables, charts, and diagrams as necessary to enhance understanding.

9.8.3.6   Examples of Design Specifications

Table 9-2. Automated Pill Dispenser Design Specifications

Parameter	Specific Values or Ranges
Name of project	Automated Pill Dispenser
Costs	$300-$500
Functions	Dispenses up to 5 types of pills; Sends reminders to patients
Performance parameters	Dispense accuracy within ±2 seconds; Battery life of at least 30 days.
Installation parameters	It should be plug-and-play with minimal setup.
Operation parameters	Operates at temperatures between 0°C and 40°C
Maintenance parameters	Requires maintenance no more than once per year
Safety parameters	Non-toxic materials; Child-proof lock
Dimensions	Maximum dimensions of 10x10x10 inches
Weight	No more than 500 grams
Special requirements	It must be user-friendly for elderly patients.
Environmental impact	Made from recyclable materials
User parameters	Intuitive interface with large buttons
Standards and codes compliance	Complies with FDA regulations for medical devices
Manufacturing specifications	Must be manufacturable using standard injection molding processes
Materials specifications	Uses BPA-free plastics; Stainless steel springs

9.8.3.7   Refining Design Specifications

As the design progresses, the team will gain a deeper understanding of the problem, and the specifications may need to be refined. This process involves:

1. **Regular Review**: Periodically review specifications to ensure they remain relevant and achievable.
2. **Stakeholder Feedback**: Continually seek input from stakeholders to verify that the design meets their evolving needs.
3. **Prototyping and Testing**: Use prototypes to test the design against specifications and identify areas for improvement.
4. **Flexibility and Adaptation**: Be prepared to adapt specifications based on new information or constraints encountered during the design process.

Design specifications are crucial for guiding the design process and ensuring that the final product meets customer needs and expectations. By translating customer needs into precise, measurable specifications, design teams can systematically address all aspects of the design and create solutions that are functional, reliable, and user-friendly. The iterative process of refining specifications based on stakeholder feedback and testing ensures that the design remains aligned with project goals and delivers a successful outcome.

9.9 Assignments

Assignment 9-1: Literature Search Assignment (Individual Assignment)

   1. Objective:

      The goal of this assignment is to conduct a comprehensive literature search on your design problem. You will gather information from various sources, filter out the less relevant ones, and summarize the five most relevant papers or sources. This exercise will help you understand the background, methodologies, and context of your design problem and identify knowledge gaps that your project can address.

   2. Instructions:

   3. Brainstorm Topics and Keywords:

      ○ Identify the main themes and sub-themes of your design problem.

      ○ Develop a comprehensive list of keywords and phrases relevant to your project.

     Response:

Main Themes:

Sub-Themes:

Keywords/Phrases:

Find Sources of Literature:
- Utilize university library databases and online resources like Google Scholar, Academia.edu, and ResearchGate.
- List at least three databases or sources you will use for your search.

Response:

Database/Source 1:

Database/Source 2:

Database/Source 3:

4. **Conduct the Search:**
   o Enter your keywords and topics into the search engines of your chosen databases.
   o Review the titles and abstracts of the search results to determine their relevance.
   o List at least ten sources you found during your search.

   Response:

   _____

   _____

   _____

   _____

   _____

   _____

   _____

   _____

   _____

   _____

5. **Filter Out Less Relevant Sources:**
   o Evaluate the relevance and quality of the sources based on their abstracts and introductions.
   o Filter out the less relevant sources and narrow down your list to the five most relevant papers or sources.

   Response:
   1. _____
   2. _____
   3. _____
   4. _____
   5. _____

6. **Summarize the Selected Papers/Sources:**
   o Write a summary of each of the five selected papers or sources. Each summary should be 500 words or less.
   o Include the following in each summary:
      ▪ Title and author(s) of the paper or source.
      ▪ Main objective or research question.
      ▪ Methodologies used.
      ▪ Key findings or conclusions.
      ▪ Relevance to your design problem.

Response:
    o **Paper 1:**
Title and Author(s):

Summary (500 words or less):

    o **Paper 2:**
Title and Author(s):

Summary (500 words or less):

      o **Paper 3:**

Title and Author(s):

_____

_____

Summary (500 words or less):

_____

_____

_____

_____

_____

_____

_____

_____

_____

_____

      o **Paper 4:**

Title and Author(s):

_____

_____

Summary (500 words or less):

_____

_____

_____

_____

_____

_____

_____

_____

_____

      ○ **Paper 5:**

Title and Author(s):

_____

_____

Summary (500 words or less):

_____

_____

_____

_____

_____

_____

_____

_____

_____

5. **Reflect on the Literature Search Process:**
   - ○ Discuss the challenges you faced during the literature search.
   - ○ Explain how the selected sources will inform your design project.
   - ○ Identify any knowledge gaps or areas that you discovered for further research.

   Response:

Challenges Faced:

_____

_____

_____

Informing the Design Project:

_____

_____

Knowledge Gaps:

_____

_____

_____

6. **Submission:**

Submit your completed assignment by the due date, including your responses to the questions and the summaries of the five selected papers or sources. Ensure that your responses are clear, concise, and well-organized.

7. **Grading Criteria:**

- Completeness and relevance of keywords and sources listed.
- Quality and relevance of the selected five papers or sources.
- Clarity, coherence, and thoroughness of the summaries.
- Reflection on the literature search process, challenges, and knowledge gaps.

By completing this assignment, you will gain valuable skills in conducting a literature search, critically evaluating sources, and integrating information to support your design project.

Assignment 9-2: Patent Search for Capstone Design Project

Objective: To conduct a comprehensive patent search related to your capstone design project using both the USPTO database and Google Patents, identify at least ten relevant patents, and use this information to inform your design process.

Instructions:

1. Familiarize yourself with patent search resources:
   - Review the USPTO's "Seven Step Strategy" for patent searching: https://www.uspto.gov/patents/search/seven-step-strategy
   - Explore the USPTO Patent Full-Text and Image Database (PatFT): https://patft.uspto.gov/
   - Visit Google Patents: https://patents.google.com/
2. Define your search:
   - Brainstorm keywords related to your project's purpose, function, and components
   - Identify synonyms and alternative terms for each keyword
   - Consider the problem your invention solves and how it solves it
3. Conducting an Advanced Search on the USPTO site:
   a. Go to the USPTO Patent Full-Text and Image Database (PatFT): https://patft.uspto.gov/
   b. Click on "Advanced Search" on the left-hand menu.
   c. In the Advanced Search page, you'll see a series of text boxes where you can enter your search terms. Each box represents a search field, and you can combine these fields using Boolean operators (AND, OR, NOT).
   d. Here's how to use the advanced search fields:
      - Use the field codes before your search terms to specify where to search. Some common field codes are:
         - TTL/: Title
         - ABST/: Abstract
         - ACLM/: Claims
         - ISD/: Issue Date
         - APD/: Application Date
         - IN/: Inventor Name
         - AN/: Assignee Name
      - Use quotation marks for exact phrases: "solar panel."
      - Use asterisks for truncation: electr* will find electric, electrical, electricity, etc.
      - Use Boolean operators to combine terms: AND, OR, NOT
      - Use parentheses to group terms: (solar OR photovoltaic) AND panel
   e. Example Advanced Search:
      - Let's say you're working on a project involving an energy-efficient

refrigeration system. Here's an example of how you might structure your advanced search:

In the search boxes, enter:

1. TTL/(refrige* OR cool*) AND (energy OR efficien*)
2. ABST/(compressor OR condenser) AND (variable speed OR inverter)
3. ISD/20100101->20230101

This search will look for:

1. Patents with titles containing words starting with "refrige" or "cool" AND either "energy" or words starting with "efficien"
2. AND patents with abstracts containing either "compressor" or "condenser" AND either the phrase "variable speed" or "inverter"
3. AND patents issued between January 1, 2010, and January 1, 2023

f. Click "Search" to run your query.
g. Review the results, noting relevant patents and their classification codes for further refinement of your search.
h. Modify your search terms and repeat the process as necessary to find the most relevant patents for your project.

Remember to apply similar advanced search techniques when using Google Patents, adapting to its specific search syntax and features.

1. Compile a list of at least ten relevant patents: For each patent, record:
   o Patent number
   o Title
   o Issue date
   o Inventor(s)
   o Brief description of how it relates to your project
   o Key features or technologies that could inform your design
2. Analyze the patents:
   o Identify common themes or approaches
   o Note any gaps or opportunities for innovation
   o Consider how these patents might influence your design choices
3. Prepare your report:
   o Use the template provided on Brightspace
   o For each patent, include the information from step 4 and your analysis from step 5
4. Reflect on the search process:
   o Discuss the effectiveness of the USPTO Seven-Step Strategy
   o Compare the results from USPTO PatFT and Google Patents
   o Explain how this patent search might influence your design approach

5. Share your findings:
   o Prepare a brief presentation (5-10 minutes) summarizing your key findings
   o Share this with your team members in your next team meeting
6. Submit your report:
   o Ensure all sections of the template are completed
   o Double-check that you've included at least ten relevant patents
   o Submit the completed report via Brightspace by the due date

Additional Guidelines:

- Pay attention to recent patents, as they represent the current state-of-the-art
- Consider both utility and design patents in your search
- Remember that this is an individual assignment, but the knowledge gained should benefit your entire team

Evaluation Criteria:

- Thoroughness of the patent search using both USPTO and Google Patents
- Proper application of the USPTO Seven-Step Strategy
- Relevance of selected patents to the project
- Quality of analysis and insights derived from the patents
- Clarity and completeness of the report
- Potential impact on the design process

By completing this assignment, you'll gain valuable insights into existing technologies related to your project and develop essential patent-searching skills that will be useful throughout your engineering career.

Assignment 9-3: Design Specifications Assignment (Team Assignment)

In this assignment, you will create an initial set of design specifications for your project using the provided template. As you progress through your project, you will refine these specifications to ensure they meet the needs of your stakeholders and the constraints of your design. This process will help guide your project from concept to prototype to the final design solution.

1. **Objective:**

Using the provided template, develop a comprehensive and measurable set of design specifications for your capstone project. Refine these specifications as you gain more knowledge about the design problem and potential solutions.

2. **Steps and Guidance:**

3. **Step 1: Understand the Design Problem**

   i. **Review the Problem Statement**: Start by carefully reviewing the problem statement provided by your sponsor or professor.
      - What is the problem or need that your design will address?
      - What are the primary goals and objectives of the project?

   ii. **Gather Initial Information**: Conduct initial research to understand the context and constraints of the problem.
      - What existing solutions or products address similar problems?
      - What are the limitations and shortcomings of these existing solutions?

4. **Step 2: Identify Customer Needs and Requirements**

   i. **Conduct Interviews and Surveys**: Gather information from potential users, stakeholders, and experts.
      - Develop a list of questions to ask during interviews or include in surveys.
      - Aim to understand the user's needs, preferences, and pain points.

   ii. **Analyze Feedback**: Compile and analyze the feedback collected.
      - Identify common themes and critical requirements from the data.
      - Categorize the needs into primary and secondary requirements.

5. **Step 3: Translate Needs into Design Specifications**

6. **Create Functional Specifications**: Define what the product or system must do.
   - Example: The pill dispenser must dispense five different types of pills.

7. **Establish Performance Specifications**: Specify how well the product must perform its functions.
   - Example: The pill dispenser must dispense each pill within ±2 seconds of the programmed time.

8. **Define Operational Specifications**: Detail the conditions under which the product will operate.
   - Example: The pill dispenser must operate at temperatures between 0°C and 40°C.

9. **Set Maintenance Specifications**: Outline requirements for maintaining the product.
   - o Example: The pill dispenser must require maintenance no more frequently than once per year.

10. **Step 4: Quantify the Specifications**
    i. **Assign Measurable Values**: Ensure each specification is clear, measurable, and testable.
       - o Example: Instead of "The pill dispenser must be lightweight," specify "The pill dispenser must weigh no more than 500 grams."
    ii. **Use Ranges When Necessary**: If exact values are not yet known, use reasonable ranges.
       - o Example: The device should cost between $300 and $500.

11. **Step 5: Validate Specifications with Stakeholders**
    i. **Review Specifications with Stakeholders**: Share your initial specifications with stakeholders for feedback.
       - o Adjust specifications based on their input and ensure they accurately reflect their needs and expectations.
    ii. **Document Feedback**: Keep a record of feedback received and changes made.

12. **Step 6: Document and Organize Specifications**
    i. **Use the Provided Template**: Organize your specifications using the template provided.
       - o Ensure all relevant parameters are included and clearly defined.
    ii. **Create a Comprehensive Document**: Compile all specifications into a well-organized document.
       - o Use tables, charts, and diagrams to enhance clarity.

13. **Step 7: Review and Refine Specifications**
    i. **Regularly Review**: Periodically review and update specifications to ensure they remain relevant and achievable.
    ii. **Test and Iterate**: Use prototypes to test the design against specifications and identify areas for improvement.
    iii. **Adapt as Necessary**: Be prepared to adjust specifications based on new information or constraints encountered during the design process.

14. **Template for Design Specifications:**

Parameters	Specific Values or Ranges
Name of project	Project name
Costs	Values and units
Functions	List each one if it can be quantified, including values and units.
Performance parameters	E.g., acceleration, efficiency, power, reliability; values and units
Installation parameters	Values and units
Operation parameters	Values and units
Maintenance parameters	Values and units
Safety parameters	Values and units
Dimensions	Values and units
Weight, load, pressure, stress, temperatures, humidity, etc.	Values and units
Special requirements	E.g., clean environment, values, and units
Environmental impact	Sustainability, durability, reliability, etc.; values and units
User parameters	Values and units
Standards and codes compliance	Numbers and references
Manufacturing specifications	Numbers and references
Materials specifications	References, values, and units

15. **Assignment Submission:**

   i. **Initial Design Specifications Document**: Submit your initial set of design specifications using the template provided.

   ii. **Stakeholder Feedback**: Include a summary of feedback received from stakeholders and any changes made to the specifications.

   iii. **Documentation of Process**: Provide a detailed account of how you gathered information, translated needs into specifications, and validated them.

   iv. **Refinement Plan**: Outline your plan for regularly reviewing and refining the specifications as your project progresses.

16. **Guiding Questions:**

   i. What are the primary and secondary needs of your stakeholders?

   ii. How can you measure and quantify each requirement?

   iii. What challenges did you encounter in translating needs into specifications?

   iv. How did stakeholder feedback influence your specifications?

   v. What is your plan for continuously updating and refining the specifications?

Use this assignment to establish a solid foundation for your design project. The specifications you develop will guide your project and help ensure it meets the needs and expectations of all stakeholders.

10   Generate Design Concepts

Generating solution concepts is a vital step in the engineering design process. It involves creatively thinking of various ways to solve the design problem, considering all possible solutions before narrowing down to the most viable options. Below, we detail several effective methods for generating solution concepts.

10.1 Brainstorming

Brainstorming is a classic technique used to generate a large number of ideas quickly. It encourages free thinking and creativity among team members, leading to innovative solutions. Here are the steps and best practices for effective brainstorming:

- **Gather a Diverse Team**: Include members with different backgrounds and expertise to bring a variety of perspectives.
- **Define the Problem Clearly**: Ensure everyone understands the design problem and the goals of the brainstorming session.
- **Set Ground Rules**: Do not criticize or judge ideas during the session. Encourage wild and far-fetched ideas. Initially, focus on quantity over quality.
- **Use a Facilitator**: Have someone guide the session to keep it focused and productive.
- **Record All Ideas**: Write down every idea, no matter how impractical it may seem.

For example, suppose the design team is working on an automated pill dispenser. In that case, brainstorming might yield ideas such as different dispensing mechanisms (e.g., rotating carousel, linear push mechanism), various reminder systems (e.g., visual, auditory, or tactile alerts), and multiple power sources (e.g., battery, solar, or plug-in).

10.2 Morphological Analysis

Morphological Analysis is a method used to systematically explore all possible solutions to a problem by breaking it down into key functions or attributes and generating ideas for each function. Here's how to perform morphological analysis:

- **Identify Key Functions**: Break down the design problem into its essential functions or attributes.
- **List Possible Solutions for Each Function**: Generate as many ideas as possible for how each function can be achieved.
- **Create a Morphological Matrix**: Combine different solutions for each function to create complete concepts.

For the automated pill dispenser, essential functions might include pill storage, pill dispensing, user interface, and power supply. Possible solutions for each function could be listed, and then various combinations can be explored to create different overall concepts.

10.3 TRIZ (Theory of Inventive Problem Solving)

TRIZ is a problem-solving methodology that uses patterns in patents to identify solutions. It is based on the idea that many problems can be solved using principles that have been used in other fields. The steps include:

- **Identify the Main Technical Contradiction**: Determine the primary conflict in the design problem.
- **Use TRIZ Principles**: Apply TRIZ principles to generate innovative solutions that resolve the contradiction.

For example, a contradiction in the pill dispenser might be the need for a compact design while still accommodating multiple large pills. TRIZ principles could help generate solutions that allow for compact yet expandable storage mechanisms.

10.4 SCAMPER

SCAMPER is an acronym for a set of questions that can be used to think creatively about how to improve or change an existing product or process. It stands for Substitute, Combine, Adapt, Modify, Put to another use, Eliminate, and Reverse. Here's how to apply SCAMPER:

- **Substitute**: What materials, components, or processes can be substituted?
- **Combine**: What can be combined to create a new solution?
- **Adapt**: What ideas from other fields or products can be adapted for this solution?
- **Modify**: What modifications can be made to improve the solution?
- **Put to Another Use**: How can existing solutions be used in different ways?
- **Eliminate**: What elements can be eliminated to simplify the solution?
- **Reverse**: What if the order of operations or elements were reversed?

For the pill dispenser, SCAMPER could generate ideas like substituting materials for biodegradable options, combining the dispenser with a health-tracking app, adapting dispensing mechanisms from other fields, modifying the size to be more portable, using the dispenser in veterinary medicine, eliminating unnecessary features, or reversing the order of pill storage and dispensing.

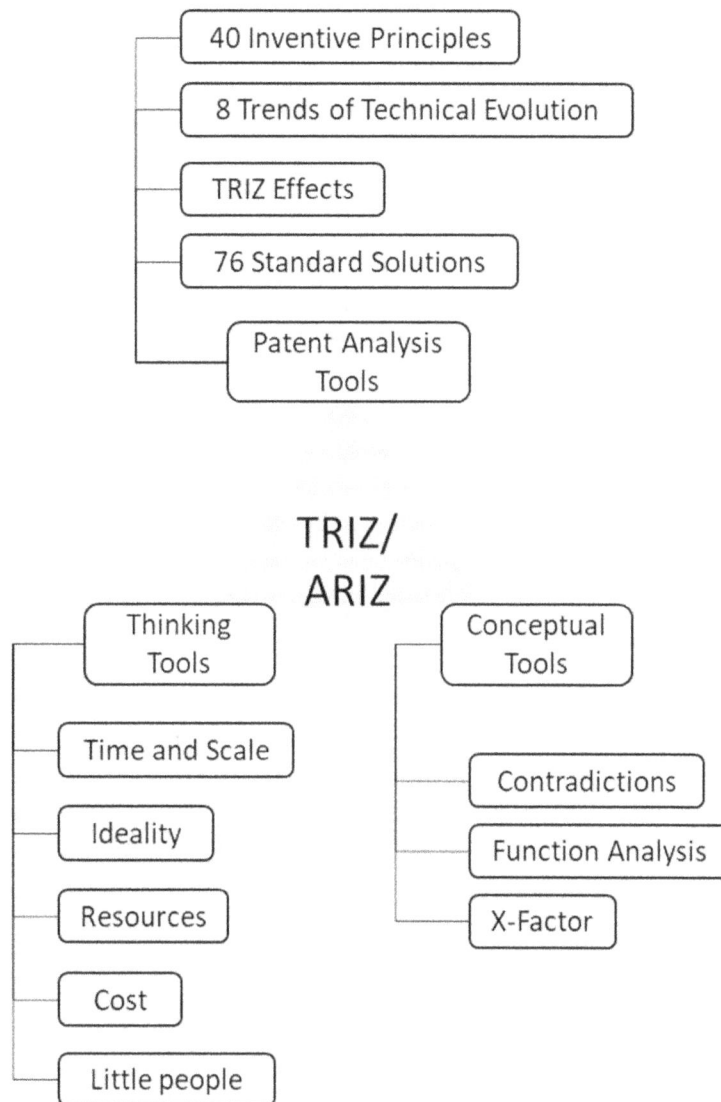

Figure 10-1. Components of the TRIZ processes.

10.5 Benchmarking

Benchmarking involves studying existing products or solutions to understand their strengths and weaknesses. It helps to identify best practices and areas for improvement. The steps include:

- **Identify Benchmarking Targets**: Select existing products or processes that are similar to the design problem.
- **Analyze Competitors**: Study how competitors solve similar problems.

- **Identify Strengths and Weaknesses**: Determine what works well and what doesn't in existing solutions.
- **Apply Insights**: Use the insights gained to generate new concepts that improve upon existing solutions.

For the automated pill dispenser, benchmarking might involve studying existing dispensers on the market, analyzing their mechanisms, user interfaces, reliability, and cost, and then identifying opportunities for innovation or improvement.

10.6 Mind Mapping

Mind Mapping is a visual tool that helps organize thoughts and ideas around a central concept. It encourages associative thinking and helps uncover relationships between different ideas. Here's how to create a mind map:

- **Start with a Central Concept**: Write the main problem or goal in the center of a page.
- **Branch Out**: Draw branches from the central concept representing different aspects or functions of the problem.
- **Add Sub-Branches**: From each branch, add sub-branches with more detailed ideas or solutions.
- **Explore Connections**: Look for connections and relationships between different branches and sub-branches.

For the pill dispenser, a mind map might start with the central concept of "automated pill dispensing" and branch out into categories like "storage mechanisms," "dispensing methods," "power sources," "user interface," and "reminder systems," each with their sub-branches detailing specific ideas. Figure 10-2. shows the mind map for the automated skimmer design project.

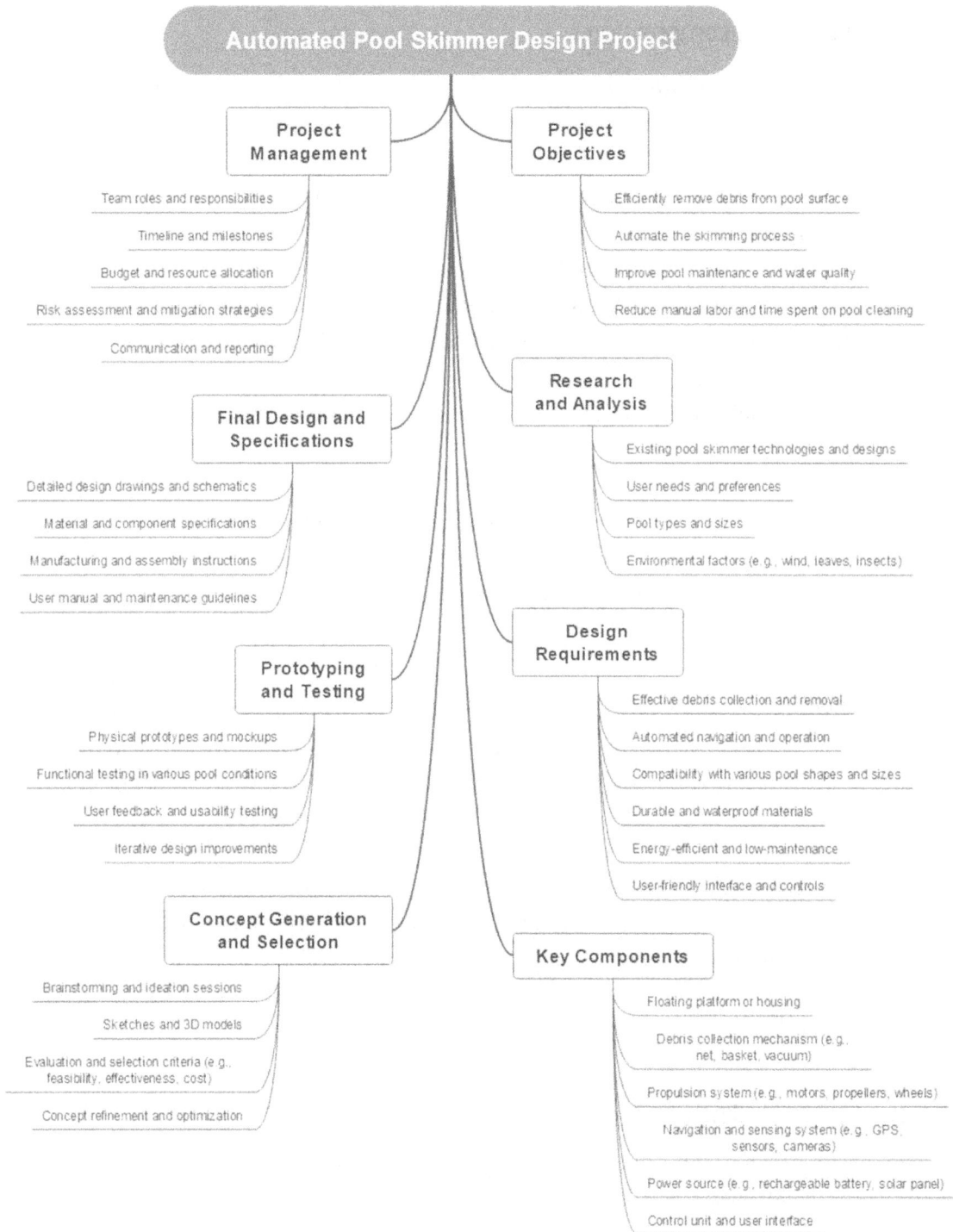

Figure 10-2. Mind map for automated pool skimmer design project

10.7 Functional Decomposition

Functional Decomposition involves breaking down the design problem into smaller, more manageable parts, each representing a function that the final design must perform. Here's how to do it:

- **Identify the Overall Function**: Define the primary function that the design must perform.
- **Break Down into Sub-Functions**: Decompose the main function into sub-functions that represent the steps needed to achieve the overall function.
- **Generate Solutions for Each Sub-Function**: Develop ideas for how to achieve each sub-function.
- **Integrate Solutions**: Combine the solutions for the sub-functions to form complete design concepts.

The overall function of the pill dispenser might be to "dispense pills on schedule." Sub-functions could include "store pills," "track time," "activate dispenser," and "alert user." Solutions for each sub-function can be generated and then integrated into complete design concepts.

Generating solution concepts is a critical phase in the engineering design process. By employing a variety of techniques—such as brainstorming, morphological analysis, TRIZ, SCAMPER, benchmarking, mind mapping, and functional decomposition—design teams can ensure that they explore a wide range of possibilities. This breadth of exploration increases the likelihood of finding innovative and effective solutions to complex design problems. The structured approach to concept generation not only fosters creativity but also ensures that the final design is robust, feasible, and aligned with the needs and constraints of the project.

10.8 Creative Thinking

10.8.1   Understanding Creativity

The dictionary definition of creativity is "the ability to transcend traditional ideas, rules, patterns, relationships, or the like, and create meaningful new ideas, forms, methods, interpretations, etc.; originality, progressiveness, or imagination." Creative thinking is essential for capstone design. But what does it mean to be creative? Do you know any creative people? Do you consider yourself to be creative? Can you learn to be creative?

Many researchers have studied creativity and creative thinking, and the search for methods and causes for creative thinking continues. While we do not fully understand what makes some people more creative than others, we recognize creativity when we encounter it. We see it as a brilliant idea or design that we have not seen before or previously thought of. When you conducted your patent search assignment, you may have noticed that some inventions were particularly creative.

You probably have had creative ideas that you can remember. When you encounter a new situation or new problem, can you think of ways to solve that problem or a workaround that seems creative to you?

10.8.2 Fostering Creative Thinking

While there is no formula or recipe for creative thinking, some behaviors or actions can facilitate original thoughts and ideas. Researchers of cognitive processes and creative thinking have suggested the following techniques.

- **Silent Brainstorming**
  Brainstorming alone and in silence can be helpful for brainstorming original ideas without distractions. Whatever environment makes one comfortable and relaxed can be beneficial to creative thinking. Some people do their most creative thinking while taking a shower! Group brainstorming dynamics are very different and can be an obstacle to creative thinking because of the processes, such as taking turns or dominating personalities who push their ideas without respect for team members.

- **Attitude**
  Attitude is the most critical factor in success in capstone design and creative design. A positive attitude towards the project and fellow team members makes all of the difference between success and failure. A positive attitude and a genuine desire to solve the design problem will enable you to think creatively and contribute to the project. A positive mindset allows you to keep an open mind about your team members' ideas and encourage them to engage fully in the design project. A positive attitude also opens the imagination and inspiration towards a plausible solution.

- **Utilizing Information Gathered**
  The information you have collected in the literature search, patent search, surveys, and problem definition will provide a foundation for original thinking and the derivation of new innovative ideas from those you have learned in your research about design problems. The design specifications you have developed will set a reasonable scope for the design problem and the solutions.

10.8.3 Barriers to Creative Thinking

In parallel to engaging in methods that foster creative thinking, one should be mindful of some behaviors that block creative thinking. The following practices should be avoided or eliminated by communication, accountability, and transparency within the design team.

- **Poor Attitude**

  A negative attitude towards the project or the team will negatively impact the team's ability to work together and achieve an excellent design solution. A poor attitude can include stereotyping, minimizing the problem, showing minimal effort, taking a lone ranger attitude, and expressing disrespect against others.

- **Tunnel Vision**

  Some design engineers get fixated on a particular solution strategy or method, which limits their thinking and contribution to creativity and negatively influences team members' creativity. When team members recognize this behavior, they should point it out to the team members and refrain from pushing only that solution.

- **Fear**

  Some students are not comfortable with the elements of capstone design, and they have to think outside of their comfort zone. They fear taking risks or entering a domain where their knowledge is uncertain. Some fear the open-ended nature of the capstone design problems where a solution is not guaranteed, or they are not sure which of the many possible solutions they should choose. They fear failure. Their fear can become a self-fulfilling prophecy. Team members should recognize this and work together to overcome fears or seek outside help if needed.

- **Poor Technique**

  Some student design engineers make poor choices in problem-solving strategies by not following the design steps carefully. For example, a poor problem definition or design specification set can be highly problematic for the design team. If the design problem is not defined well, it may create a situation where a solution cannot be achieved. If the design specifications are not fully developed or contain ill-conceived parameters or judgment, then a design solution may not be achievable.

- **Inadequate Domain Knowledge**

  Some design problems may involve aspects beyond the standard educational background of engineering students in a particular major. For example, a mechanical engineer may encounter a situation where they need to prevent electromagnetic interference in their design. They will need to seek expertise outside of their team to understand the science and work towards a solution.

- **Errors**

  Many different types of errors can cause frustration for the design team. One example of an error is incorrect assumptions or information provided by the sponsor, or perhaps a team member looked up wrong information. Errors can creep into the design process for many reasons, including a lack of attention to detail. For example, the units on some numerical value could be transcribed incorrectly, or a calculation by one of the design team members could contain errors. Errors can be devastating to a design project. All engineering work, calculations, assumptions, and data collection must be performed accurately and checked and double-checked.

As an example of engineering error, in 1999, NASA lost its Mars Climate Orbiter because the engineers failed to perform a unit conversion from engineering units to metric when communicating the critical spacecraft data before launch. The error cost NASA the $125 million spacecraft. The navigation team at NASA's Jet Propulsion Laboratory (JPL) used millimeters and meters in their calculations. At the same time, design engineers at Lockheed Martin Aeronautics in Denver provided acceleration data in feet and pounds. Lockheed Martin was responsible for the design and build of the orbiter spacecraft. JPL engineers assumed the acceleration data was provided to them in Newton-seconds. A simple error cost the mission years of setbacks, and much hard work was lost. SI units, also known as the metric system, are the international standard for communicating science, engineering, and technical data.

10.8.4 Techniques to Foster Creative Thinking

Here are some additional techniques to foster creative thinking within your design team:

- **Mind Mapping**
Mind mapping is a visual tool that helps organize thoughts and ideas around a central concept. It encourages associative thinking and helps uncover relationships between different ideas. Here's how to create a mind map:

- **Start with a Central Concept**: Write the main problem or goal in the center of a page.
- **Branch Out**: Draw branches from the central concept representing different aspects or functions of the problem.
- **Add Sub-Branches**: From each branch, add sub-branches with more detailed ideas or solutions.
- **Explore Connections**: Look for connections and relationships between different branches and sub-branches.

For example, a mind map for an automated pill dispenser might start with the central concept of "automated pill dispensing" and branch out into categories like "storage mechanisms," "dispensing methods," "power sources," "user interface," and "reminder systems," each with their sub-branches detailing specific ideas.

10.8.5   SCAMPER

SCAMPER is an acronym for a set of questions that can be used to think creatively about how to improve or change an existing product or process. It stands for Substitute, Combine, Adapt, Modify, Put to another use, Eliminate, and Reverse. Here's how to apply SCAMPER:

- **Substitute**: What materials, components, or processes can be substituted?
- **Combine**: What can be combined to create a new solution?
- **Adapt**: What ideas from other fields or products can be adapted for this solution?
- **Modify**: What modifications can be made to improve the solution?
- **Put to Another Use**: How can existing solutions be used in different ways?
- **Eliminate**: What elements can be eliminated to simplify the solution?
- **Reverse**: What if the order of operations or elements were reversed?

For the pill dispenser, SCAMPER could generate ideas like substituting materials for biodegradable options, combining the dispenser with a health-tracking app, adapting dispensing mechanisms from other fields, modifying the size to be more portable, using the dispenser in veterinary medicine, eliminating unnecessary features, or reversing the order of pill storage and dispensing.

10.8.6   Role Playing

Role-playing involves team members acting out different roles related to the design problem. This technique helps in understanding different perspectives and generating new ideas. Here's how to implement role-playing:

- **Assign Roles**: Each team member takes on a different role, such as a user, customer, sponsor, or competitor.
- **Act Out Scenarios**: Create scenarios where the roles interact with the design problem.
- **Generate Ideas**: Use insights from the role-playing exercise to generate new ideas or improve existing ones.

For instance, in the context of the pill dispenser, one team member could play the role of an elderly user, another as a pharmacist, and another as a caregiver. Acting out how each role interacts with the dispenser can reveal usability issues or innovative features.

10.8.7   Reverse Thinking

Reverse thinking involves thinking about the problem or solution in a backward manner. It helps break conventional thinking patterns and discover new ideas. Here's how to use reverse thinking:

- **Define the Problem**: Clearly state the problem or goal.
- **Reverse the Problem**: Think about the opposite of the problem or goal.
- **Generate Solutions**: Brainstorm solutions for the reversed problem.
- **Reverse Again**: Apply the ideas generated back to the original problem.

For example, instead of thinking about how to dispense pills efficiently, consider how pills could be stored inefficiently, generate solutions to those problems, and reverse the solutions to improve efficiency.

10.8.8   Six Thinking Hats

The Six Thinking Hats technique, developed by Edward de Bono, involves looking at a problem from six different perspectives. Each "hat" represents a different mode of thinking:

- **White Hat**: Focus on data and information.
- **Red Hat**: Consider emotions and feelings.
- **Black Hat**: Identify potential problems and risks.
- **Yellow Hat**: Look for benefits and feasibility.
- **Green Hat**: Generate creative ideas and alternatives.
- **Blue Hat**: Manage the thinking process and keep it on track.

By systematically wearing each hat, the design team can explore different aspects of the problem and generate a well-rounded set of ideas.

Creative thinking is essential in generating innovative and effective solutions to design problems. By understanding and fostering creative thinking and being aware of and avoiding barriers, engineering students can greatly enhance their capstone design projects. Utilizing techniques such as silent brainstorming, mind mapping, SCAMPER, role-playing, reverse thinking, and the Six Thinking Hats method, design teams can ensure that they explore a wide range of possibilities and arrive at the best possible solutions.

10.9 Techniques for Creative Thinking and Brainstorming

Creative thinking is crucial for generating innovative design solutions. Recent research in cognitive-based creative thinking has shown that creative thinking skills can be improved through training based on established methods. We will introduce some of those techniques here to enhance your design process.

10.9.1   Technique 1: Silent Brainstorming

Brainstorming is typically carried out as a group activity, but it can also be beneficial when done individually. Silent brainstorming allows individuals to generate ideas without the boundaries, procedures, and restrictions present during group sessions. It eliminates the influence of views and criticisms expressed by team members. An individual brainstorming session does not have to be very long—five to ten minutes of silent brainstorming and writing down ideas is often sufficient. Once unique ideas have been captured, they can be included in the group brainstorming session.

1. **Steps for Silent Brainstorming:**
   i. **Set a Timer**: Allocate five to ten minutes for the session.
   ii. **Find a Quiet Space**: Choose a location free from distractions.
   iii. **Write Down Ideas**: Jot down any ideas that come to mind regarding the design problem.
   iv. **Review Ideas**: After the session, review your ideas and select the most promising ones.
   v. **Share with Team**: Present your ideas in the next group brainstorming session for further discussion and development.

10.9.2   Technique 2: Evolutionary Thinking – The TRIZ Method

Genrich Altshuller and his colleagues developed the TRIZ (Theory of Inventive Problem Solving) method by analyzing thousands of patents. TRIZ focuses on deriving new design solutions by studying prior solutions and patents. The method uses "40 Inventive Principles" to guide the creative process.

Steps for Using TRIZ:
1. **Identify the Problem**: Clearly define the design problem.
2. **Analyze Prior Solutions**: Study existing patents and solutions related to the problem.
3. **Apply Inventive Principles**: Use the 40 Inventive Principles to generate new ideas. For example, one principle is "Segmentation" (dividing an object into independent parts), which can be used to rethink how a design can be modularized.
4. **Develop Solutions**: Create new design concepts based on the principles applied.
5. **Evaluate**: Assess the feasibility and effectiveness of the generated solutions.

10.9.3   Technique 3: Random Input

Lateral thinking helps break fixed patterns of thinking and introduces new ways of solving problems. Random input involves using an arbitrary word, object, or image to provoke different thinking and generate new ideas.

Steps for Random Input:

1. **Select Random Input**: Choose a word from a book, an object around you, or an image.
2. **Associate with Problem**: Relate the random input to the design problem. For instance, if you pick the word "tree," think about how the branching structure of a tree could inspire new design elements.
3. **Generate Ideas**: Use the associations to brainstorm new solutions.
4. **Document**: Write down all ideas, regardless of how far-fetched they may seem.
5. **Evaluate**: Review the ideas to find practical and innovative solutions.

10.9.4   Technique 4: SCAMPER

SCAMPER stands for Substitute, Combine, Adapt, Modify, Put to other uses, Eliminate, and Rearrange. It is a structured approach to questioning that can help generate new ideas during brainstorming.

Steps for SCAMPER:

1. **Substitute**:
   - What different processes can be used?
   - What materials can be incorporated or substituted?
   - What other sources of power could be applied?
   - What users or groups can be included or excluded instead?
   - What other methods can be used?
2. **Combine**:
   - What can be combined?
   - What purposes can be combined?
   - How can you integrate modules or sections?
   - How can you combine applications or purposes?
3. **Adapt**:
   - What else is similar to this?
   - What other thoughts does this suggest?
   - Can you adapt other solutions from the past?
4. **Modify**:
   - How can you add a new twist?
   - How can you change the meaning, color, shape, or form?
   - How can you expand or shrink the geometry?
   - How can you change the function or motion?
5. **Put to Other Uses**:
   - What other uses could it have?

      o   How can it be put to other uses if changed?

6. **Eliminate**:
   - o   What can you eliminate?
   - o   What functions, parts, or design specifications can you remove?

7. **Rearrange**:
   - o   What parts can you rearrange?
   - o   What other arrangements of components will work?
   - o   How can the process sequence, system layout, or methodology be different?

Example Application of SCAMPER:

For an automated pill dispenser:

- **Substitute**: Use biodegradable materials instead of plastic.
- **Combine**: Integrate the dispenser with a mobile app for reminders.
- **Adapt**: Use existing vending machine mechanisms for dispensing pills.
- **Modify**: Change the shape of the dispenser to make it more compact.
- **Put to Other Uses**: Use the dispenser for other small items like vitamins or candies.
- **Eliminate**: Remove unnecessary buttons to simplify the user interface.
- **Rearrange**: Change the layout of the pill compartments for easier access.

These techniques—Silent Brainstorming, Evolutionary Thinking (TRIZ), Random Input, and SCAMPER—provide structured approaches to foster creative thinking. By applying these methods, design teams can generate a diverse range of innovative solutions to their design problems, empowering them to tackle any challenge confidently. Engaging in these creative techniques regularly will enhance your ability to think outside the box and contribute to successful capstone design projects.

10.10 Generating Solution Concepts

The methods discussed so far will help the design team members in generating many design concepts. How many concepts are enough? When does the team stop generating design solutions? The aim is to generate a wide array of concepts before narrowing down to the most promising ones. This iterative process, while it may require patience, is fundamental to achieving an optimal solution and is a commitment to excellence.

10.10.1   Functional Decomposition

Functional decomposition, also known as "divide and conquer," involves breaking down a complex problem into smaller, more manageable parts. Each part is then solved individually, and the solutions are integrated into a comprehensive solution for the entire problem.

1. **Steps for Functional Decomposition:**
    i. **Identify Main Function**: Determine the primary function of the design.
    ii. **Break Down into Sub-functions**: Decompose the main function into smaller, simpler sub-functions.
    iii. **Further Decompose**: If sub-functions are still too complex, further decompose them until each function is manageable.
    iv. **Solve Sub-functions**: Develop solutions for each sub-function independently.
    v. **Integrate Solutions**: Combine the solutions for all sub-functions into a coherent overall solution.
2. **Example: SAE Mini-Baja Vehicle**

    The physical and functional decomposition of a Mini-Baja vehicle includes the following sub-systems:

    - **Chassis and Frame**: Provides structure and support.
    - **Powertrain**: Includes the engine and transmission.
    - **Suspension**: Ensures ride comfort and handling.
    - **Braking System**: Provides stopping power.
    - **Steering System**: Controls vehicle direction.

    Each of these sub-systems can be further decomposed and solved individually before integrating them into the final vehicle design.

10.10.2 Morphological Methods

Morphological methods focus on the shape and form of the product or process. This approach is particularly useful for physical product design, where the arrangement of parts or steps is crucial.

Steps for Morphological Methods:

i. **Identify Key Functions**: Determine the critical functions that the design must perform.
ii. **Generate Form Variations**: For each function, generate multiple variations of forms or structures that can fulfill the function.
iii. **Combine Variations**: Experiment with different combinations of the form variations to create new design concepts.
iv. **Evaluate Combinations**: Assess the feasibility and effectiveness of each combination.

Example: Automated Intelligent Pill Dispenser

A design team working on an automated pill dispenser might generate form variations for functions such as:

- **Pill Storage**: Different compartment shapes and sizes.

- **Pill Dispensing**: Various mechanisms for releasing pills.
- **User Interface**: Different layouts and input methods.

By combining these variations, the team can develop several concept designs for evaluation.

10.10.3   Axiomatic Design

Axiomatic design, developed by Dr. Nam P. Suh at MIT, is based on two main axioms: the independence axiom and the information axiom. This methodology helps ensure that functional requirements are independent and that the design minimizes information content, thus reducing complexity and improving robustness.

Steps for Axiomatic Design:

i. **Define Functional Requirements (FRs)**: Identify the essential functions that the design must fulfill.

ii. **Establish Design Parameters (DPs)**: Determine the physical variables that will achieve the functional requirements.

iii. **Map FRs to DPs**: Ensure that each design parameter independently satisfies one functional requirement.

iv. **Minimize Information Content**: Simplify the design to reduce the information content, enhancing reliability and ease of manufacturing.

10.10.4   TRIZ Method

The TRIZ (Theory of Inventive Problem Solving) method, created by Genrich Altshuller, uses a systematic approach to derive new design solutions by studying prior patents and inventions. It involves identifying contradictions and applying inventive principles to resolve them.

Steps for Using TRIZ:

i. **Identify Contradictions**: Determine conflicting requirements or parameters in the design.

ii. **Apply Inventive Principles**: Use one or more of the 40 inventive principles to resolve the contradictions.

iii. **Develop Solutions**: Create new design concepts based on the applied principles.

iv. **Evaluate Ideality**: Assess the ideality of each solution, which is defined as the ratio of benefits to the sum of costs and harms.

Example: Designing a Swimming Pool Skimmer Basket

Contradictions might include:

- **Increased Flow Area vs. Ease of Removal**: Larger flow areas improve performance but can make the basket harder to remove.
- **Durability vs. Cost**: More durable materials may increase costs.

Applying TRIZ principles such as "Segmentation" (dividing the basket into modular parts) and "Local Quality" (enhancing specific areas of the basket) can help resolve these contradictions and generate innovative solutions.

10.10.5 Combination of Methods

Often, the best approach to generating solution concepts involves a combination of the methods discussed. By integrating functional decomposition, morphological methods, axiomatic design, and TRIZ, teams can leverage the strengths of each approach to develop a comprehensive set of design solutions.

Steps for Combining Methods:

i. **Start with Functional Decomposition**: Break down the design problem into manageable sub-functions.

ii. **Apply Morphological Methods**: Generate form variations for each sub-function.

iii. **Use Axiomatic Design**: Ensure independence of functional requirements and minimize information content.

iv. **Incorporate TRIZ**: Identify and resolve contradictions using inventive principles.

v. **Integrate Solutions**: Combine the solutions from each method into a coherent overall design.

Example: Capstone Project - Autonomous Delivery Drone

1. Functional Decomposition:
   o **Navigation**: GPS and obstacle avoidance.
   o **Payload Management**: Secure loading and unloading.
   o **Power Supply**: Efficient battery management.
   o **Communication**: Real-time data transmission.

2. **Morphological Methods**:
   o **Navigation**: Different GPS module layouts and various sensor types for obstacle avoidance.
   o **Payload Management**: Multiple designs for secure loading mechanisms.
   o **Power Supply**: Various battery configurations and charging methods.
   o **Communication**: Different communication protocols and hardware arrangements.

3. **Axiomatic Design**:
   o Ensure the navigation system independently fulfills location tracking without interference.
   o Minimize information content in payload management for simplicity.

4. **TRIZ**:
   o **Navigation vs. Obstacle Avoidance**: Resolve the need for high accuracy without excessive sensor costs using "Segmentation" (modular sensor arrays) and "Local Quality" (enhanced critical areas).

      o **Battery Life vs. Weight**: Use "Parameter Change" (adjusting battery chemistry) and "Composite Materials" (lighter structural components).

5. **Integrate Solutions**:
      o Combine the best variations and principles from each method to create a final design for the autonomous delivery drone.

Generating solution concepts is a critical phase in the design process that benefits from a structured and systematic approach. By employing methods such as functional decomposition, morphological methods, axiomatic design, and TRIZ, design teams can create a diverse and innovative set of solutions. Combining these techniques allows for a comprehensive exploration of potential solutions, leading to an optimal and robust final design. This iterative and collaborative process is essential for the success of capstone design projects and beyond.

10.11 Analyzing Concepts

The process of analyzing concepts is crucial in engineering design as it allows the design team to evaluate the viability of different ideas and narrow down to the most promising solutions. This section outlines various methods and approaches for concept analysis, including go/no-go analysis, engineering analysis, modeling, and simulation.

10.11.1 Go/No-Go Analysis

The first step in concept analysis is a go/no-go analysis. This technique involves a preliminary review of each idea to decide whether it is worth further development or should be discarded. This initial filter helps streamline the list of viable concepts and focuses efforts on the most promising ones.

- **Steps for Go/No-Go Analysis:**
1. **List Concepts**: Start by listing all the generated concepts.
2. **Set Criteria**: Establish simple, high-level criteria for evaluating the concepts. These could include feasibility, cost, resource availability, alignment with project goals, and technical requirements.
3. **Evaluate Each Concept**: Assess each concept against the criteria. This can be done through discussion, voting, or a simple scoring system.
4. **Decide**: Categorize each concept as either "go" (for further development) or "no-go" (to be discarded).

By the end of this process, about 50% of the initial concepts can typically be eliminated, leaving a refined list of concepts that warrant a more detailed analysis.

10.11.2 Engineering Analysis

Engineering analysis involves applying scientific principles and mathematical models to evaluate the performance of the design concepts. This analysis helps predict

how the design will perform under various conditions and identify potential issues.
- **Steps for Engineering Analysis:**

1. **Define Parameters**: Identify the key parameters and state variables that influence the performance of the design.
2. **Develop Models**: Create mathematical models representing the physical, chemical, or biological processes involved in the design.
3. **Perform Calculations**: Use simple formulas, correlations, and hand calculations to perform preliminary analysis.
4. **Validate Models**: Compare the results of the calculations with experimental data or known benchmarks to validate the models.
5. **Iterate**: Refine the models and repeat the calculations as needed.

Engineering analysis can range from simple "back of the envelope" calculations to sophisticated simulations using advanced software tools like ANSYS, COMSOL, SIMULIA, and MATLAB. The level of detail and complexity of the analysis will depend on the design stage and the available resources.

10.11.3 Modeling

Modeling is an essential step in the design process, providing a visual and functional representation of the concept. Depending on the nature of the design and the available tools, models can be computational or physical.

10.11.3.1   Computational Modeling:

- **Geometry Creation**: Define the geometry of the design using CAD software.
- **Boundary Conditions**: Set up the boundary and initial conditions for the simulation.
- **Material Properties**: Assign material properties to different parts of the model.
- **Meshing**: Create a mesh to discretize the model for numerical analysis.
- **Simulation Setup**: Input the model into simulation software and set up the analysis parameters.

10.11.3.2   Physical Modeling:

- **Prototype Creation**: Build a physical prototype using materials like wood, plastic, or metal. 3D printing can be a valuable tool for creating prototypes quickly and accurately.
- **Testing**: Conduct experiments to test the performance of the prototype.
- **Analysis**: Analyze the test results to identify areas for improvement.
Physical models are particularly useful for understanding the form and function of the design. In contrast, computational models are better suited for complex systems that

cannot be easily built or tested.

10.11.4 5.3.4 Simulation

Simulation is the process of creating digital prototypes of the design to analyze its performance under various conditions. Simulations can be deterministic (based on physical laws) or stochastic (based on statistical sampling).

- **Steps for Simulation:**
1. **Develop Simulation Models**: Create models that represent the design and its environment.
2. **Define Scenarios**: Set up different scenarios to test the performance of the design under various conditions.
3. **Run Simulations**: Use simulation software to run the models and analyze the results.
4. **Analyze Results**: Evaluate the results to identify potential issues and areas for improvement.
5. **Iterate**: Refine the models and repeat the simulations as needed.

Simulation can be used to analyze various aspects of the design, including structural integrity, thermal performance, fluid dynamics, and electrical behavior. Advanced simulation techniques, such as finite element analysis (FEA) and computational fluid dynamics (CFD), can provide detailed insights into the design's performance.

Example: Analyzing Concepts for an Automated Pill Dispenser
- **Go/No-Go Analysis**
1. **List Concepts**: Generate a list of possible designs for the automated pill dispenser, including various dispensing mechanisms, storage configurations, and user interfaces.
2. **Set Criteria**: Criteria might include feasibility, cost, ease of use, safety, and manufacturability.
3. **Evaluate Each Concept**: Each team member evaluates the concepts against the criteria, and a group discussion is held to decide which concepts are worth further development.
4. **Decide**: Eliminate concepts that do not meet the criteria or are not favored by the majority of the team.

- **Engineering Analysis**
1. **Define Parameters**: Identify parameters such as pill size, dispensing force, storage capacity, and power consumption.
2. **Develop Models**: Create mathematical models for the dispensing mechanism,

including forces and torques involved.

3. **Perform Calculations**: Use simple equations to estimate the forces required to dispense pills and the power needed for the motor.
4. **Validate Models**: Compare the calculated forces with experimental data or literature values.
5. **Iterate**: Refine the models and repeat the calculations as needed.

- **Computational Modeling**
1. **Geometry Creation**: Design the dispenser geometry using CAD software.
2. **Boundary Conditions**: Set up boundary conditions for the moving parts of the dispenser.
3. **Material Properties**: Assign material properties to the components.
4. **Meshing**: Create a mesh for the numerical analysis.
5. **Simulation Setup**: Input the model into simulation software and set up the parameters for structural analysis.

- **Physical Modeling**
1. **Prototype Creation**: Use a 3D printer to create a physical prototype of the dispenser.
2. **Testing**: Conduct experiments to test the dispensing mechanism's performance.
3. **Analysis**: Analyze the test results to identify any issues with the mechanism.

- **Simulation**
1. **Develop Simulation Models**: Create simulation models for the dispenser, including the dispensing mechanism and storage configuration.
2. **Define Scenarios**: Set up scenarios to test the dispenser under different conditions, such as varying pill sizes and weights.
3. **Run Simulations**: Use simulation software to run the models and analyze the results.
4. **Analyze Results**: Evaluate the results to identify potential issues and areas for improvement.
5. **Iterate**: Refine the models and repeat the simulations as needed.

Analyzing design concepts is a critical step in the engineering design process, enabling the team to evaluate the viability of different ideas and narrow down to the most promising solutions. By using methods such as go/no-go analysis, engineering analysis, modeling, and simulation, the team can systematically assess each concept's performance and potential. This iterative process helps ensure that the final design is robust, feasible, and optimized for the intended application. The combination of different analysis techniques provides a comprehensive understanding of the design's strengths and weaknesses, guiding the team toward a successful final solution.

10.12 Decision Making

Decision-making is the process of choosing among two or more solution alternatives to find a practical design solution. Engineering decision-making is critical to the entire design process. For instance, in preparation for the critical design review, each design concept must be considered and compared to all of the other concepts. In fact, every step in the design process involves decision-making. The methodology used in decision-making directly affects the outcome and quality of the design solution.

Problem-solving and decision-making are related but are not synonymous. When a design solution does not meet its established design specifications or does not produce the expected or desired results or function as planned, it creates a problem that requires a solution. In general, decision-making can be considered a step in the problem-solving process. Based on education, training, and life experiences, each individual develops and employs a different set of logically ordered steps to identify and solve problems and make decisions. A formalized process and method must be utilized for problem-solving and decision-making in engineering design. Several tools for decision-making in engineering design are described next.

10.12.1   Pugh Chart

Pugh Analysis, invented by Professor Stuart Pugh of the University of Strathclyde in Glasgow, Scotland, is an evaluation matrix where alternatives or design solutions are listed in the first row, and evaluation criteria are listed in the first column. The method evaluates and prioritizes the many design solutions by scoring each design against each evaluation criterion.

10.12.1.1   Steps of a Pugh Analysis:

1. **Develop the design specifications**: These become the evaluation criteria.
2. **Generate design solution concepts**: Refer to previous discussions on concept generation.
3. **Create the evaluation matrix**: Using a simple spreadsheet, list the evaluation criteria in the first column and the design concept reference numbers across the first row.
4. **Select a reference design**: This may be one of the design concepts at random or a preferred design that the team intuitively likes best.
5. **Evaluate each design concept**: For each evaluation criterion (design specification), assess each design concept against the reference design, using scores of worse (-1), same (0), or better (+1).
6. **Record the ratings**: Enter the scores in the corresponding row and column cells. Sum the results and select the alternative with the highest score.

7. **Iterate if necessary**: If no clear best idea emerges, look for a hybrid new concept that captures the best features of the competing alternatives and rescore.

Criteria	1	2	3	4	5	6	7	8	9	10	11	12	13	14	15	16	17	18	19	20	21	22	23	24	25	26	27	28	29	30
Drag	S	S	S	-	S	+	-	+	-	+	-	-	-	-	+	-	S	S	S	S	S	-	+	+	-	S	S	-	-	S
Safety	-	-	+	S	+	+	S	S	S	-	+	+	+	+	S	S	S	S	S	S	+	S	+	-	S	+	-	+	-	+
Hydrodynamic	S	S	S	-	S	-	+	+	-	+	-	+	+	+	+	-	-	-	+	+	-	S	S	+	-	-	S	+	S	S
Manufacturability	+	+	-	+	+	+	+	-	+	-	+	-	+	+	-	+	-	+	-	S	S	-	+	-	-	S	-	+	+	-
Cost	S	S	S	+	+	+	-	+	-	S	S	-	-	-	S	S	S	S	S	S	S	-	+	-	-	-	S	-	S	-
Durability	+	+	-	-	-	-	+	-	+	-	+	+	+	+	S	S	S	S	S	+	-	-	+	-	-	+	-	S	S	S
Weight	S	S	S	S	+	+	-	+	-	S	S	-	-	-	S	S	S	S	S	S	S	S	S	-	S	S	S	-	S	-
Functionality	-	+	-	-	-	+	S	S	S	+	-	+	+	+	+	-	-	-	+	+	+	+	-	+	+	+	+	+	+	+
Aesthetic	S	S	S	S	S	-	S	S	S	+	-	+	+	+	+	-	-	-	+	+	+	-	+	+	S	S	S	+	-	+
# of pluses (+)	2	3	1	2	4	6	3	4	2	4	3	5	6	6	4	1	0	1	3	5	2	2	5	4	2	3	1	5	2	3
# of minuses (-)	2	1	3	4	2	3	3	2	4	3	4	4	3	3	1	4	4	3	1	0	2	5	2	4	5	2	3	3	3	3

Figure 10-3. Example of a Pugh Analysis Worksheet.

Figure 10-3. shows an example of a Pugh worksheet for an underwater jet scooter. The design solution concepts are numbered 1 through 30 on the top row, and the evaluation criteria (design specifications) are listed in the first column. Each concept is evaluated with respect to case 1, and the results are recorded in the corresponding row and column. After evaluating all concepts against all criteria, sum the number of pluses and minuses. Repeat the analysis by selecting the highest value for the count of plusses and the lowest count of minuses for the same. In this example, design concepts 6, 13, and 14 have the highest count of pluses (6), and the lowest number of minuses is (3) for each of the three concepts. The team decides which of the three top concepts will be used as the reference design in the next iteration.

This evaluation matrix, also known as a decision matrix, can be used to evaluate and prioritize any list of options and, as such, is a decision-making tool. The criteria can also be weighted to achieve a more precise decision method. The weighted criteria method is a step in the quality function deployment method described in the next section.

10.12.1.2 Tutorial on the Pugh Analysis Method for Comparing Multiple Design Concepts

Pugh Analysis, also known as the Pugh Method or Decision-Matrix Method, is a qualitative technique used in engineering for concept selection, decision-making, and design improvement. Developed by Stuart Pugh in the 1980s, this method provides a systematic approach to evaluating and comparing multiple design concepts against a set of criteria.

The primary goals of Pugh Analysis are to a) Identify the best concept among several alternatives, b) Generate new, hybrid concepts by combining the strengths of different designs, and c) Improve existing designs by highlighting their weaknesses and strengths.

Pugh Analysis is instrumental in the early stages of product development when detailed quantitative data may not be available, and decisions need to be made based on expert judgment and qualitative assessments.

10.12.1.2.1   The Pugh Analysis Process

The Pugh Analysis process consists of several key steps:

i.   Define the selection criteria. Begin by establishing a set of criteria against which the design concepts will be evaluated. These criteria should be derived from customer requirements, engineering specifications, and other relevant factors.

ii. Select a datum concept. Choose one concept as the baseline or datum against which all other concepts will be compared. This is typically an existing design or concept that seems most promising at first glance.

iii. Prepare the Pugh Matrix. Create a matrix with the selection criteria listed in rows and the design concepts in columns. The datum concept is usually placed in the leftmost column.

iv. Compare concepts For each criterion, compare each concept to the datum and assign a score: '+' if the concept is better than the datum '-' if the concept is worse than the datum 'S' if the concept is the same as the datum

v.   Sum the scores Calculate the total number of '+', '-', and 'S' scores for each concept.

vi. Analyze the results. Evaluate the strengths and weaknesses of each concept based on the scores.

vii. Improve concepts Use the analysis to generate new concepts or improve existing ones by combining positive features from different designs.

viii. Iterate and repeat the process with the improved concepts until a clear winner emerges or further analysis is no longer beneficial.

10.12.1.2.2   Advantages and Limitations of Pugh Analysis

Advantages:
- Provides a structured approach to decision-making
- Encourages team discussion and consensus-building
- Helps identify strengths and weaknesses of different concepts
- Facilitates the generation of new, improved concepts
- Can be used with limited quantitative data

Limitations:
- Relies on subjective judgments, which can introduce bias
- Does not account for the relative importance of different criteria
- May oversimplify complex design problems
- Can be time-consuming for a large number of concepts or criteria

10.12.1.2.3   Case Study: Designing a Tree Climbing Machine

To illustrate the Pugh Analysis method, let's apply it to the design of a tree-climbing machine.

10.12.1.2.3.1   Customer Requirements

First, we need to establish the customer requirements for the tree-climbing machine. These might include:

i.   Safe for the operator
ii.   Minimizes damage to the tree
iii.   Able to climb trees of various diameters
iv.   Easy to operate
v.   Portable and lightweight
vi.   Capable of reaching high altitudes (at least 30 meters)
vii.   Stable during operation
viii. Affordable for small tree service businesses
ix.   Low maintenance
x.   Weather-resistant

10.12.1.2.3.2   Engineering Requirements

Next, we translate these customer requirements into more specific engineering requirements:

i.   Safety: Meets or exceeds relevant safety standards (e.g., OSHA regulations)
ii.   Tree protection: Contact pressure on bark < 0.5 MPa
iii.   Tree diameter range: Suitable for trees 20-150 cm in diameter
iv.   Usability: Requires < 4 hours of training for proficiency
v.   Portability: Total weight < 25 kg
vi.   Climbing height: Maximum height ≥ 30 meters
vii.   Stability: Maintains position with wind speeds up to 30 km/h
viii. Cost: Manufacturing cost < $5,000
ix.   Maintenance: Service interval > 500 hours of operation
x.   Durability: Operates in temperatures from -10°C to 40°C and up to 95% humidity

10.12.1.2.3.3   Generating Design Concepts

Let's consider four design concepts for the tree-climbing machine:

A. Rope and Harness System: A traditional climbing method using ropes, harnesses, and ascending devices. B. Telescoping Boom Lift: A vehicle-mounted extendable boom with a bucket for the operator. C. Treaded Climber: A motorized device that grips the tree trunk and crawls up using treads. D. Robotic Arm Climber: A multijointed robotic arm that wraps around the tree and climbs by repositioning its grippers.

10.12.1.2.3.4 Applying Pugh Analysis

Now, we'll create a Pugh Matrix to compare these concepts. The rope and Harness System (Concept A) will be the datum.

Table 10-1. Sample Pugh Analysis for a Tree-Climbing Machine.

Criteria	A (Datum)	B	C	D
Safety	S	+	S	+
Tree protection	S	-	-	S
Tree diameter range	S	-	+	+
Ease of operation	S	+	+	S
Portability	S	-	+	-
Climbing height	S	+	S	S
Stability	S	+	+	+
Affordability	S	-	-	-
Low maintenance	S	-	-	-
Weather resistance	S	S	+	+
Total '+'	0	4	5	4
Total '-'	0	5	3	3
Total 'S'	10	1	2	3

10.12.1.2.3.5 Analysis of results:

- Concept B (Telescoping Boom Lift) has mixed results, with good safety and operation but poor portability and affordability.
- Concept C (Treaded Climber) shows the most positive scores, excelling in tree diameter range, ease of operation, and portability.
- Concept D (Robotic Arm Climber) performs well in safety and stability but has issues with portability and affordability.

Based on this analysis, Concept C (Treaded Climber) appears to be the most promising design. However, we can identify areas for improvement:
1. Enhance tree protection to minimize bark damage
2. Improve affordability through design optimization
3. Develop a better maintenance plan to increase service intervals

We can now iterate on the Treaded Climber design, incorporating these improvements and potentially borrowing features from other concepts (e.g., safety features from the Robotic Arm Climber).

10.12.1.2.3.6   Best Practices and Tips

To make the most of Pugh's Analysis, consider the following best practices:

i.   Involve a diverse team: Include members from different disciplines to ensure a well-rounded evaluation.

ii. Carefully select the datum: Choose a datum that is well understood by the team and represents a solid benchmark.

iii. Use clear, specific criteria: Ensure that all team members have a shared understanding of each criterion.

iv. Avoid weighting criteria initially: Start with unweighted criteria to prevent bias, then consider weighting in later iterations if necessary.

v.   Encourage discussion: Use the Pugh Matrix as a tool to facilitate team discussions rather than as a rigid decision-making formula.

vi. Document reasoning: Record the rationale behind each evaluation to aid in future iterations and decision-making.

vii. Iterate and refine: Don't expect to find the perfect solution in one round. Use multiple iterations to improve concepts and generate new ideas.

viii. Combine with other methods: Consider using Pugh Analysis in conjunction with other decision-making tools, such as QFD or TRIZ, for a more comprehensive approach.

Pugh Analysis is a powerful tool for comparing and selecting design concepts, particularly in the early stages of product development. Providing a structured approach to evaluation and encouraging team discussion helps engineers and designers make informed decisions and generate improved concepts.

In our tree climbing machine example, we saw how Pugh Analysis can be applied to compare diverse design concepts and identify areas for improvement. By iterating on the process and refining our designs, we can work towards an optimal solution that best meets the customer and engineering requirements.

While Pugh Analysis has its limitations, such as relying on subjective judgments and potentially oversimplifying complex problems, it remains a valuable technique in the engineer's toolkit. When used in conjunction with other design and decision-making methods, it can significantly contribute to successful product development and innovation.

10.12.2   Quality Function Deployment (QFD)

Quality Function Deployment (QFD) is a structured approach to product development that focuses on translating customer needs and wants into specific technical requirements. Developed in Japan in the late 1960s, QFD has become a widely used tool in various industries for improving product quality, reducing development time, and enhancing customer satisfaction.

The primary goals of QFD are: a) To ensure that customer requirements are the driving force behind product development, b) To prioritize design efforts based on customer needs, c) To translate customer needs into measurable engineering characteristics, d) To identify and resolve potential conflicts early in the development process e) To foster cross-functional collaboration and communication.

QFD is often associated with the "House of Quality," a visual tool that forms the basis of the QFD process. However, a full QFD implementation typically involves four phases, which we will explore in the next section.

10.12.2.1   The Four Phases of QFD

A complete QFD process consists of four interlinked phases, each building upon the previous one:

10.12.2.1.1   Phase 1: Product Planning (House of Quality)

- Translates customer requirements into technical characteristics
- Identifies relationships between customer needs and product features
- Prioritizes technical characteristics based on customer importance

10.12.2.1.2   Phase 2: Part Deployment

- Translates technical characteristics into part characteristics
- Identifies critical parts and components
- Establishes target values for part characteristics

10.12.2.1.3   Phase 3: Process Planning

- Translates part characteristics into key process operations
- Identifies critical process parameters
- Establishes process control plans

10.12.2.1.4   Phase 4: Production Planning

- Translates key process operations into production requirements
- Identifies critical production metrics and controls
- Establishes quality assurance measures

While all four phases are important for a comprehensive QFD implementation, this tutorial will focus primarily on Phase 1 (House of Quality), as it forms the foundation of the QFD process and is the most commonly used phase in practice.

10.12.2.2   The House of Quality

The House of Quality is a matrix-based tool that forms the core of the QFD process. It visually represents the relationships between customer requirements and engineering characteristics, as well as the interactions between different engineering characteristics.

The House of Quality consists of several interconnected sections:

i.   Customer Requirements (WHATs): A list of customer needs and wants
ii. Engineering Characteristics (HOWs): Technical measures that address customer requirements
iii. Relationship Matrix: Shows how engineering characteristics relate to customer requirements
iv. Correlation Matrix (Roof): Indicates interactions between engineering characteristics
v.   Competitive Assessment: Compares the product against competitors
vi. Target Values: Specific goals for each engineering characteristic g) Importance Ratings: Prioritizes customer requirements and engineering characteristics

Figure 14-1. shows the process for QFD analysis in general.

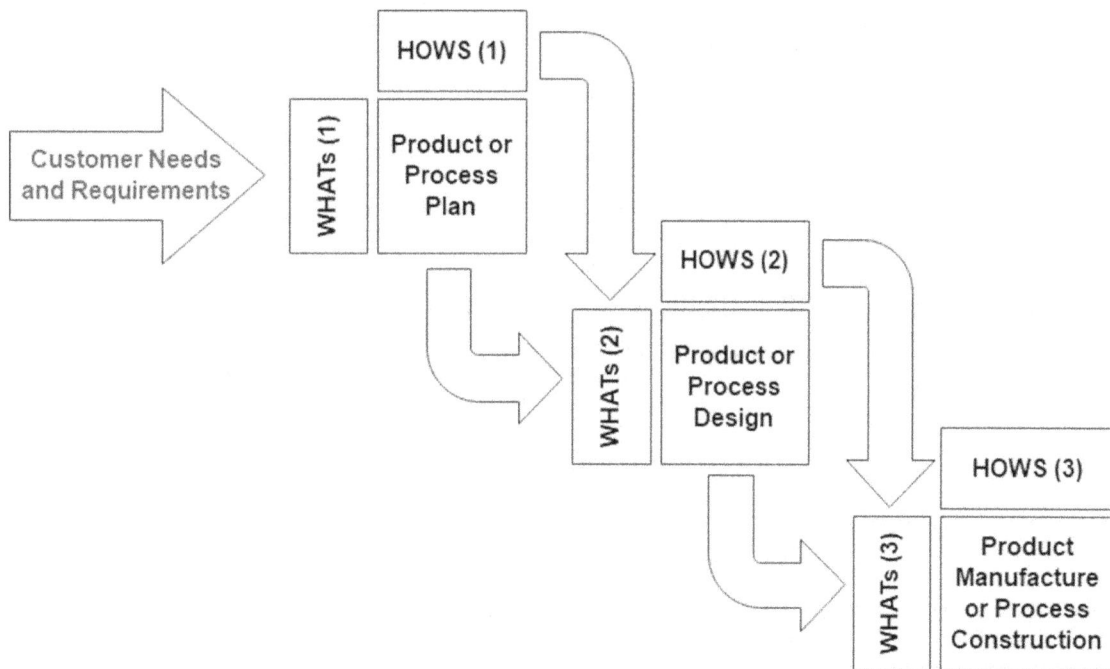

Figure 10-4. QFD Process.

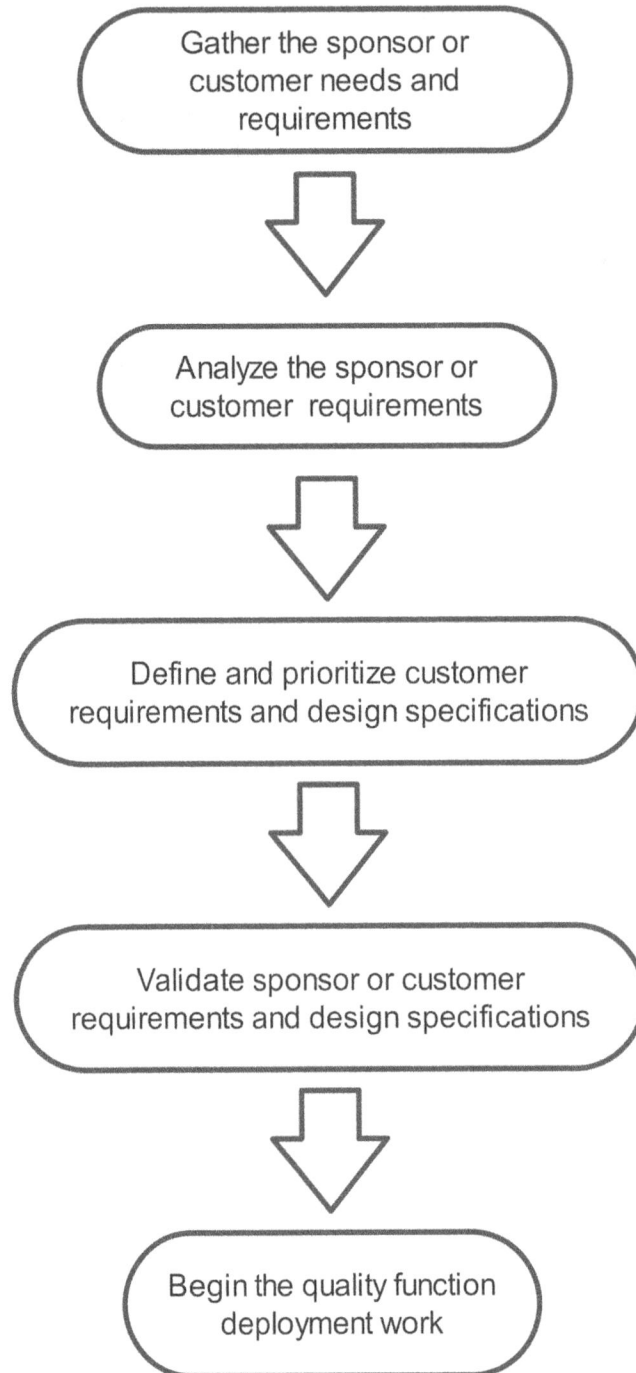

Figure 10-5. Assess customer needs and requirements.

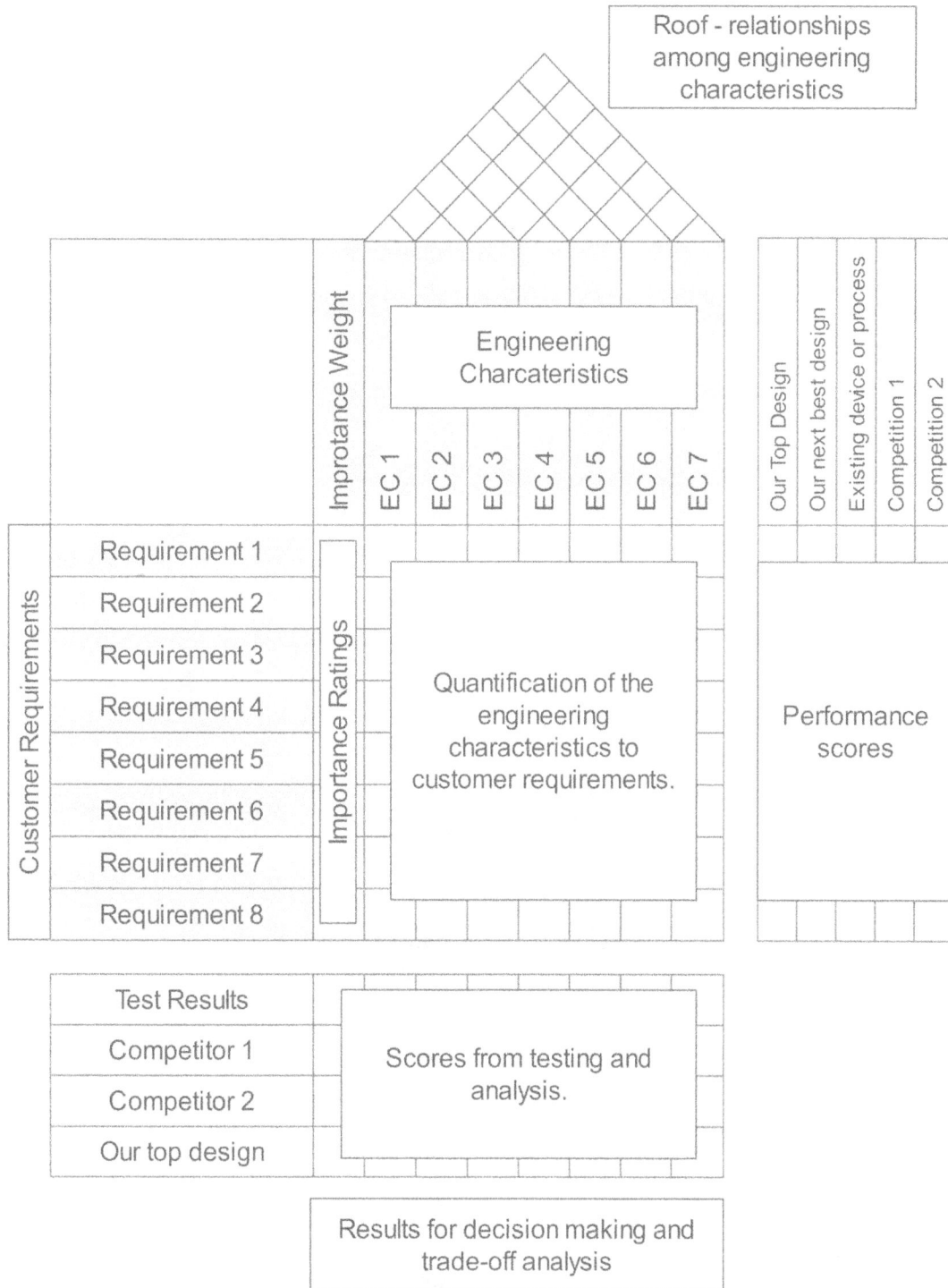

Figure 10-6. QFD overview.

10.12.2.3   Step-by-Step Guide to Creating a House of Quality

10.12.2.3.1   Step 1: Identify Customer Requirements
- Gather customer input through surveys, interviews, focus groups, etc.
- Organize and prioritize customer needs

10.12.2.3.2   Step 2: Develop Engineering Characteristics
- Identify measurable technical characteristics that address customer needs
- Ensure characteristics are specific and quantifiable

10.12.2.3.3   Step 3: Create the Relationship Matrix
- Assess how each engineering characteristic relates to each customer requirement
- Use symbols or numbers to indicate strong, moderate, or weak relationships

10.12.2.3.4   Step 4: Build the Correlation Matrix
- Identify positive or negative interactions between engineering characteristics
- Use symbols to indicate strong positive, positive, negative, or strong negative correlations

10.12.2.3.5   Step 5: Perform Competitive Assessment
- Evaluate how well your product and competitors meet customer requirements
- Use a numeric scale (e.g., 1-5) to rate performance

10.12.2.3.6   Step 6: Set Target Values
- Establish specific, measurable goals for each engineering characteristic
- Consider technical feasibility and competitive benchmarks

10.12.2.3.7   Step 7: Calculate Importance Ratings
- Determine the relative importance of customer requirements
- Calculate the technical importance of engineering characteristics

10.12.2.3.8   Step 8: Analyze and Interpret Results
- Identify key areas for improvement and innovation
- Prioritize development efforts based on importance ratings and competitive assessment

10.12.2.4   Benefits and Limitations of QFD

Benefits:
- Customer-focused product development
- Improved cross-functional communication

- Early identification of design conflicts and trade-offs
- Reduced development time and costs
- Systematic approach to decision-making
- Enhanced product quality and customer satisfaction

Limitations:

- Can be time-consuming and resource-intensive
- Requires significant data collection and analysis
- May oversimplify complex relationships
- Subjective assessments can introduce bias
- Limited ability to capture rapidly changing customer needs

10.12.2.5 Case Study: Designing an Autonomous Pool Cleaning Robot

To illustrate the QFD process, let's apply it to the design of an autonomous, solar-powered robot for collecting debris in swimming pools.

10.12.2.5.1 Customer Requirements

Through customer research, we've identified the following requirements:

i. Effective debris collection
ii. Easy to use
iii. Low maintenance
iv. Long-lasting battery life
v. Safe for swimmers
vi. Eco-friendly operation
vii. Suitable for various pool sizes
viii. Quiet operation
ix. Attractive design
x. Affordable price

10.12.2.5.2 Engineering Characteristics

Based on the customer requirements, we've identified these engineering characteristics:

i. Debris collection rate (kg/hour)
ii. Filter capacity (liters)
iii. Battery capacity (Watt-hours)
iv. Solar panel efficiency (%)
v. Noise level (dB)
vi. Weight (kg)
vii. Maximum pool size (square meters)
viii. Number of user controls
ix. Maintenance interval (hours)

x.   Manufacturing cost ($)

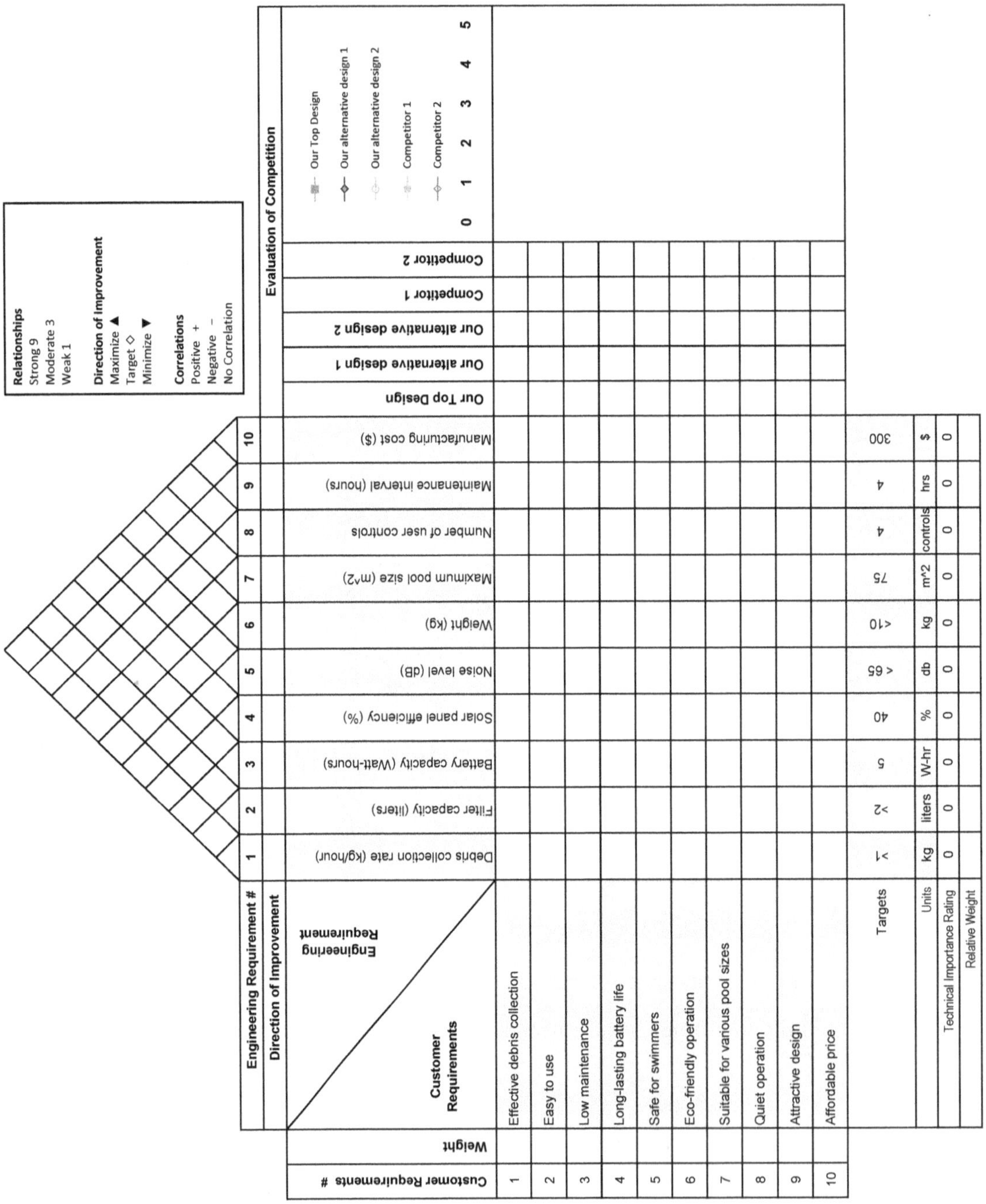

Relationships
Strong 9
Moderate 3
Weak 1

Direction of Improvement
Maximize ▲
Target ◇
Minimize ▼

Correlations
Positive +
Negative –
No Correlation

Evaluation of Competition

- ▦ Our Top Design
- ◆ Our alternative design 1
- ╪ Our alternative design 2
- ✳ Competitor 1
- ◇ Competitor 2

0   1   2   3   4   5

Engineering Requirement #		Direction of Improvement											Evaluation of Competition

Engineering Requirements (numbered 1–10):
1. Debris collection rate (kg/hour)
2. Filter capacity (liters)
3. Battery capacity (Watt-hours)
4. Solar panel efficiency (%)
5. Noise level (dB)
6. Weight (kg)
7. Maximum pool size (m^2)
8. Number of user controls
9. Maintenance interval (hours)
10. Manufacturing cost ($)

Evaluation columns: Our Top Design, Our alternative design 1, Our alternative design 2, Competitor 1, Competitor 2

Customer Requirements

#	Customer Requirements	Weight
1	Effective debris collection	
2	Easy to use	
3	Low maintenance	
4	Long-lasting battery life	
5	Safe for swimmers	
6	Eco-friendly operation	
7	Suitable for various pool sizes	
8	Quiet operation	
9	Attractive design	
10	Affordable price	

	1	2	3	4	5	6	7	8	9	10
Targets	>1	>2	5	40	<65	<10	75	4	4	300
Units	kg	liters	W-hr	%	db	kg	m^2	controls	hrs	$
Technical Importance Rating	0	0	0	0	0	0	0	0	0	0
Relative Weight										

Figure 10-7. QFD Chart with customer requirements and engineering requirements filled out for the example problem of automated pool skimmer design.

Relationships
Strong 9
Moderate 3
Weak 1

Direction of Improvement
Maximize ▲
Target ◇
Minimize ▼

Correlations
Positive +
Negative −
No Correlation

Evaluation of Competition
- Our Top Design
- Our alternative design 1
- Our alternative design 2
- Competitor 1
- Competitor 2

(scale 0 1 2 3 4 5)

Customer Requirements #	Customer Requirements	1 Debris collection rate (kg/hour)	2 Filter capacity (liters)	3 Battery capacity (Watt-hours)	4 Solar panel efficiency (%)	5 Noise level (dB)	6 Weight (kg)	7 Maximum pool size (m^2)	8 Number of user controls	9 Maintenance interval (hours)	10 Manufacturing cost ($)
1	Effective debris collection	9	9	3			1	3		3	
2	Easy to use	1					3	3	9		
3	Low maintenance		3		3				1	9	1
4	Long-lasting battery life			9	9					3	1
5	Safe for swimmers	1				3	3				1
6	Eco-friendly operation		3	3	9	3	9		3	3	1
7	Suitable for various pool sizes	3		3			3	9			3
8	Quiet operation					9	3				3
9	Attractive design				1		3		1		3
10	Affordable price	1	1	3	3	1	3		1	3	9
	Targets	>1	>2	5	40	< 65	<10	75	4	4	300
	Units	kg	liters	W-hr	%	db	kg	m^2	controls	hrs	$
	Technical Importance Rating	0	0	0	0	0	0	0	0	0	0
	Relative Weight	0	0	0	0	0	0	0	0	0	0

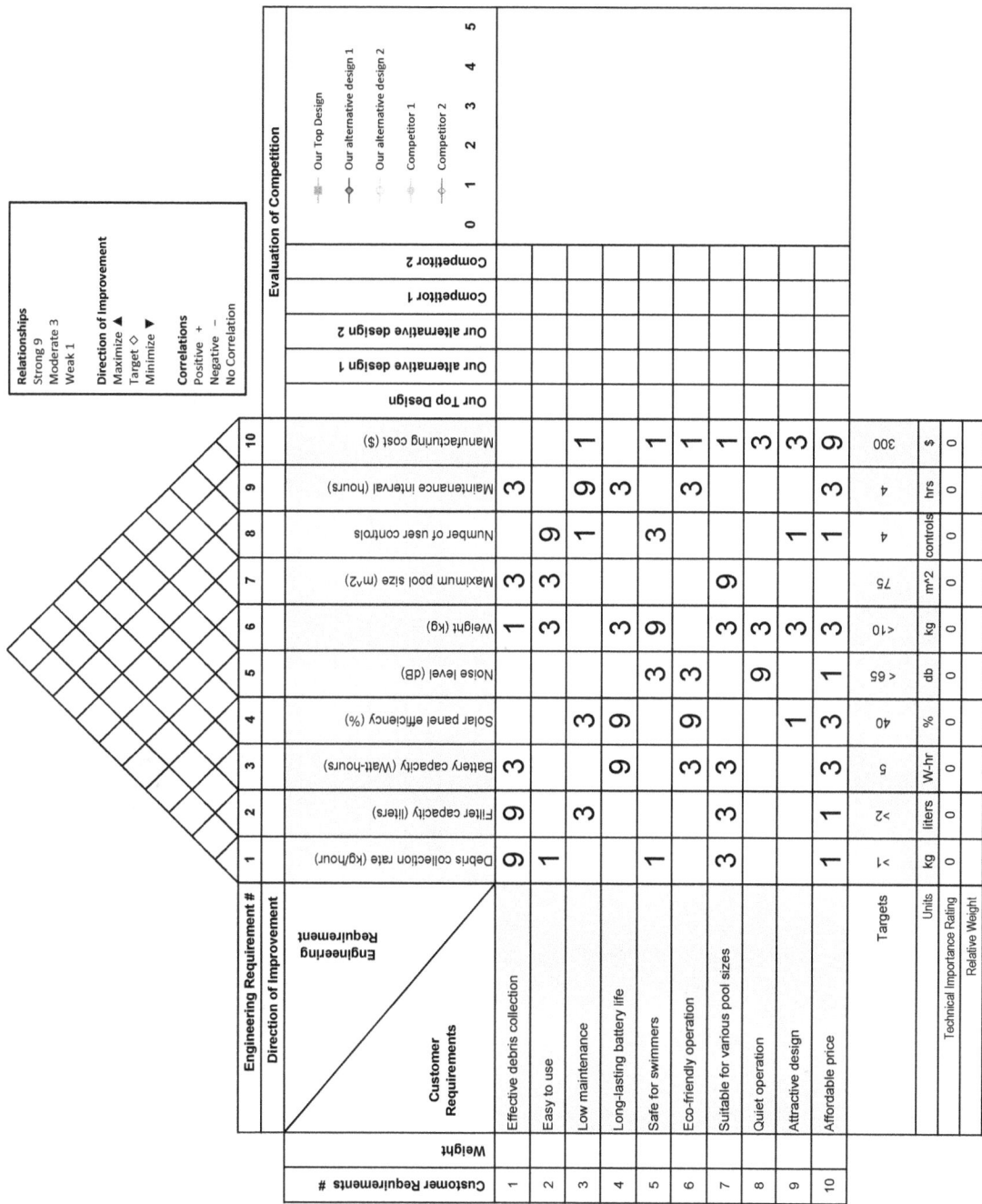

Figure 10-8. QFD Chart with the relationship matrix completed.

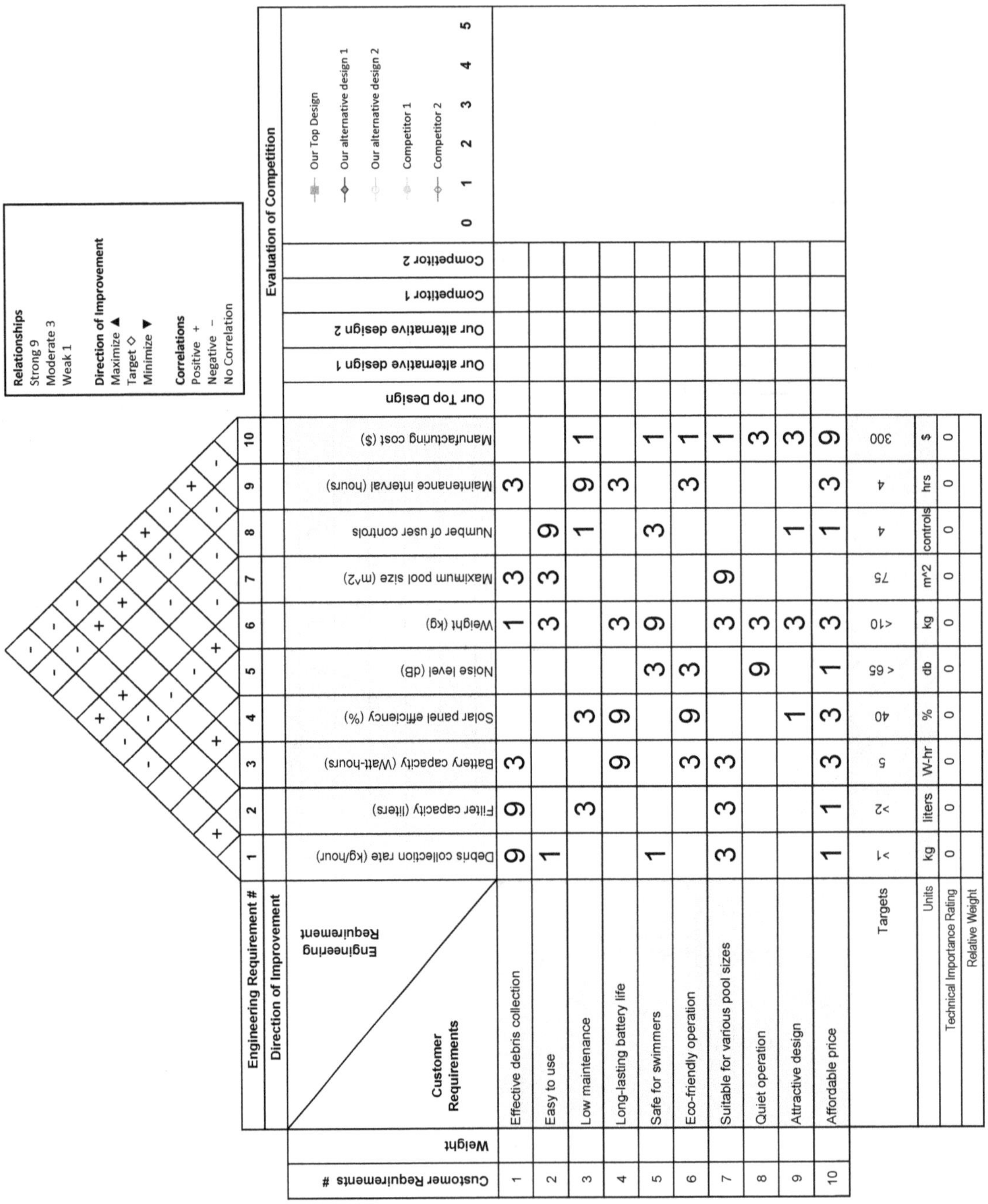

Relationships
Strong 9
Moderate 3
Weak 1

Direction of Improvement
Maximize ▲
Target ◇
Minimize ▼

Correlations
Positive +
Negative –
No Correlation

Customer Requirements	1. Debris collection rate (kg/hour)	2. Filter capacity (liters)	3. Battery capacity (Watt-hours)	4. Solar panel efficiency (%)	5. Noise level (dB)	6. Weight (kg)	7. Maximum pool size (m^2)	8. Number of user controls	9. Maintenance interval (hours)	10. Manufacturing cost ($)
1. Effective debris collection	9	9	3			1	3		3	
2. Easy to use	1					3	3	9		
3. Low maintenance		3		3				1	9	1
4. Long-lasting battery life			9	9		3			3	
5. Safe for swimmers	1				3	3				1
6. Eco-friendly operation		3	3	9	3	9			3	1
7. Suitable for various pool sizes	3		3			3	9	3		1
8. Quiet operation					9	3				3
9. Attractive design				1	1	3		1		3
10. Affordable price	1	1	3	3	1	3		1	3	9
Targets	>1	>2	5	40	<65	<10	75	4	4	300
Units	kg	liters	W-hr	%	db	kg	m^2	controls	hrs	$
Technical Importance Rating	0	0	0	0	0	0	0	0	0	0
Relative Weight	0									

Evaluation of Competition

Our Top Design
Our alternative design 1
Our alternative design 2
Competitor 1
Competitor 2

0   1   2   3   4   5

Competitor 2
Competitor 1
Our alternative design 2
Our alternative design 1
Our Top Design

Figure 10-9. QFD Chart with the correlation matrix completed.

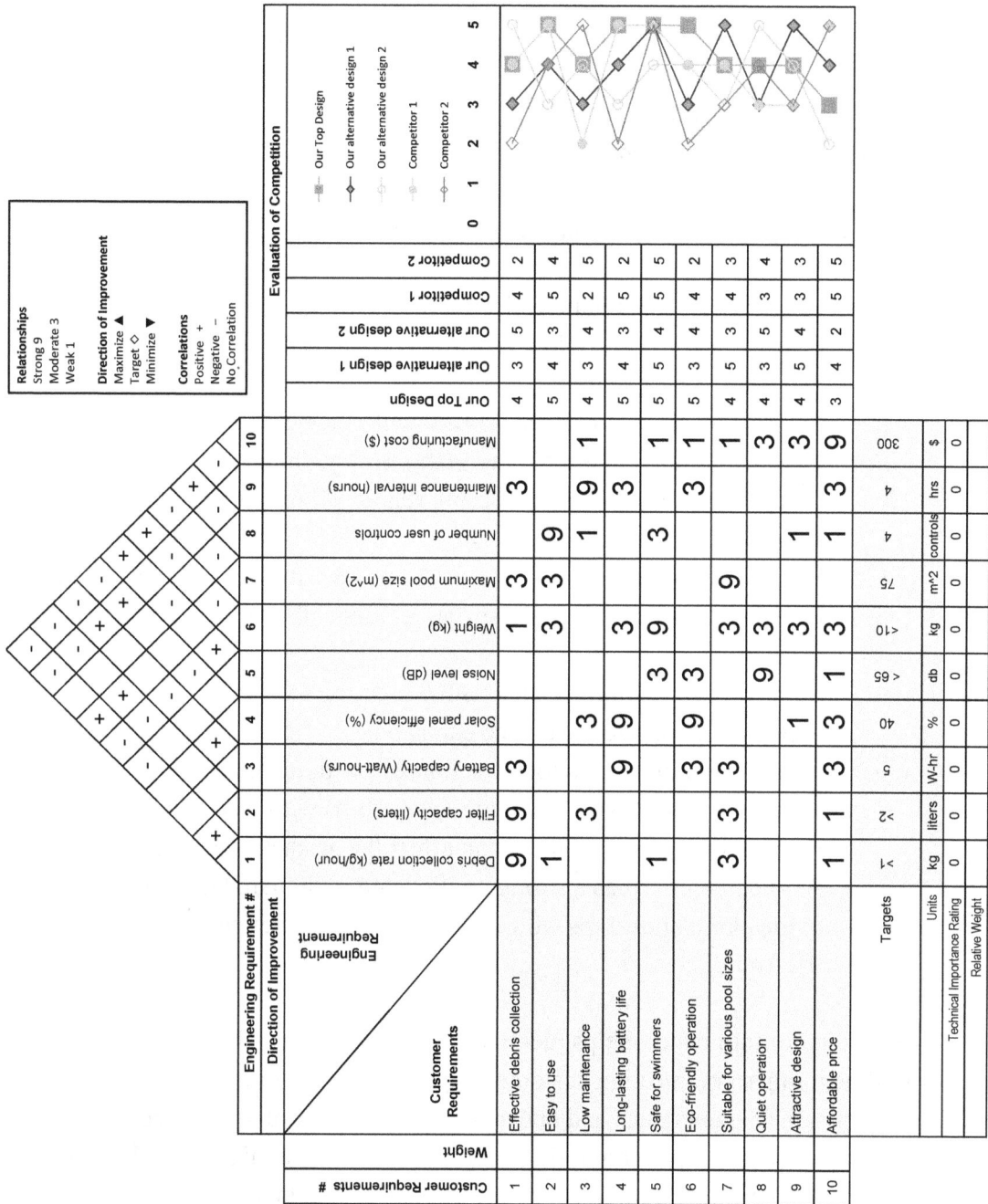

Figure 10-10. QFD Chart with the competitive analysis completed.

10.12.2.5.3   Relationship Matrix

We'll use the following numerical values to indicate relationships:
Strong relationship (9 points)
Moderate relationship (3 points)
Weak relationship (1 point)

Figure 10-9. shows the relationship matrix completed in the QFD chart.

Explanation of some key relationships:
  i.   Effective debris collection has strong relationships with debris collection rate and filter capacity.
  ii.   Easy to use has a strong relationship with the number of user controls.
  iii. Low maintenance has a strong relationship with the maintenance interval.
  iv.   Long-lasting battery life has strong relationships with battery capacity and solar panel efficiency.
  v.   Safe for swimmers has a strong relationship with the weight of the device.
  vi.   Eco-friendly operation has a strong relationship with solar panel efficiency.
  vii. Suitability for various pool sizes has a strong relationship with the maximum pool size the device can handle.
  viii. Quiet operation has a strong relationship with the noise level.
  ix.   Attractive design has moderate relationships with weight and manufacturing cost, as these can affect the aesthetics and materials used.
  x.   Affordable price has a strong relationship with manufacturing cost and moderate relationships with several other characteristics that can impact the overall cost.

This relationship matrix helps identify which engineering characteristics have the most significant impact on meeting customer requirements. It guides the design team in prioritizing certain characteristics and understanding the trade-offs involved in different design decisions. For example, improving debris collection rate might negatively impact ease of use due to increased weight, so the team needs to find an optimal balance.

10.12.2.5.4   Correlation Matrix

We'll use the following symbols for correlations:
• Positive correlation (+)
• Negative correlation (-)

Figure 10-10. shows the correlation matrix completed in the QFD chart.

Engineering Characteristics:

A. Debris collection rate (kg/hour)

B. Filter capacity (liters)

C. Battery capacity (Watt-hours)

D. Solar panel efficiency (%)

E. Noise level (dB)

F. Weight (kg)

G. Maximum pool size (square meters)

H. Number of user controls

I. Maintenance interval (hours)

J. Manufacturing cost ($)

Key correlations explained:

i.   Debris collection rate (A) positively correlates with filter capacity (B) and maximum pool size (G), but negatively with weight (F) and noise level (E).

ii. Battery capacity (C) positively correlates with solar panel efficiency (D) and maintenance interval (I), but negatively with weight (F) and manufacturing cost (J).

iii. Weight (F) negatively correlates with several characteristics, as increased weight can impact performance and usability.

iv. Noise level (E) positively correlates with weight (F) and manufacturing cost (J), but negatively with the number of user controls (H).

v.   Manufacturing cost (J) negatively correlates with many characteristics, as improving performance often increases costs.

This simplified correlation matrix helps identify potential conflicts and synergies in the design, allowing the team to make informed decisions about trade-offs and prioritize improvements that have the most positive overall impact on the product.

10.12.2.5.5   Competitive Assessment

We'll evaluate our concept against four existing products on a scale of 1-5, where 5 is excellent and 1 is poor.

Figure 10-10. shows the competitive assessment matrix completed in the QFD chart. This competitive assessment matrix provides insights into how our concept compares to existing products in the market. Here are some observations:

i.   Our concept excels in ease of use, battery life, and eco-friendly operation.

ii. We're competitive in debris collection, but Competitor B has an edge.

iii. We have room for improvement in affordability compared to Competitors C and

D.

iv.   Our concept is well-rounded, scoring at least average (3) or above in all categories.

v.   Each competitor has strengths and weaknesses, e.g., Competitor D is strong in low maintenance and affordability but weak in debris collection and battery life.

This assessment helps identify areas where our concept has a competitive advantage and areas where improvement might be needed to stand out in the market. It can guide design decisions and prioritization of features in the development process.

10.12.2.5.6   Target Values and Difficulty

For each engineering characteristic, we'll set target values and assess the difficulty of achieving them on a scale of 1-5, where 5 is very difficult and 1 is easy.

Figure 10-10. shows the QFD chart with updated target values and difficulty entered (technical importance rating).

Explanation of some key target values and their difficulties:

i.   Debris collection rate: A high target of 2.5 kg/hour is set to ensure effective cleaning, but it's difficult to achieve while maintaining other constraints like weight and power consumption.

ii.   Filter capacity: A 3-liter capacity is relatively easy to achieve and balances effectiveness with size constraints.

iii.   Battery capacity: 200 Wh should provide good operating time, but achieving this capacity in a compact, waterproof design is moderately difficult.

iv.   Solar panel efficiency: 22% efficiency is very difficult to achieve, as it requires advanced and potentially expensive solar cell technology.

v.   Noise level: Keeping the noise below 50 dB is challenging, especially when balancing with debris collection rate and motor power.

vi.   Weight: Keeping the weight below 7 kg while incorporating all necessary components is difficult and may require advanced materials.

vii.   Maximum pool size: Covering 100 square meters is moderately difficult, balancing battery life and debris collection rate.

viii.   Number of user controls: Limiting to 3 controls is relatively easy but requires thoughtful user interface design.

ix.   Maintenance interval: A 500-hour interval between maintenance is difficult to achieve, requiring robust design and high-quality components.

x.   Manufacturing cost: Keeping the cost below $300 is very difficult, especially when considering the advanced features and quality requirements.

This information helps the design team understand which targets might require more resources, innovation, or compromise to achieve. It can guide the prioritization of development efforts and help in making trade-off decisions during the design process.

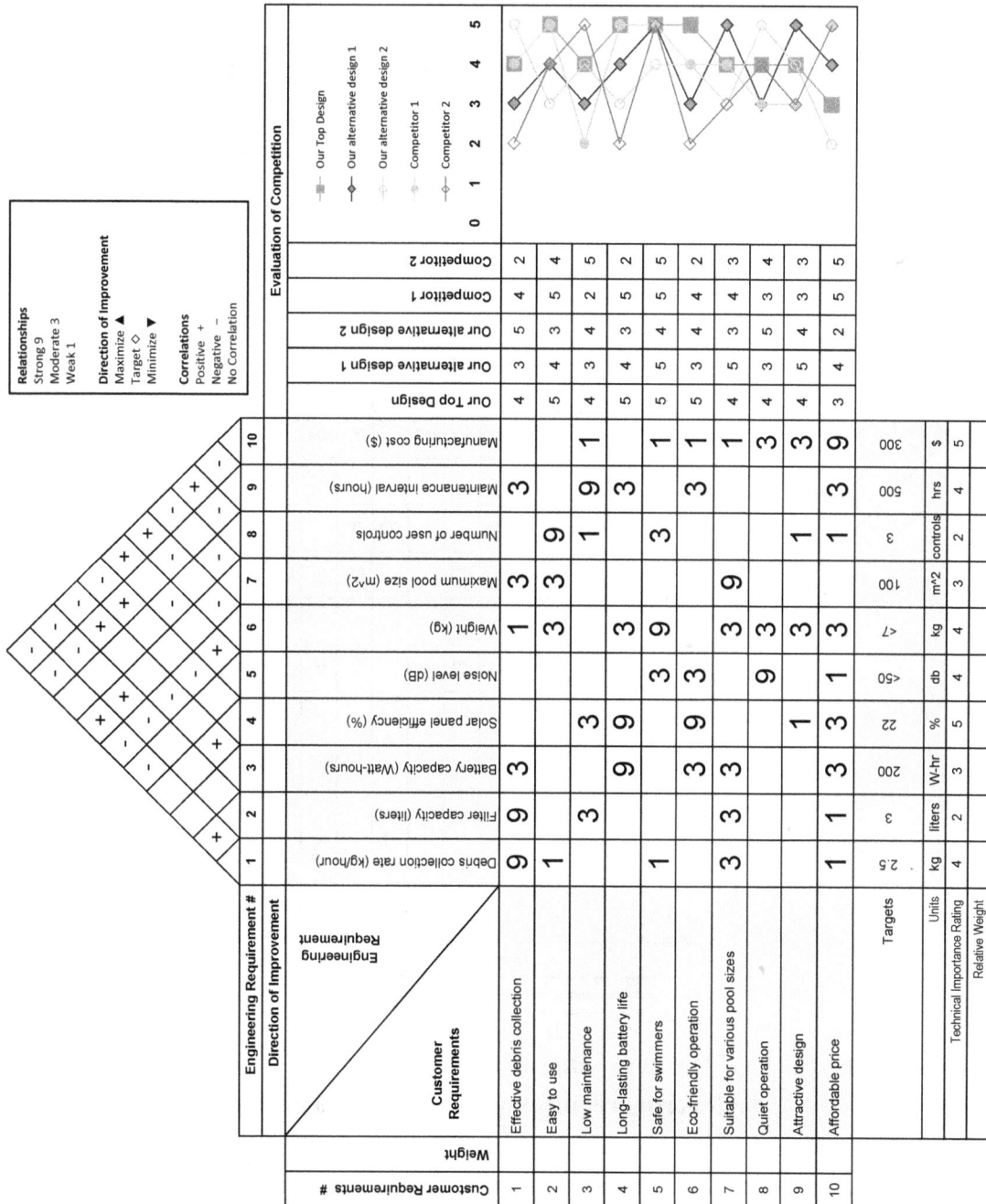

Relationships
Strong 9
Moderate 3
Weak 1

Direction of Improvement
Maximize ▲
Target ◇
Minimize ▼

Correlations
Positive +
Negative −
No Correlation

Customer Req #	Customer Requirements	Weight	1. Debris collection rate (kg/hour)	2. Filter capacity (liters)	3. Battery capacity (Watt-hours)	4. Solar panel efficiency (%)	5. Noise level (dB)	6. Weight (kg)	7. Maximum pool size (m^2)	8. Number of user controls	9. Maintenance interval (hours)	10. Manufacturing cost ($)	Our Top Design	Our alternative design 1	Our alternative design 2	Competitor 1	Competitor 2
1	Effective debris collection		9	9	3			1	3		3		4	3	5	4	2
2	Easy to use		1					3	3	9		1	5	4	3	5	4
3	Low maintenance			3		3				1	9		4	3	4	2	5
4	Long-lasting battery life				9	9		3			3	1	5	4	3	5	2
5	Safe for swimmers		1				3	9				1	5	5	4	5	5
6	Eco-friendly operation			3	3	9	3			3	3	1	5	3	4	4	2
7	Suitable for various pool sizes		3	3	3			3	9				4	5	3	4	3
8	Quiet operation						9	3		1		3	4	3	5	3	4
9	Attractive design					1		3		1		3	4	5	4	3	3
10	Affordable price		1	1	3	3	1	3		1	3	9	3	4	2	5	5
	Targets		2.5	3	200	22	<50	<7	100	3	500	300					
	Units		kg	liters	W-hr	%	db	kg	m^2	controls	hrs	$					
	Technical Importance Rating		4	2	3	5	4	4	3	2	4	5					
	Relative Weight																

Evaluation of Competition legend:
- ■ Our Top Design
- ◆ Our alternative design 1
- ○ Our alternative design 2
- ◇ Competitor 1
- ◇ Competitor 2

Figure 10-11. The QFD Chart with the target values and difficulty was updated and completed.

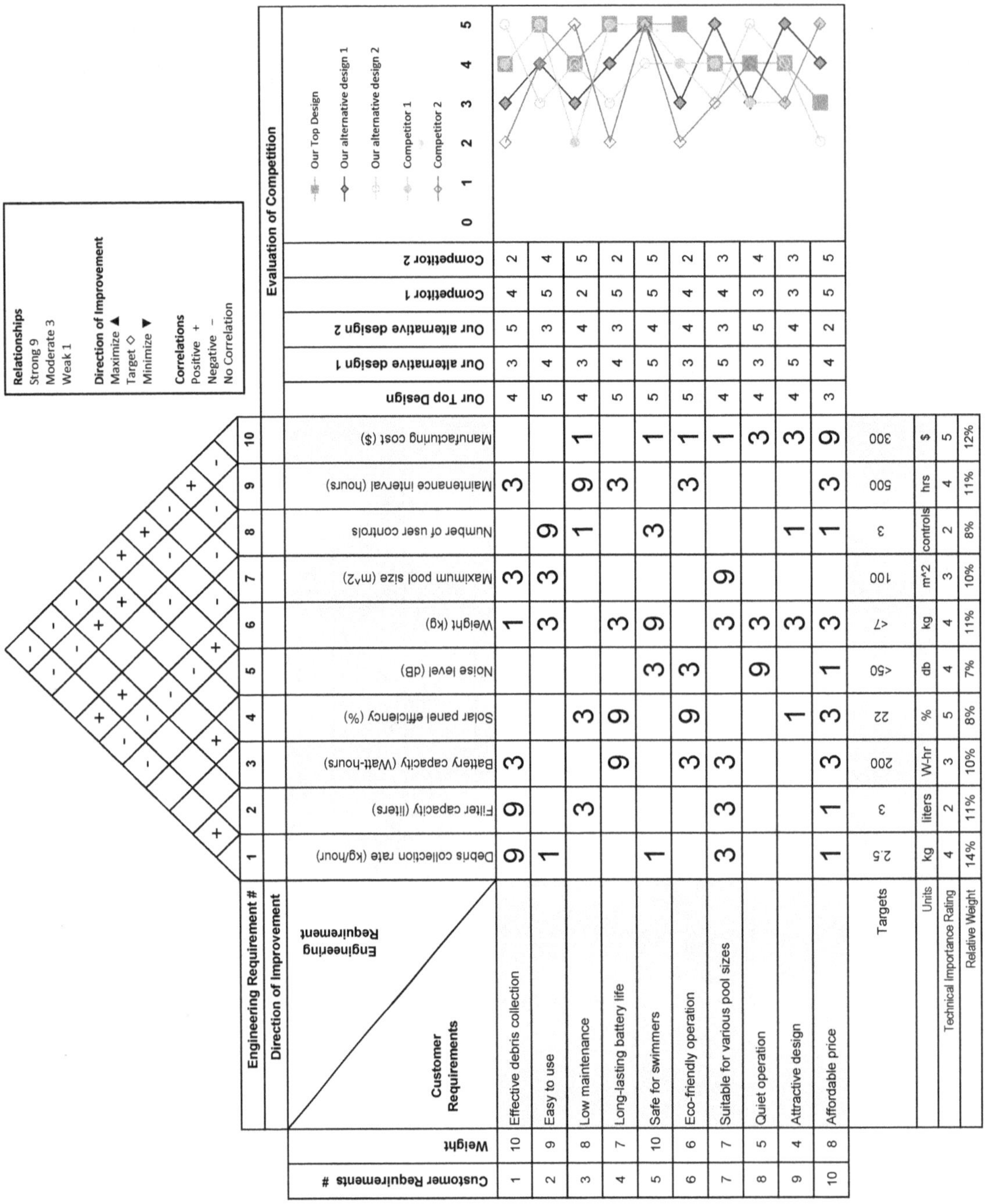

Legend

Relationships
Strong 9
Moderate 3
Weak 1

Direction of Improvement
Maximize ▲
Target ◇
Minimize ▼

Correlations
Positive +
Negative –
No Correlation

QFD Chart — Relationship Matrix

CR #	Weight	Customer Requirements	1. Debris collection rate (kg/hour)	2. Filter capacity (liters)	3. Battery capacity (Watt-hours)	4. Solar panel efficiency (%)	5. Noise level (dB)	6. Weight (kg)	7. Maximum pool size (m^2)	8. Number of user controls	9. Maintenance interval (hours)	10. Manufacturing cost ($)
1	10	Effective debris collection	9	9	3			1	3		3	
2	9	Easy to use	1					3	3	9		
3	8	Low maintenance		3		3				1	9	
4	7	Long-lasting battery life			9	9					3	1
5	10	Safe for swimmers	1				3	3				
6	6	Eco-friendly operation			3	9	3	9		3	3	1
7	7	Suitable for various pool sizes	3	3	3			3	9			1
8	5	Quiet operation				1	9	3				3
9	4	Attractive design						3		1		3
10	8	Affordable price	1	1	3	3	1	3		1	3	9
		Targets	2.5	3	200	22	<50	<7	100	3	500	300
		Units	kg	liters	W-hr	%	db	kg	m^2	controls	hrs	$
		Technical Importance Rating	4	2	3	5	4	4	3	2	4	5
		Relative Weight	14%	11%	10%	8%	7%	11%	10%	8%	11%	12%

Evaluation of Competition

Customer Requirement	Our Top Design	Our alternative design 1	Our alternative design 2	Competitor 1	Competitor 2
Effective debris collection	4	3	5	4	2
Easy to use	5	4	3	5	4
Low maintenance	4	3	4	2	5
Long-lasting battery life	5	4	3	5	2
Safe for swimmers	5	5	4	5	5
Eco-friendly operation	5	3	4	4	2
Suitable for various pool sizes	4	5	3	4	3
Quiet operation	4	3	5	3	4
Attractive design	4	5	4	3	3
Affordable price	3	4	2	5	5

Evaluation of Competition — line plot legend:
- Our Top Design
- Our alternative design 1
- Our alternative design 2
- Competitor 1
- Competitor 2

(scale 0 1 2 3 4 5)

Figure 10-12. A QFD Chart with the weights for customer requirements and engineering requirements has been completed.

10.12.2.5.7 Importance Ratings

We'll calculate the importance of each customer requirement and engineering characteristic based on the relationship matrix and competitive assessment.

Figure 10-10. shows the completed QFD chart.

Explanation of ratings:
i.   Effective debris collection (10): This is the primary function of the device, so it's given the highest importance.
ii. Easy to use (9): High importance as it directly affects customer satisfaction and adoption.
iii. Low maintenance (8): Important for long-term customer satisfaction and reducing ongoing costs.
iv. Long-lasting battery life (7): Important for convenience but slightly less critical than other factors.
v.   Safe for swimmers (10): Safety is paramount, so this is given the highest importance.
vi. Eco-friendly operation (6): Important for marketing and environmental concerns but not as critical as core functionality.
vii. Suitable for various pool sizes (7): This is important for market reach, but some compromise is acceptable.
viii. Quiet operation (5): Desirable but not critical to the core function.
ix. Attractive design (4): The least important factor, as functionality is prioritized over aesthetics for this product.
x.   Affordable price (8): This is very important for market competitiveness and adoption, but some premium for quality is acceptable.

These importance ratings help prioritize design efforts and can be used in conjunction with the relationship matrix to calculate the overall importance of each engineering characteristic. This guides the development team in focusing their efforts on the most critical aspects of the product from the customer's perspective.

It is also a good time at this step to assign weights to customer requirements and engineering requirements. Figure 10-10. shows the completed chart.

Explanation of the relative weights for engineering requirements:
i.   Debris collection rate (13.5%): The highest weight, reflecting its critical importance to the product's primary function.
ii. Manufacturing cost (12.1%): This is the second-highest weight, emphasizing the importance of achieving a competitive price point.
iii. Filter capacity, Weight, and Maintenance interval (10.8% each): These

three characteristics share the third-highest weight, indicating their significant impact on product performance and user satisfaction.

iv. Battery capacity and Maximum pool size (9.5% each): These characteristics have a notable impact on the product's versatility and convenience.

v. Solar panel efficiency and Number of user controls (8.1% each): While important, these characteristics have slightly lower weights as they are supportive features rather than core functionalities.

vi. Noise level (6.8%): This is the lowest weight, as it's a desirable feature but less critical than other performance aspects.

These percentage weights provide a clear picture of how to allocate resources and effort during the development process. For example, the team should spend about twice as much time and resources on improving the debris collection rate compared to reducing noise levels. This prioritization helps ensure that development efforts are aligned with the characteristics that will have the most significant impact on the product's success in the market.

10.12.2.5.8   Comparing Design Concepts

Based on the QFD analysis, we'll compare five design concepts for the autonomous pool-cleaning robot:

Concept A: Catamaran-style float with central collection basket Concept B: Circular design with perimeter suction Concept C: Submersible robot with surface debris skimmer Concept D: Floating disc with extendable collection arms Concept E: Modular design with detachable collection units

We'll evaluate each concept against the engineering characteristics and customer requirements identified in the QFD process.

10.12.2.6   Implementing QFD Results

After completing the QFD analysis and concept comparison, the next steps in the product development process include:

i. Detailed design of the selected concept
ii. Prototyping and testing
iii. Design refinement based on test results
iv. Production planning
v. Market launch and customer feedback collection

Throughout these stages, the QFD results should be used to guide decision-making and ensure that the final product aligns with customer needs and technical requirements.

10.12.2.7   Advanced QFD Techniques

While the House of Quality is the most common QFD tool, several advanced techniques can enhance the QFD process:

 i. Blitz QFD: A streamlined approach for rapid product development

 ii. Modern QFD: Incorporates new tools and methods for complex systems

 iii. Voice of Customer (VOC) analysis: Advanced techniques for capturing and analyzing customer needs

 iv. Analytical Hierarchy Process (AHP): A method for prioritizing customer requirements

 v. Kano Model integration: Classifying customer requirements into basic, performance, and excitement categories

These advanced techniques can be particularly useful for complex products or when dealing with conflicting customer requirements.

Quality Function Deployment is a robust methodology for translating customer needs into actionable engineering requirements. By systematically analyzing the relationships between customer wants and technical characteristics, QFD helps development teams create products that truly meet customer expectations.

In our case study of the autonomous pool cleaning robot, we saw how QFD can be applied to guide the design process, from identifying customer requirements to comparing design concepts. The structured approach of QFD ensures that all aspects of the product are considered and that development efforts are focused on the most important features from the customer's perspective.

While QFD can be time-consuming and complex, especially for large-scale projects, its benefits in terms of improved product quality, reduced development time, and enhanced customer satisfaction make it a valuable tool for any product development team. By fostering cross-functional collaboration and maintaining a strong focus on customer needs throughout the development process, QFD helps companies create successful products in an increasingly competitive marketplace.

As with any tool, QFD's effectiveness depends on how well it is implemented and integrated into the overall product development process. Teams should be prepared to invest time in data collection, analysis, and interpretation and use the insights gained from QFD to inform decision-making at all stages of product development.

By mastering the QFD process and adapting it to their specific needs, companies can gain a significant competitive advantage in product development and innovation.

10.12.3   Decision Trees

Decision trees are graphical representations of possible solutions to a decision based on certain conditions. They help in visualizing the consequences of various choices, allowing the team to make informed decisions.

10.12.3.1   Steps to Create a Decision Tree:

1. **Identify the Decision**: Define the problem or decision to be made.
2. **List Possible Alternatives**: Identify all possible courses of action.
3. **Evaluate Consequences**: Determine the outcomes of each alternative.
4. **Calculate Probabilities**: Assign probabilities to each possible outcome.
5. **Analyze the Tree**: Use the decision tree to evaluate the expected values of each alternative.

10.12.4   Multi-Criteria Decision Analysis (MCDA)

Multi-Criteria Decision Analysis (MCDA) is a method for evaluating multiple conflicting criteria in decision-making. It is particularly useful in complex decision-making scenarios where various factors must be considered.

10.12.4.1   Steps of MCDA:

1. **Define the Problem**: Clearly state the problem and objectives.
2. **Identify Criteria**: List all criteria relevant to the decision.
3. **Assign Weights**: Determine the relative importance of each criterion.
4. **Evaluate Alternatives**: Score each alternative against the criteria.
5. **Aggregate Scores**: Combine the scores to determine the best alternative.
6. **Example: Using Pugh Analysis for Concept Selection**

Consider a team tasked with designing an underwater jet scooter. The team has generated 30 design concepts and developed the following evaluation criteria: speed, cost, maneuverability, durability, ease of use, and energy efficiency.

1. **Develop Design Specifications**: Establish that the scooter must achieve a speed of 5 knots, cost less than $500, and be highly maneuverable, durable, easy to use, and energy-efficient.
2. **Generate Design Concepts**: List the 30 concepts generated by the team.
3. **Create the Evaluation Matrix**: Use a spreadsheet to list the evaluation criteria and design concepts.
4. **Select a Reference Design**: Choose design concept 1 as the reference.
5. **Evaluate Each Design Concept**: Score each concept against the reference design using +1, 0, or -1.
6. **Record Ratings**: Enter the scores and sum the results.

7. **Iterate if Necessary**: If no clear best idea emerges, refine the concepts and repeat the evaluation.

After conducting the Pugh Analysis, the team found that design concepts 6, 13, and 14 have the highest scores. They decide to combine the best features of these concepts into a new hybrid design and evaluate it again, resulting in an optimal solution.

Decision-making is integral to the engineering design process. By using structured methods such as Pugh Analysis, Quality Function Deployment, Decision Trees, and Multi-Criteria Decision Analysis, design teams can systematically evaluate and select the best design concepts. These tools help ensure that the final design meets customer requirements, is feasible to produce, and performs well under real-world conditions. Structured decision-making leads to better-informed choices, ultimately resulting in higher-quality design solutions.

10.13 Assignments

Assignment 10-1: Developing Customer Requirements and Importance Weights

   Objective: As a team, create a comprehensive list of customer requirements for your capstone design project, along with associated importance weights. This assignment will guide you through the process of gathering, organizing, and prioritizing customer requirements, which will serve as the foundation for your project.

   Instructions:

1. Individual Brainstorming (30 minutes)
   - In your workbook, create a section titled "Customer Requirements Brainstorming."
   - Independently, write down as many potential customer requirements as you can think of for your capstone design project.
   - Consider various aspects such as functionality, usability, performance, safety, reliability, aesthetics, and cost.
   - Be specific and clear in your descriptions.

2. Team Discussion and Consolidation (60 minutes)
   - As a team, discuss and share the customer requirements each member has identified.

- o In your workbook, create a section titled "Consolidated Customer Requirements."
- o Combine similar requirements and eliminate duplicates to create a comprehensive list of customer requirements.
- o Ensure that each requirement is clearly written and understood by all team members.

3. Importance Weight Assignment (45 minutes)
   - o In your workbook, create a section titled "Importance Weights."

- o   As a team, assign importance weights to each customer requirement using a scale of 1 to 5 (1 being least important, five being most important).
- o   Consider factors such as the impact on the project's success, customer satisfaction, and alignment with the project's goals.
- o   Discuss and reach a consensus on the importance of weights for each requirement.

4. Electronic Template Completion (60 minutes)
   - o   Access the "Customer Requirements Template" provided on Brightspace.

- o As a team, transfer the consolidated customer requirements and their corresponding importance weights from your workbooks to the electronic template.
- o Ensure that the information is accurately and neatly entered into the template.
- o Organize the requirements in a logical manner, such as by category or priority.

5. Review and Finalization (30 minutes)
   - o As a team, review the completed electronic template to ensure all requirements are included and the importance weights are accurately assigned.
   - o Make any necessary revisions or adjustments based on team feedback.
   - o Save the final version of the electronic template on Brightspace.

Workbook Requirements:
- Each student must maintain a neat and well-organized workbook throughout the

assignment.

- Write legibly and clearly to ensure readability.
- Include the sections "Customer Requirements Brainstorming," "Consolidated Customer Requirements," and "Importance Weights" in your workbook.
- Individual work in the workbook will be assessed at the end of the semester. Collaboration and Electronic Submission:
- Actively collaborate with your team members throughout the assignment.
- Use your individual written responses in the workbook to contribute to the team's consolidated list of customer requirements and importance weights.
- Complete the electronic template provided on Brightspace as a team, ensuring accuracy and clarity.
- Submit the final electronic template on Brightspace by the specified deadline.

Remember, the quality and comprehensiveness of your customer requirements and importance weights will have a significant impact on the success of your capstone design project. Take the time to carefully consider and discuss each requirement as a team.

If you have any questions or need further guidance, please don't hesitate to reach out to your instructor or teaching assistants.

Assignment 10-2. Develop Engineering Requirements, Target Values, and Importance Weights

Objective: As a team, create a comprehensive list of engineering requirements for your capstone design project, along with their target values and importance weights. This assignment will guide you through the process of translating customer requirements into measurable engineering specifications, setting target values, and prioritizing them based on their importance to the project's success.

Instructions:

1. Review Customer Requirements (30 minutes)
   - As a team, review the consolidated list of customer requirements and their importance weights from the previous assignment.
   - Below write your responses and how each customer requirement may be converted into an engineering requirement. Explain.
   - Discuss and ensure that all team members have a clear understanding of each customer requirement.

2. Identify Engineering Requirements (60 minutes)
    o For each customer requirement, identify the corresponding engineering requirement(s) that will enable you to meet that requirement.
    o Translate customer requirements into engineering requirements.
    o Be specific, measurable, and technically feasible in your engineering requirements.
    o Example: If a customer requirement is "Easy to use," an engineering requirement could be "User interface response time < 1 second."

3. Set Target Values (45 minutes)
   - For each engineering requirement, determine a target value that quantifies the desired performance or specification.
   - Use relevant industry standards, benchmarks, or customer expectations to guide your target value selection.
   - Example: For the engineering requirement "User interface response time < 1 second," the target value would be "0.5 seconds."

4. Assign Importance Weights (45 minutes)
   o As a team, assign importance weights to each engineering requirement using a scale of 1 to 5 (1 being least important, five being most important).
   o Consider factors such as the impact on meeting customer requirements, technical feasibility, and project constraints.
   o Through group discussion and analysis, determine the relative importance of each engineering requirement.

5. Electronic Assignment Completion (60 minutes)
    - o Access the "Engineering Requirements Template" provided on Brightspace.
    - o As a team, transfer your written engineering requirements, target values, and importance weights from your workbooks to the electronic template.
    - o Ensure that the information is accurately and neatly entered into the template.
    - o Organize the requirements in a logical manner, such as by subsystem or functional area.

6. Review and Finalization (30 minutes)
    - o As a team, review the completed electronic template to ensure all engineering requirements, target values, and importance weights are included and accurate.
    - o Make any necessary revisions or adjustments based on team feedback.
    - o Upload the final version of the electronic template on Brightspace.

Workbook Requirements:
- Each student must maintain a neat and well-organized workbook throughout the assignment.
- Write legibly and clearly to ensure readability.
- Individual work in the workbook will be assessed at the end of the semester.

Collaboration and Electronic Submission:
- Actively collaborate with your team members throughout the assignment.
- Use your individual written responses in the workbook to contribute to the team's consolidated list of engineering requirements, target values, and importance weights.
- Complete the electronic template provided on Brightspace as a team, ensuring accuracy and clarity.
- Submit the final electronic template on Brightspace by the specified deadline.

Remember, well-defined engineering requirements, target values, and importance weights are crucial for guiding your design process and ensuring that your final product meets the desired specifications and customer expectations.

If you have any questions or need further guidance, please don't hesitate to reach out to your instructor or teaching assistants.

Assignment 10-3: Brainstorming Design Solutions – Class Exercise
1. **Objective**
   This assignment's objective is to facilitate a brainstorming session where students generate multiple design solutions for their capstone engineering project. This exercise will help students explore various ideas, encourage creativity, and prepare for a more detailed design concept generation phase.
2. **Materials Needed**
   - Color Post-it Notes (enough for each student to have at least 10)
   - Pens or pencils
   - Large table or flat surface for sorting and grouping ideas
3. **Instructions**
   Follow these step-by-step instructions to complete the brainstorming exercise during a 30-minute class session.
4. **Step 1: Individual Brainstorming (10 minutes)**
   1. **Preparation**:
      - Each student should have at least ten color Post-it notes and a pen or pencil.
   2. **Brainstorming**:
      - Students will individually brainstorm design solutions for their capstone project.
      - Write each idea on a separate Post-it note. Use words, sketches, or a combination of both to describe the idea.
      - Aim to generate at least ten unique design concepts within the 10-minute time frame.
      - Do not critique or dismiss any ideas during this phase; all ideas are valuable.
5. **Step 2: Sorting and Grouping Ideas (10 minutes)**
   - **Place Post-it Notes on the Table**:
      1. At the end of the brainstorming session, students will place their Post-it notes on a large table or flat surface.
      2. Ensure all notes are visible and easily accessible for everyone to see.
   - **Sort and Group Ideas**:
      1. As a team, spend 10 minutes sorting and grouping the Post-it notes into clusters based on common themes or similarities.
      2. Discuss the ideas as you sort them, but focus on identifying patterns and grouping similar concepts.
      3. Create clusters of related ideas and label each cluster if necessary.
6. **Step 3: Documenting Design Concepts (10 minutes)**
   1. **Record the Clusters**:

  - o Once the ideas are grouped, document the clusters by taking notes or photographing the Post-it notes on the table.
  - o Ensure that each cluster is clearly documented with a description of the common theme and individual ideas within the cluster.

2. **Prepare for Detailed Design Concept Generation**:
  - o Discuss the documented clusters and decide on the most promising design concepts to explore further.
  - o Each student should select one or more clusters to focus on for the next phase of detailed design concept generation.

3. **Set the Stage for Pugh Analysis**:
  - o Explain that in the next class, each student will generate 30 detailed design concepts based on the selected clusters.
  - o These concepts will then be evaluated using a Pugh analysis to determine the best solutions.

7. **Submission Requirements**

1. **Documentation of Brainstorming Session**:
  - o Submit a summary of the brainstorming session, including photographs or scanned images of the Post-it notes and clusters.
  - o Provide a brief description of each cluster and the individual ideas within it.

2. **Preparation for Next Phase**:
  - o Ensure that each student is prepared to generate 30 detailed design concepts for the next class session.
  - o Document any initial thoughts or plans for the Pugh analysis.

By following these steps, students will engage in a productive brainstorming session, generating a wide range of design solutions and preparing for a more detailed evaluation of their ideas. This exercise encourages creativity, collaboration, and effective documentation, all of which are essential skills in the engineering design process.

Assignment 10-4: Individual Concept Generation and Pugh Analysis

Objective: Generate a minimum of 30 unique design concepts for your team's capstone design project. Document your concepts through sketches and written descriptions in the provided logbook pages within this capstone design workbook. Use formal concept generation methods to aid in the ideation process. Perform an individual Pugh analysis to select your top five concepts. As a team, combine your top concepts and conduct a team-based Pugh analysis to identify the most promising ideas for further development.

Instructions:

1. Individual Concept Generation a. Set aside dedicated time for concept generation, ensuring you have at least an hour of uninterrupted focus. b. In the designated "Design Project Concepts" section of your capstone design workbook, begin generating ideas. c. Use formal concept generation methods to help generate ideas. Some methods to consider include:
   - Brainstorming: Engage in a free-flowing, non-judgmental ideation session.
   - Mind Mapping: Create a visual diagram to explore and connect related ideas.
   - SCAMPER: Apply the SCAMPER checklist (Substitute, Combine, Adapt, Modify, Put to another use, Eliminate, Reverse) to generate variations of existing concepts.
   - Morphological Analysis: Break down the problem into key parameters and generate combinations of solutions for each parameter.
   - Analogies and Biomimicry: Draw inspiration from nature or other fields to generate novel concepts. d. Document the formal methods used for each concept in the provided space within the workbook. e. For each generated concept, create a clear sketch that illustrates the key features and components of the design on the designated sketch pages. f. Accompany each sketch with a written description that explains the concept in sufficient detail. Include how the design addresses the project requirements, its unique features, and any potential advantages or disadvantages. g. Aim to generate a minimum of 30 distinct concepts. Push yourself to think creatively and explore diverse ideas. h. Number each concept sequentially for easy reference.

2. Individual Concept Evaluation a. In the "Individual Pugh Analysis" section of your workbook, set up a Pugh analysis matrix with your 30 concepts listed in the first column. b. Establish evaluation criteria based on the project requirements and objectives. Consider factors such as:
   - Functionality and performance
   - User experience and ergonomics

- o Manufacturing feasibility and cost
- o Sustainability and environmental impact
- o Aesthetics and visual appeal c. Choose one concept as the reference (datum) against which all other concepts will be compared. d. Evaluate each concept against the reference for each criterion using the following scale:
- o "+": Better than the reference
- o "S": Same as the reference
- o "-": Worse than the reference e. Tally the scores for each concept, assigning a weight to each "+" and "-" based on the importance of the criterion. f. Select your top five concepts based on the highest scores and most promising evaluations.

3. Team Concept Consolidation a. Gather as a team and share your individual top five concepts. b. Discuss the merits and drawbacks of each concept, considering the insights gained from your individual Pugh analyses. c. Identify common themes, innovative features, and potential synergies among the team's concepts. d. Combine and refine the concepts to create a consolidated set of team concepts. Aim for a manageable number of concepts (e.g., 10-15) that represent the best ideas from the team.

4. Team Pugh Analysis a. In the "Team Pugh Analysis" section of your team's workbook, set up a Pugh analysis matrix with the consolidated team concepts listed in the first column. b. Use the same evaluation criteria and scoring system as in the individual Pugh analysis. c. As a team, evaluate each concept against the reference, discussing and reaching a consensus on the scores. d. Calculate the weighted scores and rank the concepts based on their overall performance. e. Select the top 3-5 concepts that show the most promise for further development and prototyping.

5. Documentation and Submission a. Ensure that your concept sketches, descriptions, formal concept generation methods used, and Pugh analysis are clearly documented in the designated sections of your capstone design workbook. b. Submit your completed individual pages of the capstone design workbook as a single PDF file on the designated platform (e.g., learning management system, project repository). c. As a team, compile the consolidated team concepts and Pugh analysis into the team section of the workbook and submit it along with the individual files.

Remember, the goal of this assignment is to generate a wide range of creative ideas using formal concept generation methods and systematically evaluate them to identify the most promising concepts for your team's design project. Encourage open-minded thinking, collaboration, and constructive feedback within your team. The selected top concepts will form the basis for further refinement, prototyping, and testing in the following stages of the project.

Date	Notes

Topic	Designer

Date	Notes

Topic	Designer

Date	Notes

Topic	Designer

Date	Notes

Topic	Designer

Date	Notes

Topic	Designer

Date	Notes

Topic	Designer

Date	Notes

Topic	Designer

Date	Notes

Topic	Designer

Date	Notes

Topic	Designer

Date	Notes

Topic	Designer

Date	Notes

Topic	Designer

Date	Notes

Topic	Designer

Date	Notes

Topic	Designer

Date	Notes

Topic	Designer

Date	Notes

Topic	Designer

Date	Notes

Topic	Designer

Date	Notes

Topic	Designer

Date	Notes

Topic	Designer

Date	Notes

Topic	Designer

Date	Notes

Topic	Designer

Date	Notes

Topic	Designer

Date	Notes

Topic	Designer

Date	Notes

Topic	Designer

Date	Notes

Topic	Designer

Date	Notes

Topic	Designer

Date	Notes

Topic	Designer

Date	Notes

Topic	Designer

Date	Notes

Topic	Designer

Date	Notes

Topic	Designer

Date	Notes

Topic	Designer

Date	Notes

Topic	Designer

Date	Notes

Topic	Designer

Date	Notes

Topic	Designer

Date	Notes

Topic	Designer

Date	Notes

Topic	Designer

Date	Notes

Topic	Designer

Date	Notes

Topic	Designer

Date	Notes

Topic	Designer

Date	Notes

Topic	Designer

Date	Notes

Topic	Designer

11 Critical Design Review (CDR)

A Critical Design Review (CDR) is an essential technical assessment in the design process, aiming to ensure that the design project team selects a superior design idea before moving to the prototyping phase. This review involves a detailed presentation and evaluation of the top design concepts, and it is typically scheduled after the completion of key analyses such as Pugh and QFD (Quality Function Deployment). The CDR mimics industry processes but is tailored to meet academic timelines and constraints.

1. **Purpose and Importance**

   The CDR serves several critical purposes:

   i. **Validation of Design Concepts:** Ensures that the selected design concepts meet the predefined design specifications and effectively solve the design problem.

   ii. **Feedback and Improvement:** This provides a platform for receiving feedback from a diverse audience, including mentors, sponsors, and technical experts, which helps refine the design.

   iii. **Risk Mitigation:** Identifying potential issues early in the design process reduces the risk of costly changes later.

   iv. **Documentation and Accountability:** Creates a formal record of the design decisions and the rationale behind them, which is crucial for future reference and accountability.

2. **Participants**

   The participants in a CDR typically include:

   - **Design Project Team:** The primary presenters who showcase their top design concepts.
   - **Capstone Class:** Fellow students who provide peer feedback.
   - **Mentors and Sponsors:** Industry experts and sponsors who offer professional insights and critiques.

- **Technical Experts:** Specialists who evaluate the technical feasibility and robustness of the designs.

3. **Presentation Structure**

   A successful CDR presentation is structured around the following key elements:

   i.   **Introduction:** This section provides an overview of the design problem, objectives, and criteria for selecting the top concepts.

   ii. **Design Concepts:** Detailed presentation of the top two to four design concepts identified from the QFD analysis. Each concept should be explained in terms of how it meets the design specifications and solves the problem.

   iii. **Analysis and Justification:** Presentation of the Pugh and QFD analyses that led to the selection of the top concepts, including any trade-offs and decision matrices used.

   iv. **Technical Feasibility:** Discussion on the technical aspects of the designs, including any simulations, calculations, and prototypes developed to validate the concepts.

   v.   **Risk Assessment:** Identification of potential risks and challenges associated with each design concept, along with proposed mitigation strategies.

   vi. **Feedback and Q&A:** This session is for receiving feedback from the audience and addressing any questions or concerns.

4. **Process and Guidelines**

   The CDR process involves the following steps:

   i.   **Preparation:** The design team prepares a concise 15-minute presentation focusing on the top design concepts and their validation. Visual aids such as slides, prototypes, and simulations are crucial for effective communication.

   ii. **Briefing Session:** Prior to the formal review, a briefing session may be held to ensure that all participants have a common understanding of the design requirements and the review process.

   iii. **Review Meeting:** A formal review meeting is conducted, during which the design team presents their concepts, followed by a structured Q&A session. The meeting should have a well-planned agenda and a checklist of items to be reviewed.

   iv. **Documentation:** Minutes of the review meeting are documented, including the decisions made, action items assigned, and any feedback received. This documentation is crucial for tracking progress and accountability.

5. **Checklist for CDR**

A typical CDR checklist includes the following items:

- **Technical Requirements:** Verification that the design meets all technical specifications, including performance, reliability, and safety criteria.
- **Cost and Resources:** Assessment of the design's cost-effectiveness and the resources required for development and production.
- **Manufacturability:** Evaluation of the design's feasibility for manufacturing, including the use of standard components and ease of assembly.
- **Risk and Reliability:** Identification of potential risks and the design's reliability under various operating conditions.

By adhering to these guidelines and processes, the CDR ensures a thorough evaluation of the design concepts, facilitating the selection of the most viable and superior design for further development and prototyping.

11.1 Guidelines for CDR Project Presentation

The Critical Design Review (CDR) is a pivotal moment in the engineering design process, providing an opportunity for the team to present their top design concepts and receive valuable feedback. To ensure a successful CDR presentation, the following guidelines should be adhered to:

1. **Introduction and Definition of the Problem**
   - **Objective:** Clearly state the problem your project aims to solve.
   - **Context:** Provide background information and context to help the audience understand the significance of the problem.
   - **Problem Statement:** Formulate a precise and concise problem statement.
   - **Importance:** Explain why solving this problem is important to the sponsor and stakeholders.
2. **Design Specifications and Their Relative Significance**
   - **Specifications:** List all the design specifications your project must meet.
   - **Criteria:** Explain the criteria used to define these specifications, including any industry standards, sponsor requirements, and constraints.
   - **Prioritization:** Rank the specifications in order of their importance and discuss any trade-offs made during the design process.
3. **Basis for Reducing Design Solution Concepts**
   - **Initial Concepts:** Describe the initial pool of design solution concepts generated.
   - **Screening Process:** Explain the methodology used to narrow down these concepts, such as feasibility studies, preliminary analysis, and stakeholder feedback.

- **Selection Criteria:** Detail the criteria used to select the top two to four concepts presented during the CDR, highlighting why these were deemed the most promising.

4. Presentation of Solution Concepts
- **Concept Overview:** Provide a high-level overview of each solution concept.
- **Technical Details:** Dive into the technical aspects of each design, including:
  - Key features and innovations
  - Design drawings and schematics
  - Material selection and justification
  - Performance analysis and expected outcomes
- **Sponsor Requirements:** Show how each concept meets the sponsor's requirements and aligns with their expectations.

5. QFD Analysis and Comparative Assessment
- **QFD Methodology:** Briefly explain the Quality Function Deployment (QFD) process and its role in your project.
- **QFD Results:** Present the QFD analysis, showing how each design concept was evaluated against the design specifications.
- **Comparison:** Compare the top design concepts to each other and existing competitive solutions. Use charts, matrices, and graphs to illustrate comparisons clearly.
- **Assessment:** Discuss the strengths and weaknesses of each concept based on the QFD analysis.

6. Audience Interaction
- **Question Time:** Allocate sufficient time for the audience to ask questions and provide feedback. Plan for at least 15-20 minutes of Q&A.
- **Engagement:** Encourage audience interaction by preparing questions and prompts that can stimulate discussion and critique.

7. Documentation and Feedback
- **Note-Taking:** Designate a team member to take detailed notes on the questions asked and comments provided by the audience.
- **Feedback Integration:** Plan a follow-up meeting to discuss the feedback received and how it can be integrated into the design process.
- **Documentation:** Ensure all feedback is documented and included in the project records for future reference.

11.1.1   Additional Tips for a Successful CDR Presentation

- **Practice:** Rehearse the presentation multiple times to ensure clarity and confidence.
- **Clarity:** Use clear and concise language, avoiding technical jargon unless absolutely necessary.

- **Visual Aids:** Utilize slides, prototypes, and other visual aids effectively to enhance understanding.
- **Time Management:** Stick to the allocated time for each section to ensure the presentation remains on schedule.
- **Professionalism:** Maintain a professional demeanor throughout the presentation and Q&A session.

By following these guidelines, your engineering team can effectively communicate your design concepts, demonstrate their technical robustness, and engage with the audience to gather valuable feedback that will enhance the final design.

11.2 Giving and Receiving Critique

11.2.1 Importance of Critique in the CDR Process

A Critical Design Review (CDR) is not just a presentation; it is an interactive process aimed at improving the design through constructive feedback. The audience for the CDR typically includes fellow students, sponsors, mentors, teaching assistants, and professors. Their diverse perspectives are invaluable in identifying potential issues, suggesting improvements, and validating the design concepts presented.

11.2.2 Guidelines for Giving Critique

1. **Be Constructive and Specific:**
   - Focus on specific aspects of the design rather than general comments. For example, instead of saying, "The design needs improvement," specify which part of the design needs improvement and why it needs improvement.
   - Provide actionable suggestions that the team can implement.
2. **Balance Positive and Negative Feedback:**
   - Highlight the strengths of the design to acknowledge the team's efforts.
   - Address weaknesses with the aim of improvement, not criticism.
3. **Be Respectful and Professional:**
   - Critique should be delivered respectfully, considering the effort and hard work of the presenting team.
   - Avoid personal comments and focus on the design and technical aspects.
4. **Ask Questions:**
   - Pose questions that can lead the team to think deeper about their design choices and potential improvements.
   - Questions should be clear and to the point to avoid confusion.
5. **Provide Rationale:**
   - When giving suggestions or pointing out flaws, explain the reasoning

behind your feedback. This helps the presenting team understand the context and importance of the feedback.

11.2.3   Guidelines for Receiving Critique

1. **Listen Actively:**
   - Pay close attention to the feedback without interrupting. This shows respect for the audience's input and ensures you fully understand their points.
   - Take notes during the feedback session to capture all comments and suggestions.
2. **Stay Open-Minded:**
   - Be open to suggestions and criticisms, even if they are unexpected or challenging.
   - Avoid becoming defensive; remember that the feedback is aimed at improving your design.
3. **Clarify and Ask Questions:**
   - If a piece of feedback is unclear, politely ask for clarification.
   - Engage with the feedback by asking follow-up questions to understand the audience's perspective better.
4. **Acknowledge and Appreciate:**
   - Thank the audience for their feedback, showing appreciation for their time and effort in providing constructive critique.
5. **Reflect and Act:**
   - After the CDR, review the notes taken during the feedback session.
   - Discuss the feedback with your team to decide on actionable steps to improve the design.

11.2.4   Handling Critique in Different Class Sizes

11.2.4.1   Small Classes (< 40 students):

- **Interactive Comments:**
  - Allow for real-time comments and questions during or at the end of the presentation. This interactive approach can provide immediate insights and foster a dynamic discussion.
  - Designate one or two team members as recorders to capture all questions, comments, and suggestions from the audience.
  - Provide your e-mail contact information to the audience for additional feedback after the presentation.

11.2.4.2   Large Classes (> 60 students):

- **Time Constraints:**
    - o Each team may only have 10 to 15 minutes for their presentation, leaving limited time for interactive comments.
    - o Focus on presenting the most critical aspects of your design concisely.
- **Feedback Collection:**
    - o The professor should utilize a class survey instrument, such as Google Forms, to collect structured feedback from all class members.
    - o The survey should include sections for strengths, weaknesses, and suggestions for each team's design.
    - o Completing the survey can be part of the grading for the class, ensuring participation from all students.
- **Post-Presentation Feedback:**
    - o Compile the feedback from the surveys and provide it to each team after all comments have been collected. This ensures that each team receives comprehensive feedback without the constraints of limited presentation time.

By following these guidelines for giving and receiving critique, the CDR process becomes a valuable learning experience that significantly enhances the quality of the final design.

11.3   Incorporating Feedback

Incorporating feedback effectively is a crucial step in the design process, particularly after a Critical Design Review (CDR). This process involves not only analyzing the feedback received but also making informed decisions about which suggestions to implement to enhance the design. Here is a detailed guide on how to incorporate feedback after a CDR.

11.3.1   1. Reviewing Feedback

11.3.1.1   Collect and Compile Feedback:

- Gather all questions, answers, comments, and suggestions from the CDR session.
- Include feedback received during the presentation, post-presentation discussions, and through follow-up emails or survey forms.

11.3.1.2   Categorize Feedback:

- Sort the feedback into categories such as technical feasibility, design aesthetics,

user experience, manufacturability, and cost.
- Identify whether the feedback is positive reinforcement, constructive criticism, or neutral suggestions.

11.3.1.3   Assess Relevance:

- Evaluate the applicability of each piece of feedback to the project goals and constraints.
- Consider the source of the feedback—whether it comes from a sponsor, mentor, professor, or fellow student—and its relevance to their area of expertise.

11.3.2   2. Deliberation and Decision-Making

11.3.2.1   Team Discussion:

- Hold a regular team meeting dedicated to discussing the feedback received.
- Invite the sponsor to participate in this meeting to ensure their perspective is considered in the decision-making process.
- Discuss each piece of feedback in detail, weighing the potential benefits and drawbacks of implementing it.

11.3.2.2   Prioritizing Feedback:

- Prioritize feedback based on its potential impact on the project and feasibility of implementation.
- Use criteria such as improvement in performance, cost reduction, ease of implementation, and alignment with project objectives.

11.3.2.3   Agreement on Changes:

- Reach a consensus among team members and the sponsor on which feedback to incorporate.
- Document the rationale for each decision, noting why certain suggestions were accepted or rejected.

11.3.3   Implementing Changes

11.3.3.1   Planning the Implementation:

- Develop a detailed plan for implementing the accepted feedback.
- Assign tasks and responsibilities to team members, setting clear deadlines and milestones.

11.3.3.2   Design Modifications:

- Make the necessary changes to the design concept or specific parts of it.
- Ensure that all modifications align with the overall project requirements and constraints.

11.3.3.3   Prototype and Test:

- If applicable, create prototypes to test the modifications.
- Conduct tests to validate that the changes achieve the desired improvements without introducing new issues.

11.3.4   Documentation and Follow-Up

11.3.4.1   Documenting Changes:

- Update the design report to reflect the changes made based on the feedback.
- Reference the comments and suggestions with citations to the individuals who provided them, ensuring proper acknowledgment.

11.3.4.2   In-Depth Discussions:

- For significant, design-changing feedback, schedule meetings with the individuals who provided it to discuss their suggestions in more depth.
- This helps ensure a thorough understanding of the feedback and its implications.

11.3.4.3   Follow-Up Actions:

- Follow up on all significant comments and suggestions, especially those that have led to major design changes.
- Keep the communication lines open with the feedback providers, updating them on how their suggestions were implemented and the outcomes of those changes.

11.3.4.4   Continuous Improvement:

- Treat feedback incorporation as an ongoing process rather than a one-time event.
- Regularly review and update the design based on new feedback and findings as the project progresses.

11.3.5   Reflect and Learn

11.3.5.1   Team Reflection:

- At the next team meeting, reflect on the entire feedback incorporation process.
- Discuss what was learned from the feedback and how it has improved the design.

11.3.5.2   Sponsor Involvement:

- Engage with the sponsor to review the changes made and gather any additional insights or feedback they may have.
- Ensure that the sponsor is satisfied with the direction and progress of the project post-feedback incorporation.

By following these steps, design teams can effectively incorporate feedback from the CDR, ensuring that their final design is robust, well-validated, and meets the needs and expectations of all stakeholders. This iterative process not only improves the design but also fosters a culture of continuous learning and improvement within the team.

11.4 Select Top Concept

Selecting the top design concept to implement for a proof of concept (PoC) is a critical decision in the engineering design process. This selection process follows the completion of key analyses such as the Pugh method, Quality Function Deployment (QFD), and the Critical Design Review (CDR). Here is a detailed guide on how to systematically choose the best design concept for PoC implementation.

11.4.1   Recap of Pugh, QFD, and CDR

Pugh Method:
- A comparative analysis tool used to evaluate multiple design concepts against a baseline. Concepts are scored based on criteria such as cost, feasibility, and performance.
- The Pugh matrix helps narrow down the design options by highlighting each concept's strengths and weaknesses relative to the baseline.

Quality Function Deployment (QFD):
- A structured approach to translating customer needs (the "what") into technical requirements (the "how").
- QFD matrices (e.g., House of Quality) prioritize design features based on their impact on customer satisfaction and competitive benchmarking.

Critical Design Review (CDR):
- A formal presentation and review process where design teams present their top concepts to an audience of stakeholders for feedback.
- The CDR involves detailed technical presentations, risk assessments, and Q&A sessions to evaluate the feasibility and alignment of the design concepts with project goals.

11.4.2 Criteria for Selecting the Top Design Concept

11.4.2.1 Technical Feasibility:

- **Engineering Analysis:** Ensure that the design can be realistically implemented using current engineering capabilities and technologies.
- **Prototyping Capability:** Assess whether the design can be prototyped effectively with available resources.

11.4.2.2 Performance Metrics:

- **Design Specifications:** Evaluate how well each design meets or exceeds the required specifications.
- **Reliability and Durability:** Consider the expected reliability and lifespan of the design under operating conditions.

11.4.2.3 Cost and Resource Allocation:

- **Initial Costs:** Analyze the upfront costs for development, prototyping, and testing.
- **Long-Term Costs:** Consider long-term production costs, maintenance, and potential scalability.

11.4.2.4 Risk Assessment:

- **Technical Risks:** Identify any technical uncertainties or potential failures and their mitigation strategies.
- **Market Risks:** Evaluate potential market acceptance and competitive risks.

11.4.2.5 Alignment with Project Goals:

- **Sponsor Requirements:** Ensure the design aligns with the sponsor's needs and expectations.
- **Strategic Fit:** Confirm that the design supports the overall strategic objectives of the project.

11.4.3 Integrating Results from Pugh, QFD, and CDR

11.4.3.1 Synthesis of Pugh Matrix Results:

- Summarize the scores and rankings from the Pugh matrix.
- Identify any patterns or consistently high-scoring concepts across different criteria.

11.4.3.2   QFD Analysis Review:

- Review the QFD matrix to see which design features are most critical to customer satisfaction.
- Cross-reference these features with the concepts evaluated in the Pugh matrix.

11.4.3.3   Feedback from CDR:

- Compile and categorize the feedback received during the CDR.
- Pay special attention to any recurring themes or significant comments from experts and stakeholders.

11.4.4   Decision-Making Process

11.4.4.1   Team Deliberation:

- Hold a dedicated meeting with the design team to discuss the findings from the Pugh matrix, QFD analysis, and CDR feedback.
- Use a structured decision-making framework to facilitate objective discussions.

11.4.4.2   Weighted Scoring:

- Develop a weighted scoring system that incorporates all evaluation criteria (technical feasibility, performance, cost, risk, alignment with goals).
- Assign weights based on the relative importance of each criterion to the project.

11.4.4.3   Scoring and Ranking:

- Score each design concept against the weighted criteria.
- Rank the concepts based on their total scores to identify the top contender.

11.4.4.4   Sensitivity Analysis:

- Conduct a sensitivity analysis to test the robustness of the decision.
- Adjust weights and scores to see if the top-ranked design remains consistent under different scenarios.

11.4.5   Final Selection and Justification

11.4.5.1   Final Review:

- Conduct a final review session with the team, including key stakeholders and sponsors, to present the top-ranked design concept.
- Discuss the rationale behind the selection, supported by data from Pugh, QFD, and CDR.

11.4.5.2   Justification Report:

- Prepare a detailed report documenting the selection process, including:
  - ○ Overview of the initial design concepts.
  - ○ Summary of the Pugh matrix and QFD analysis.
  - ○ Key feedback from the CDR.
  - ○ Weighted scoring results and sensitivity analysis.
  - ○ Final decision and justification.

11.4.5.3   Approval and Next Steps:

- Seek formal approval from the sponsor and key stakeholders.
- Outline the next steps for PoC implementation, including detailed planning for prototyping, testing, and validation.

11.4.6   Continuous Evaluation and Feedback Loop

11.4.6.1   Monitor and Adapt:

- Continuously monitor the progress of the PoC implementation.
- Be prepared to adapt and make iterative improvements based on ongoing feedback and testing results.

11.4.6.2   Feedback Integration:

- Implement a feedback loop to incorporate insights from the PoC phase back into the design process.
- Ensure that lessons learned are documented and applied to future design iterations.

By following this comprehensive approach, design teams can ensure that they select the most viable and promising design concept for PoC implementation. This process not only enhances the likelihood of project success but also fosters a culture of systematic evaluation and continuous improvement within the team.

11.5 Assignments

Assignment 11-1: Prepare a 15-minute Presentation for a Critical Design Review

Objective:

This assignment's goal is to prepare and deliver a 15-minute presentation for your Critical Design Review (CDR). Your presentation should effectively communicate your design concepts, technical analysis, and key findings. You must also engage the audience and facilitate an interactive Q&A session.

Guidelines for Presentation Preparation:

1. Presentation Structure:
- **Total Time:** 15 minutes
  - **Presentation:** 10 minutes
  - **Q&A:** 3 minutes
  - **Setup:** 2 minutes

2. Slide Guidelines:
- **Number of Slides:** Aim for 10-12 slides. This allows roughly one minute per slide.
- **Slide Content:**
  - **Title Slide:** Project title, team members, date.
  - **Introduction (1 slide):** Briefly introduce the problem statement and objectives.
  - **Design Specifications (1 slide):** Outline key design specifications and their significance.
  - **Concept Development (2-3 slides):** Describe the initial pool of design solutions and the process of narrowing them down.
  - **Top Design Concepts (3-4 slides):** Provide detailed technical descriptions of your top 2-4 design concepts, including:
    - Design features
    - Technical drawings or schematics
    - Material selection and justification
    - Performance analysis
  - **QFD Analysis (1-2 slides):** Present the QFD analysis results and compare the top design concepts.
  - **Conclusion (1 slide):** Summarize the key points and state your preferred design concept for the PoC.
  - **Q&A Prompt (1 slide):** Prepare a slide to encourage questions and interactions from the audience.

3. Content Preparation:

- **Clarity and Conciseness:** Ensure that each slide is clear and concise. Avoid cluttering slides with too much information.
- **Visual Aids:** Use diagrams, charts, and images to support your points and make the presentation more engaging.
- **Rehearsal:** Practice your presentation multiple times to ensure smooth delivery and adherence to the time limit.
- **Engagement:** Plan how to engage the audience. Prepare questions to ask the audience and think about how to encourage participation.

4. Technical Setup:

- **Laptop and AV Compatibility:**
  - Ensure your laptop is fully charged.
  - Check compatibility with the classroom's audio-visual system.
  - Bring necessary adapters and cables.
- **Classroom Check:**
  - Reserve time to practice in the classroom.
  - Verify that all equipment works properly.
  - Test your presentation on the classroom AV system.
  - Make sure your slides are displayed correctly and the sound works (if applicable).

5. Presentation Delivery:

- **Setup Time:** Use the first 2 minutes to set up your equipment and ensure everything is working.
- **Introduction:** Start with a solid introduction to grab the audience's attention.
- **Main Content:** Clearly and logically present your design concepts, analyses, and findings.
- **Q&A Session:**
  - Allocate 3 minutes at the end for questions and answers.
  - Encourage the audience to ask questions by prompting them with engaging content.
  - Designate one team member to take notes on the questions and feedback received.
- **Conclusion:** Conclude with a summary slide and thank the audience for their attention and feedback.

6. Additional Tips:
- **Timing:** Keep track of your time to ensure you stay within the 10-minute limit for the presentation.
- **Engagement:** Make eye contact with the audience, use gestures, and vary your tone to maintain interest.
- **Team Coordination:** Decide in advance which team members will present which parts of the presentation to ensure smooth transitions.
- **Submission of Slide Deck:** Upload your final slide deck at least 24 hours before your scheduled presentation.
- **Practice Session:** Schedule a practice session in the classroom to test your setup and receive feedback from your instructor or peers.

By following these guidelines, you will be well-prepared to deliver an effective and engaging CDR presentation that clearly communicates your design concepts and technical analyses while also fostering interaction and feedback from your audience.

12 Financial Analysis

Engineering design is a multifaceted discipline requiring not only technical proficiency but also a robust understanding of financial and economic principles. In the context of capstone engineering projects, financial and economic analysis becomes particularly crucial as it bridges the gap between theoretical designs and practical implementations. This chapter aims to elucidate the essential concepts, methodologies, and applications of financial and economic analysis in capstone engineering design projects, providing students with the necessary tools to evaluate and justify their design choices.

12.1 Importance of Financial and Economic Analysis

Financial and economic analysis in engineering design is indispensable for several reasons:

1. **Decision Making**: Engineers frequently face choices among alternative designs, materials, and processes. Financial analysis helps in making informed decisions by comparing the economic implications of each alternative.
2. **Project Viability**: Assessing a project's economic feasibility ensures that the design meets technical specifications and provides value for money. This is crucial for gaining stakeholder approval and securing funding.
3. **Resource Allocation**: Effective financial analysis aids in optimal resource allocation, ensuring that limited resources are utilized efficiently.
4. **Risk Management**: Identifying and evaluating financial risks associated with engineering projects helps in devising strategies to mitigate these risks, thereby enhancing the project's robustness and reliability.

12.2 Key Concepts in Financial and Economic Analysis

To conduct a thorough financial and economic analysis, several key concepts must be understood:

1. **Time Value of Money**: This fundamental principle states that a dollar today is worth more than a dollar in the future due to its earning potential. This is the

basis for discounting future cash flows to their present value.

2. **Net Present Value (NPV)**: NPV is the sum of the present values of incoming and outgoing cash flows over some time. It measures a project's profitability.

3. **Internal Rate of Return (IRR)**: The IRR is the discount rate that makes a project's NPV zero. It represents the project's potential return on investment.

4. **Payback Period**: This is the time required to recover the initial investment from the net cash inflows. It is a simple measure of investment risk.

5. **Benefit-Cost Ratio (BCR)**: BCR is the ratio of the present value of benefits to the present value of costs. A BCR greater than one indicates a financially viable project.

12.3 Methodologies for Economic Decision Making

Several methodologies are employed to conduct financial and economic analyses in engineering design:

1. **Present-Worth Analysis**: This method involves discounting all costs and benefits to the present time to evaluate the net present worth of different alternatives. It is particularly useful when comparing projects with different lifespans.

2. **Annual Cost Analysis**: This method converts all cash flows over time into an equivalent annual cost or benefit, facilitating the comparison of alternatives with different time periods.

3. **Capitalized Cost Analysis**: This is a special case of present-worth analysis for projects that are intended to exist in perpetuity. It is often used for public infrastructure projects.

4. **Benefit-Cost Ratio Analysis**: This involves comparing the ratio of benefits to costs for different projects to determine the most economically efficient option

.

12.4 Application in Capstone Engineering Design

In capstone engineering design projects, the application of financial and economic analysis is essential from the conceptual phase to the final implementation:

1. **Conceptual Design Phase**: Initial cost estimates are developed using analogies with previous projects or parametric methods. These estimates help screen and select feasible design concepts.

2. **Embodiment Design Phase**: More detailed cost analysis is conducted, including material, labor, and overhead costs. This phase may involve the use of software tools for cost modeling and optimization.

3. **Detail Design Phase**: A comprehensive economic evaluation is performed, considering life-cycle costs, including manufacturing, operation, maintenance, and disposal costs. This ensures that the final design is not only technically

sound but also economically viable.

Financial and economic analysis is a critical component of the engineering design process, particularly in capstone projects where real-world applicability and economic justification are paramount. By mastering these analytical techniques, engineering students can enhance their ability to deliver successful, sustainable, and economically viable projects.

12.5 Estimating Project Costs

Estimating project costs is a critical aspect of engineering capstone design, ensuring that projects are financially viable and resources are effectively allocated. This detailed guide covers the essential elements of cost estimation, including accounting for student time, resource usage, and external consultation costs. We will convert student hours into dollar amounts using an average hourly rate of $25 for senior engineering students. Additionally, we will consider costs associated with using resources such as 3D printing, materials testing, mechanical shops, consulting time, industry sponsors, and professors.

12.5.1   1. Student Time and Hours

Student time is one of the most significant cost factors in a capstone project. Senior engineering students' efforts must be quantified and converted into monetary terms to reflect the project's true cost. Effective time management and precise estimation of student hours are essential for accurate budgeting and resource allocation.

12.5.1.1   Example Calculation:

If a team of four students works on a project for 15 weeks, dedicating 10 hours per week per student, the total hours are calculated as follows:

Total hours=4 students×15 weeks×10 hours/week=600 hours

Converting hours to dollars:

Total cost=600 hours×$25/hour=$15,000

This calculation should be adjusted based on actual project schedules and time commitments.

To provide more detailed tracking, consider categorizing student hours into specific activities such as:

- **Research and Concept Development**: Time spent on initial research, brainstorming, and developing design concepts.

- **Design and Modeling**: Hours dedicated to creating detailed designs and models using CAD software.
- **Prototyping and Testing**: Time spent on building prototypes, conducting tests, and analyzing results.
- **Documentation and Reporting**: Efforts related to preparing reports, presentations, and documentation of the design process.
- **Meetings and Collaboration**: Hours spent in team meetings, consultations with advisors, and collaboration sessions.

By tracking hours in these categories, teams can better understand where their time is being allocated and identify areas for improvement in efficiency.

12.5.2   Resource Utilization

Resource utilization is another critical aspect of cost estimation in capstone projects. Different resources have varying costs, and accurate estimation requires an understanding of both direct and indirect costs associated with resource usage.

12.5.2.1   3D Printing Time

3D printing is often used for prototyping in engineering projects. Costs include machine time and materials.

- **Machine Time**: Typically, 3D printing services charge by the hour. For example, if the rate is $50 per hour and the project requires 20 hours of printing:

Total cost=20 hours×$50/hour=$1,000

- **Materials**: Costs vary depending on the material used (e.g., PLA, ABS, metal). Assuming the material cost is $30 per kg, and the project uses 5 kg:

Total material cost=5 kg×$30/kg=$150

Considerations for reducing 3D printing costs:

- **Optimizing Design for 3D Printing**: Simplifying designs to reduce print time and material usage.
- **Utilizing In-House Facilities**: University-owned 3D printers may have lower rates than commercial services.

12.5.2.2   Materials Testing Machines

Material testing, such as tensile and impact tests, is crucial for validating designs. Accurate cost estimation includes machine usage fees and any additional consumables.

- **Machine Usage**: If the cost is $100 per hour and the project requires 10 hours of testing: Total cost=10 hours×$100/hour=$1,000\text{Total cost} = 10 \text{hours} \times \$100/\text{hour} = \$1,000Total cost=10 hours×$100/hour=$1,000

- **Consumables and Specimens**: Costs associated with test specimens and any consumable materials used during testing should be included. For example, if specimens cost $50 each and 20 specimens are tested:

Total specimen cost=20 specimens×$50/specimen=$1,000

12.5.2.3   Mechanical Shop

Mechanical shops provide essential services like machining, welding, and assembly. Estimating these costs requires understanding both labor rates and equipment usage fees.

- **Labor and Equipment**: If shop labor is billed at $75 per hour and the project requires 40 hours of work:

Total cost=40 hours×$75/hour=$3,000

Additional considerations include:

- **Material Costs**: Costs of raw materials used in machining and fabrication.
- **Setup Fees**: One-time fees associated with setting up machines for specific tasks.

12.5.2.4   Consulting Time

Consulting with practicing engineers provides valuable insights and expertise. Estimating these costs involves understanding the consulting rates and required consultation hours.

- **Consulting Fees**: Assuming an hourly consulting fee of $150 and 20 hours of consultation:

Total cost=20 hours×$150/hour=$3,000

- **Types of Consultation**: Costs may vary based on the type of consultation, such as design review, project management advice, or technical troubleshooting.

12.5.2.5   Industry Sponsors

Industry sponsors often provide financial support or in-kind contributions such as materials and access to facilities. Estimating these contributions accurately involves:

- **Direct Financial Contributions**: If a sponsor provides $5,000 in funding, this reduces the total project cost.
- **In-kind Contributions**: Estimating the value of materials or services provided. For example, if a sponsor provides $2,000 worth of materials: In-kind contribution=$2,000\text{In-kind contribution} = \$2,000In-kind contribution=$2,000

Incorporating these contributions into the budget helps provide a clearer picture of the project's financial landscape.

12.5.2.6   Professors

Professors' time, although often not directly billed, should be accounted for in terms of the guidance and supervision provided.

- **Estimating Time Value**: Assuming a professor spends 2 hours per week over 15 weeks at an estimated hourly rate of $100:
Total cost=2 hours/week×15 weeks×$100/hour=$3,000\text{Total cost} = 2 \text{ hours/week} \times 15 \text{ weeks} \times \$100/\text{hour} = \$3,000Total cost=2 hours/week×15 weeks×$100/hour=$3,000
- **Types of Supervision**: Costs may vary based on the level of involvement, such as weekly meetings, design reviews, or detailed feedback on reports.

12.5.3   Example Cost Breakdown

Combining all these elements, we can estimate the total project cost. Here's a hypothetical example:

Student Time: $15,000

3D Printing:
- Machine Time: $1,000
- Materials: $150 Total 3D Printing=$1,150

Materials Testing:
- Machine Usage: $1,000
- Specimens: $1,000 Total Materials Testing=$2,000

Mechanical Shop: $3,000

Consulting: $3,000

Industry Sponsorship:
- Financial Contribution: -$5,000
- In-kind Contributions: $2,000
- Total Industry Sponsorship=-$7,000

Professors' Supervision: $3,000

Total Estimated Cost:

$15,000+$1,150+$2,000+$3,000+$3,000−$3,000+$3,000=$24,150

12.5.4   4Cost Management Strategies

To manage and potentially reduce project costs, consider the following strategies:

1. **Efficient Time Management**: Ensure that student hours are used effectively through detailed planning and regular progress reviews. Implement time-tracking tools and methods to monitor student contributions and identify inefficiencies.
2. **Resource Optimization**: Use shared resources where possible, negotiate rates for services, and leverage university facilities. Coordinate with other project teams to share costs for common resources, such as 3D printers or testing machines.
3. **Sponsorship and Grants**: Actively seek industry sponsorships and academic grants to offset costs. Develop strong proposals that highlight the project's benefits to attract potential sponsors and funding opportunities.
4. **In-Kind Contributions**: Maximize the use of in-kind contributions from industry partners. Engage with industry sponsors early to secure material donations, access to specialized equipment, and expert consultations.
5. **Budgeting and Forecasting**: Regularly update the project budget and forecast future costs. Monitor actual expenditures against the budget to identify variances and take corrective actions as needed.
6. **Contingency Planning**: Allocate a contingency budget for unexpected costs. Typically, a 10-15% contingency is included to cover unforeseen expenses and ensure project completion without financial strain.

Estimating project costs in engineering capstone design projects involves careful accounting of various elements, from student time to resource usage and external consultations. By understanding and managing these costs effectively, student teams can ensure their projects are both financially viable and successful, providing a valuable learning experience that mirrors real-world engineering practice. Accurate cost estimation not only aids in project planning and resource allocation but also enhances the team's ability to deliver high-quality, economically feasible solutions.

12.6 Tracking and Reporting Estimated Costs

Effective cost tracking and reporting are vital for the success of capstone engineering design projects. These activities ensure that projects remain within budget, help identify potential financial issues early, and provide transparency to all stakeholders.

This section discusses the importance of tracking and reporting estimated costs, methods for doing so, and best practices for maintaining accurate and up-to-date financial records.

12.6.1   1. Importance of Tracking and Reporting Costs

Tracking and reporting costs in capstone projects serve several purposes:
1. **Budget Management**: Ensures the project does not exceed its allocated budget and helps in making informed financial decisions.
2. **Resource Allocation**: Helps in efficient allocation and utilization of resources, preventing wastage and ensuring optimal use.
3. **Transparency**: Provides a clear financial picture to all stakeholders, including faculty, industry sponsors, and team members.
4. **Performance Measurement**: This allows for evaluating financial performance against the project plan and facilitating adjustments as needed.
5. **Documentation**: Creates a historical record that can be useful for future projects and for understanding cost dynamics.

12.6.2   Methods for Tracking Costs

To track costs effectively, teams should implement a structured approach that includes the following components:

12.6.2.1   Initial Budget Creation

At the outset, create a detailed budget that outlines all expected costs. This budget should cover:
- **Student Hours**: Estimate the total number of hours each student will work and convert this to a dollar amount. For example:
  Total student cost=Total hours×\$25/hour\text{Total student cost} = \text{Total hours} \times \$25/\text{hour}Total student cost=Total hours×\$25/hour
- **Materials and Supplies**: List all required materials and their estimated costs.
- **Equipment Usage**: Include costs for using equipment such as 3D printers, testing machines, and mechanical shop tools.
- **Consulting Fees**: Account for any fees associated with consulting engineers, industry experts, or professors.
- **Miscellaneous Costs**: Include any other costs that might arise during the project, such as travel or software licenses.

12.6.2.2   Ongoing Cost Tracking

Use a tracking system to record all expenses as they occur. This can be done using:
- **Spreadsheets**: Maintain a detailed spreadsheet with columns for date,

description, category, amount, and running total.
- **Project Management Software**: Tools like Microsoft Project, Asana, or Trello can be used to track expenses alongside project tasks.

12.6.2.3   Weekly Progress Reports

Include a section in weekly progress reports dedicated to financial updates. This section should:
- **Summarize Expenditures**: Provide a summary of expenditures for the week.
- **Compare to Budget**: Show how current spending compares to the initial budget.
- **Forecast Future Spending**: Update forecasts based on current spending patterns.

12.6.2.4   Preliminary and Final Design Reports

Incorporate detailed cost reports in both the preliminary design report and the final design report. These reports should include:
- **Initial Budget Summary**: A summary of the initial budget.
- **Cost Tracking Details**: Detailed records of all costs incurred, categorized by type (e.g., materials, labor, equipment).
- **Variance Analysis**: An analysis of variances between the budgeted and actual costs, explaining significant differences.

12.6.2.5   Record Keeping

Keep copies of all purchase requests, invoices, and receipts. Upload these documents to a shared drive accessible to all team members. This ensures transparency and provides a backup in case of discrepancies.

12.6.3   Best Practices for Cost Tracking and Reporting

Implementing best practices can enhance the accuracy and efficiency of cost tracking and reporting:

12.6.3.1   Regular Updates

Ensure that the cost-tracking spreadsheet or software is updated regularly, ideally after each transaction. This reduces the likelihood of errors and ensures real-time financial visibility.

12.6.3.2   Detailed Descriptions

Record detailed descriptions for each expense. This helps in understanding the nature of expenditures and facilitates easier auditing and reporting.

12.6.3.3   Categorization

Categorize expenses into predefined categories such as labor, materials, equipment, and consulting. This categorization helps analyze spending patterns and identify areas of concern.

12.6.3.4   Approval Processes

Implement an approval process for significant expenditures. This ensures that all major costs are reviewed and approved by a faculty advisor or project manager before being incurred.

12.6.3.5   Regular Audits

Conduct regular internal audits to review the accuracy of cost tracking and ensure compliance with the budget. This can be done weekly or monthly, depending on the project's complexity.

12.6.3.6   Communication

Maintain open communication within the team regarding financial matters. Regular discussions about budget status and financial concerns can prevent misunderstandings and foster a collaborative approach to cost management.

12.6.4   Reporting in Preliminary and Final Design Reports

12.6.4.1   Preliminary Design Report

The preliminary design report should provide an initial financial overview, including:

- **Budget Summary**: An overview of the initial budget, including all estimated costs.
- **Funding Sources**: Information on funding sources, such as grants, sponsorships, and team contributions.
- **Cost Breakdown**: A detailed breakdown of anticipated costs, categorized by type.

12.6.4.2   Final Design Report

The final design report should provide a comprehensive financial analysis, including:

- **Final Cost Summary**: A summary of all costs incurred during the project.
- **Budget vs. Actual Comparison**: A comparison of budgeted costs to actual expenditures, highlighting any variances.
- **Detailed Expense Report**: A detailed report of all expenses, categorized and

itemized.
- **Lessons Learned**: An analysis of what was learned from the cost tracking process, including any challenges faced and how they were addressed.

12.6.5 Reporting in Weekly Progress Reports

Weekly progress reports should include a financial section with the following:
- **Weekly Expenditures**: A summary of all expenses incurred during the week.
- **Cumulative Spending**: The total spending to date, compared to the budget.
- **Budget Forecast**: An updated forecast of total project costs based on current spending trends.
- **Issues and Risks**: Identification of any financial issues or risks that could impact the project, along with proposed mitigation strategies.

12.6.6 Document Management and Shared Drives

Maintaining accurate records is crucial for financial transparency and accountability. Best practices for document management include:

12.6.6.1 Centralized Storage

Use a shared drive (e.g., Google Drive, Dropbox) to store all financial documents. This ensures that all team members have access to the latest information.

12.6.6.2 Organized Folders

Organize documents into folders based on categories such as invoices, receipts, purchase requests, and financial reports. This makes it easier to locate specific documents when needed.

12.6.6.3 Regular Backups

Ensure that the shared drive is regularly backed up to prevent data loss. This can be automated using backup software or services.

12.6.6.4 Access Control

Control access to the shared drive to ensure that only authorized team members can modify financial documents. This helps prevent accidental deletions or unauthorized changes.

Tracking and reporting estimated costs in student capstone engineering design projects are essential for maintaining financial control, ensuring transparency, and achieving project success. By implementing structured cost-tracking methods, regularly updating financial records, and maintaining open communication within the team, student teams can effectively manage their project finances. Additionally, incorporating detailed financial reports in preliminary and final design reports and maintaining organized document management systems ensures that all stakeholders are informed and that the project remains on track financially.

12.8 Assignments

Assignment 12-1: Create a Financial Analysis Document for Your Capstone Engineering Design Project

1. **Objective**

   Create a comprehensive financial analysis document for your capstone engineering design project. This document will be updated throughout the project's duration and included in both the preliminary and final design reports. The aim is to track and report all estimated and actual costs, ensuring the project remains within budget and providing transparency to all stakeholders.

2. **Instructions**

   Follow these step-by-step instructions to create, maintain, and report on your project's financial analysis:

3. **Step 1: Initial Budget Creation**

1. **Define Cost Categories**:
   - **Student Hours**: Estimate the total number of hours each team member will work.
   - **Materials and Supplies**: List all required materials with their estimated costs.
   - **Equipment Usage**: Include costs for using equipment such as 3D printers, testing machines, and mechanical shop tools.
   - **Consulting Fees**: Account for any fees associated with consulting engineers, industry experts, or professors.
   - **Miscellaneous Costs**: Include any other potential costs, such as travel or software licenses.

2. **Estimate Costs**:
   - Calculate the total student hours and convert them to dollars using the rate of $25 per hour. Total student cost=Total hours×$25/hour\text{Total student cost} = \text{Total hours} \times \$25/\text{hour}Total student cost=Total hours×$25/hour
   - Research and list the estimated costs for all materials, equipment usage, consulting fees, and miscellaneous costs.

3. **Create an Initial Budget Spreadsheet**:
   - Use a spreadsheet tool (e.g., Microsoft Excel, Google Sheets) to create a budget table.
   - Organize the table into columns for Date, Description, Category, Estimated Cost, Actual Cost, and Variance.

Example:

Date	Description	Category	Estimated Cost	Actual Cost	Variance
01/09/2024	Student Labor	Labor	$15,000	$0	$0
01/09/2024	3D Printing Materials	Materials	$150	$0	$0
01/09/2024	3D Printing Machine Time	Equipment	$1,000	$0	$0
01/09/2024	Consulting Fees	Consulting	$3,000	$0	$0
01/09/2024	Mechanical Shop Usage	Equipment	$3,000	$0	$0
01/09/2024	Miscellaneous Expenses	Miscellaneous	$2,000	$0	$0

4. **Step 2: Ongoing Cost Tracking**
1. **Record All Expenses**:
   - Update the spreadsheet regularly (at least once a week) with actual expenses as they occur.
   - Include detailed descriptions for each expense to clarify its purpose.
2. **Track Variances**:
   - Calculate the variance by subtracting the actual cost from the estimated cost.
   - Highlight significant variances and investigate the reasons behind them.
3. **Monitor the Budget**:
   - Keep an eye on the total spending compared to the budget to ensure you are on track.
   - Adjust the budget, if necessary, based on new information or changes in project scope.
5. **Step 3: Weekly Progress Reports**
6. **Create a Financial Section in Weekly Reports**:
   - Summarize the week's expenditures, including the total spent to date.
   - Compare current spending to the initial budget and update the forecast for future spending.
   - Identify any financial issues or risks and propose mitigation strategies.

Example:

Financial Update (Week Ending MM/DD/YYYY)

- **Weekly Expenditures**: $1,200
- **Cumulative Spending**: $1,200
- **Budget vs. Actual**:
  - Budgeted: $24,150
  - Actual: $1,200
- **Variance**: $22,950 remaining

- **Issues and Risks**:
  - None identified this week.

7. **Step 4: Preliminary Design Report**
1. **Include Financial Analysis in the Preliminary Design Report**:
   - Provide an overview of the initial budget, including all estimated costs.
   - Present a summary of expenditures to date, highlighting any significant variances.
   - Discuss any financial risks or challenges identified during the initial phases of the project.

   Example Outline:

   ### Financial Analysis

   #### Initial Budget Summary
   - **Total Budget**: $24,150
   - **Breakdown by Category**:
     - Labor: $15,000
     - Materials: $150
     - Equipment: $4,000
     - Consulting: $3,000
     - Miscellaneous: $2,000

   #### Expenditures to Date
   - **Total Spent**: $1,200
   - **Breakdown by Category**:
     - Labor: $1,000
     - Materials: $0
     - Equipment: $200
     - Consulting: $0
     - Miscellaneous: $0

   #### Financial Risks and Challenges
   - **Potential Issues**:
     - Higher-than-expected equipment usage costs.
     - Delays in material procurement affecting the budget.

8. **Step 5: Final Design Report**
1. **Include Comprehensive Financial Analysis in the Final Design Report**:
   - Provide a final summary of all costs incurred, categorized by type.

- o   Compare budgeted costs to actual expenditures, highlighting any significant variances.
- o   Discuss lessons learned regarding financial management and provide recommendations for future projects.

Example Outline:

Financial Analysis

Final Cost Summary
- **Total Budget**: $24,150
- **Total Spent**: $23,500
- **Variance**: $650 remaining

Detailed Expense Report
- **Labor**:
  - Budgeted: $15,000
  - Actual: $14,800
  - Variance: $200 under budget

- **Materials**:
  - Budgeted: $150
  - Actual: $180
  - Variance: $30 over budget

- **Equipment**:
  - Budgeted: $4,000
  - Actual: $4,200
  - Variance: $200 over budget

- **Consulting**:
  - Budgeted: $3,000
  - Actual: $3,000
  - Variance: $0

- **Miscellaneous**:
  - Budgeted: $2,000
  - Actual: $1,320
  - Variance: $680 under budget

Lessons Learned
- **Financial Management**:
  - Accurate initial estimates are crucial for effective budget management.
  - Regular updates and tracking prevent significant deviations from the budget.
  - Clear communication within the team regarding financial status and changes is essential.

Recommendations for Future Projects
- Establish a more detailed initial budget with potential contingency plans.
- Implement stricter controls on resource usage and material procurement.
- Enhance collaboration with sponsors and advisors to identify cost-saving opportunities.

9. **Step 6: Record Keeping and Shared Drive Management**
   1. **Organize Financial Documents**:
      - Store all financial documents, including purchase requests, invoices, and receipts, in a centralized shared drive (e.g., Google Drive, Dropbox).
      - Organize documents into clearly labeled folders (e.g., Invoices, Receipts, Purchase Requests, Financial Reports).
   2. **Ensure Accessibility**:
      - Ensure all team members have access to the shared drive and can view and upload documents.
      - Implement access controls to prevent unauthorized modifications.
   3. **Regular Backups**:
      - Schedule regular backups of the shared drive to prevent data loss.
      - Use automated backup solutions if available.

10. **Submission Requirements**
    1. **Initial Budget**: Submit the initial budget spreadsheet by the end of Week 2.
    2. **Weekly Progress Reports**: Submit weekly progress reports every Monday, including the financial section.
    3. **Preliminary Design Report**: By the end of Week 8, submit the preliminary design report, including the financial analysis.
    4. **Final Design Report**: By the project deadline, submit the final design report, including the comprehensive financial analysis.

By following these steps, you will create a thorough and accurate financial analysis document that will help you effectively manage and report the financial aspects of your capstone engineering design project.

13 Create a Proof of Concept

13.1 Introduction

Creating a fresh, brilliant solution concept for a design problem is an exciting venture for student teams. However, before building and testing their solution ideas, they must pace their design and developmental process and validate their solution ideas. Therefore, the team needs to have a comprehensive project plan that includes a proof of concept (POC) design. Creating a POC can mitigate errors and prevent overlooking critical aspects of their project that would cause them to fail.

Moreover, suppose the team wants their sponsor to accept their ideas and design solution proposals. In that case, project teams need to prove that their ideas are practical, functional, viable, and worth the additional human and financial resources needed to complete the project.

13.1.1 Importance of Proof of Concept

A POC is a crucial step for capstone design projects. It manifests by presenting the proposed design solution and its potential viability for further development. A POC typically involves a small-scale visualization or prototype to verify the idea's real-life functioning. It's not about delivering the final design concept but demonstrating and proving its feasibility. Through the POC, the design team can prove that building the proposed solution, product, process, or method is achievable.

The POC also allows the sponsor to see and understand the idea's potential, giving them a glimpse of what the team intends to develop further. This way, the design team can ensure their solution supports the sponsors' needs and the overall requirements of the capstone course.

13.1.2   Steps to Develop a Proof of Concept

1. Define Objectives and Scope:
- ☐ Clearly outline what the POC aims to achieve.
- ☐ Define the scope to focus on the essential aspects of the design that need validation.

2. Develop Key Design Features:
- ☐ Identify and detail the top ideas and their proposed functionality.
- ☐ Highlight specific design features and their practicality.

3. Create a Prototype:
- ☐ Develop a small-scale visualization or prototype to verify the real-life functioning of the idea.
- ☐ The prototype should include the critical features of the proposed solution.

4. Present and Validate:
- ☐ Present the POC to sponsors, mentors, and other stakeholders.
- ☐ Use illustrations, visuals, and interactive elements to explain the design concept thoroughly.

5. Feedback and Iteration:
- ☐ Gather feedback from the audience and refine the design accordingly.
- ☐ Identify and address any issues or obstacles encountered during the POC phase.

13.1.3   POC vs. Prototype

Designers often use POCs and prototypes interchangeably. However, they serve different purposes:

- **POC:** Demonstrates the feasibility of the idea. It shows whether the product or process can be built.
- **Prototype:** Physically presents the essential functions of the design concept. It provides an interactive working model that demonstrates how the design solution works as specified in the POC.

13.1.4   Benefits of a Proof of Concept

1. Error Mitigation:
- Identifies unknowns and obstacles early in the development process.
- Reduces the probability of project failure by addressing issues before full-scale implementation.

2. Resource Allocation:
- Justifies the expenditure of human and financial resources.
- Helps sponsors and stakeholders make informed decisions about project investment.

3. Design Validation:
- Ensures that the design solution meets the requirements and specifications.
- Provides a basis for scaling up the design for manufacturability and production.

4. Stakeholder Confidence:
- Builds confidence among sponsors and mentors in the viability of the design solution.
- Demonstrates the design team's commitment to a structured and thorough development process.

13.1.5   Conducting a Successful POC

1. Detailed Planning:
- Create a comprehensive project plan outlining the POC objectives, scope, and deliverables.
- Assign tasks and responsibilities within the team to ensure efficient execution.

2. Effective Presentation:
- Use clear and concise communication to present the POC.
- Include detailed illustrations, visuals, and interactive elements to engage the audience.

3. Collaborative Feedback:
- Encourage open and constructive feedback from sponsors and stakeholders.
- Use the feedback to refine the design and address any potential issues.

4. Documentation and Reporting:
- Document the POC process, including objectives, methods, results, and feedback.
- Provide a detailed report to sponsors and stakeholders for review and approval.

13.1.6   Outcome of POC

1. Approval and Advancement:
- If the POC is successful, the design team can proceed with the full-scale development and testing of the design solution.
- The POC serves as a solid foundation for further development, ensuring that the project is on the right track.

2. Re-evaluation and Revision:
- If the POC is not successful, the design team must reconsider their choice of design solution and strategy.
- The team should revisit the initial problem statement, specifications, and

feedback to identify alternative solutions.

Having a proof of concept and prototype is essential for student design teams to refine their ideas and begin their product or process development process. The POC helps the design team identify unknowns and obstacles they may face in further developing the proposed solution. Uncovering these obstacles during the POC phase mitigates problems later during the build and testing of the design solution, reducing the probability of project failure. While the POC does not guarantee the smooth progression of the project, it significantly increases the chances of success by providing a validated, practical, and feasible design foundation.

13.1.7   How to Create a Proof of Concept

Creating a proof of concept (POC) involves several fundamental steps that design project teams should follow to demonstrate that their design solution meets sponsor requirements and is feasible. The POC process provides a way to validate ideas and ensure that they align with stakeholders' needs and expectations.

13.1.7.1   Demonstrate the Design Solution Meets Sponsor Requirements

The initial step in the POC process is to confirm that the design solution meets the requirements of the sponsor. This involves several key activities:

1. **Establish the Need for the Product or Process**:
   o **Market Surveys**: Conduct surveys to understand the market demand and identify gaps that the proposed solution can fill.
   o **Sponsor Reviews**: Engage with sponsors to get their insights and feedback on the proposed solution.
   o **Focus Groups**: Organize sessions with potential users to gather detailed feedback and preferences.
   o **End-User Interviews**: Conduct one-on-one interviews with end-users to understand their likes, dislikes, and desires regarding the proposed solution.
2. **Document Sponsors' Feedback**:
   o Develop thoughtful questions to understand the sponsors' perspectives.
   o Document sponsors' feelings, intuitions, and perspectives, as these will guide the design team in refining the solution.
   o Incorporate this feedback into the planning and progression of the design work.

Having this information will guide the design team in streamlining the design solution and fine-tuning the design specifications to meet the sponsor's requirements better.

13.1.7.2   Generate an Improved Design Solution

Using the insights gained from sponsor and user feedback, the design team should aim to improve the original design concepts:

1. **Brainstorm Improvements**:
   - Use the feedback from interviews and focus groups to brainstorm potential improvements to the design.
   - Integrate these ideas into the Quality Function Deployment (QFD) analysis to reassess and rate the design concepts.
2. **Revise QFD Charts**:
   - Update the QFD chart with new information collected from interviews, reflecting changes in sponsors' needs and requirements.
   - Reevaluate the engineering criteria to ensure they comply with the updated requirements.

This step ensures that the design evolves in response to real-world feedback and aligns more closely with stakeholder expectations.

13.1.7.3   Build a Prototype and Test It

Once an improved design solution is developed, the next step is to create a prototype and test it:

1. **Create a Prototype**:
   - Develop a prototype based on the updated requirements, criteria, and design specifications.
   - Ensure the prototype reflects the features and design solutions decided upon.
2. **Test with Stakeholders**:
   - Have sponsors and individuals from the interview groups test the prototype.
   - Gather feedback on their experience, satisfaction, and any issues they encounter.
3. **Efficiency in Prototyping**:
   - Build the prototype quickly and efficiently, using methods such as 3D printing or modeling in process design software.
   - Ensure the prototype is functional, even if it is not complete or perfect, as the goal is to validate the concept.

The feedback gathered during testing helps in verifying and validating the design choices, allowing the team to make necessary adjustments before finalizing the design.

13.1.7.4   Collect Test Data, Analyze, and Document

The final step in the POC process involves collecting and analyzing data from the prototype testing:

1. **Collect Data**:
   o Gather detailed feedback from the test group on their experiences, satisfaction, reactions, complaints, and other pertinent details.
2. **Analyze Data**:
   o Analyze the collected data to validate the design and verify the feasibility of the solution.
   o Identify areas for improvement based on the feedback and make necessary adjustments to the design.
3. **Document Findings**:
   o Document all findings, feedback, and actions taken in the team's project management plan.
   o Use this documentation to guide future iterations of the design and ensure continuous improvement.

By following these steps, the design team can develop a robust proof of concept that demonstrates the viability and effectiveness of their design solution, ensuring it meets the requirements of sponsors and stakeholders.

13.2 Demonstrate the Design Solution Meets Sponsor Requirements

Ensuring that a design solution meets sponsor requirements is a critical step in the engineering design process. This validation phase verifies that the final product aligns with the initial specifications and expectations set by the sponsor. It involves a comprehensive assessment of the design's performance, functionality, and adherence to predefined criteria.

13.2.1   Key Steps to Demonstrate Compliance

1. **Understanding Sponsor Requirements**

   o **Initial Documentation**: At the outset, compile a detailed list of sponsor requirements. This includes functional specifications, performance targets, cost constraints, and any other specific needs or preferences.
   o **Product Design Specification (PDS)**: Develop a comprehensive PDS that outlines all critical requirements. This document serves as a control and

reference throughout the design process.

2. **Design Reviews and Iterative Feedback**

   o **Regular Reviews**: Conduct design reviews at various stages of the project. These reviews involve the sponsor and other stakeholders to ensure continuous alignment with requirements.
   o **Iterative Improvements**: Use feedback from these reviews to make necessary adjustments. This iterative process helps refine the design to better meet sponsor expectations.

3. **Verification and Validation**

   o **Prototype Testing**: Build and test prototypes to validate that the design meets the functional and performance criteria specified in the PDS. Verification ensures the product behaves as expected under real-world conditions.
   o **Performance Metrics**: Establish and measure key performance indicators (KPIs) to objectively assess the design's compliance with the requirements. This includes mechanical, electrical, and thermal load tests, among others.

4. **Documentation and Reporting**

   o **Detailed Reporting**: Prepare comprehensive reports documenting how the design meets each sponsor's requirement. Include test results, performance data, and any deviations along with their resolutions.
   o **Final Review Documentation**: Ensure that all findings and adjustments are documented and reviewed in the final design review. This serves as a formal approval stage before production or implementation.

13.2.2   Critical Aspects to Address

1. **Functional Requirements**

   o **Core Functions**: Demonstrate that the design performs the primary functions as intended. Use specific tests to validate critical functionalities.
   o **User Interaction**: Ensure the design is user-friendly and meets the sponsor's usability expectations. This can be validated through user testing and feedback sessions.

2. **Performance Requirements**

    ○ **Reliability and Durability**: Test the product for reliability under various conditions to ensure it will perform consistently over its expected lifespan.

    ○ **Safety and Compliance**: Verify that the design meets all safety standards and regulatory requirements. This includes conducting thorough safety analyses and compliance checks.

3. **Cost and Resource Management**

    ○ **Budget Adherence**: Ensure the design process stays within the budget constraints set by the sponsor. This involves careful cost tracking and value engineering.

    ○ **Resource Allocation**: Efficiently manage resources to meet project timelines and deliverables. Document any deviations and their impact on the project.

4. **Aesthetic and Ergonomic Considerations**

    ○ **Design Appeal**: Ensure the design meets the aesthetic criteria specified by the sponsor. This includes material selection, color schemes, and overall visual appeal.

    ○ **Ergonomics**: Validate that the product is ergonomically designed for ease of use and comfort. This can be assessed through user trials and ergonomic evaluations.

Demonstrating that a design solution meets sponsor requirements is a multifaceted process that involves thorough planning, continuous feedback, rigorous testing, and detailed documentation. By following these structured steps, design teams can ensure that their solutions not only meet but exceed sponsor expectations, leading to successful project outcomes and satisfied stakeholders.

13.3 POC Presentation

The Proof of Concept (POC) presentation marks a critical milestone in the capstone design project. It is the culmination of the first phase of the project, occurring at the end of the first semester in a two-semester course sequence. This presentation allows the design team to showcase their initial design efforts to peers, sponsors, mentors, and professors, setting the stage for the next phase of development.

13.3.1   Purpose of the POC Presentation

The primary aim of the POC presentation is to demonstrate that the proposed design solution is practical, valuable for the target audience, and achievable. The presentation is a formal event where the design team must convincingly communicate several critical aspects of their project:

1. **Customer Requirements**: Detailed explanation of the customer's needs, problems, and dislikes that the design solution aims to address.
2. **Design Features and Benefits**: Presentation of specific features that solve the identified problems, highlighting the benefits and advantages of the design.
3. **Development Process**: This section describes the product or process development stages, including design iterations and decision-making processes.
4. **Project Management**: This section illustrates project management strategies using tools such as Gantt charts to depict timelines, milestones, and resource allocation.

13.3.2   Key Components of the Presentation

1. **Customer Requirements and Problem Statement**
   - Clearly outline the customer's needs and the specific problems the design aims to solve.
   - Include any initial feedback or dislikes from the customer regarding early design concepts.
2. **Design Solution**
   - Present the POC design solution, emphasizing how it addresses the customer requirements and problems.
   - Highlight critical features and demonstrate how these features provide tangible benefits.
3. **Development and Management Process**
   - Discuss the overall development process, including significant phases such as ideation, prototyping, and testing.
   - Use visual aids such as Gantt charts to illustrate the project timeline, milestones achieved, and future plans.
4. **Success Criteria and Evaluation Metrics**
   - Define the criteria for success and how the design solution meets these criteria.
   - Present the evaluation metrics used to assess the design's performance, such as functionality tests, user feedback, and financial viability .
5. **Future Plans**
   - Outline the next steps post-POC presentation, including plans for

further development, testing, and final implementation.

  o   Identify any additional resources needed, such as materials, funding, or expertise.

13.3.3   Importance of the POC

The POC presentation is not only a demonstration of the design solution but also a critical checkpoint for receiving feedback and approval to proceed to the next phase of the project. It helps in:

- **Validating the Design**: Ensuring that the design is on the right track and meets the customer's needs.
- **Identifying Improvements**: Gathering feedback from peers, mentors, and sponsors to refine and improve the design.
- **Securing Support**: Convincing stakeholders of the design's viability is essential for securing further support and resources.
- **Conclusion**

In summary, the POC presentation is a pivotal event in the capstone design project. It provides a platform for the design team to validate their work, receive constructive feedback, and secure approval to advance their project. Through detailed presentation of customer requirements, design solutions, development processes, and project management, the team demonstrates the feasibility and value of their proposed design, setting the stage for successful project completion.

13.4 Viability and Usability

The viability and usability of the Proof of Concept (POC) are crucial in determining a project's probability of success and assessing whether the fundamental assumptions about the proposed design solutions are sound. This phase aims to identify the key features and functionalities of the design solution before significant resources are committed. It involves creating a prototype as a preliminary attempt to develop a working model that can potentially be used in real-world applications.

13.4.1   Purpose of the POC Step

The primary objectives of the POC step are:

1. **Feature and Functionality Identification**: Determine the essential features and functionalities of the design solution.
2. **Resource Optimization**: Avoid excessive expenditure of time, money, and materials on unviable solutions.
3. **Assumption Testing**: Validate or refute the initial assumptions about the

design solution, including functionality, feasibility, and user needs.

13.4.2 Prototype Development

Developing a prototype during the POC phase serves as a practical means to test and refine the design solution. The prototype should emulate most of the functionalities and performance characteristics of the final product. However, it may lack efficiency, aesthetics, durability, and the actual size and materials of the end solution.

1. **Error Identification**: The prototype helps in identifying and rectifying wrong assumptions, mistakes, and errors early in the design process.
2. **Problem Solving**: It exposes unknowns, stumbling blocks, and unresolved issues that could impede the progress of the project.
3. **Functionality Demonstration**: The prototype should demonstrate the core functionalities and performance capabilities of the final design solution, even if it is not fully optimized.

13.4.3 Benefits of Prototyping

Prototyping during the POC phase offers several benefits:

1. **Knowledge Sharing**: Facilitates the exchange of design solution knowledge among team members, enhancing collaboration and innovation.
2. **Design Technique Exploration**: Provides an opportunity to experiment with various design techniques and approaches.
3. **Validation Evidence**: Offers tangible evidence of the solution concept's validity to sponsors and other stakeholders.

13.4.4 Stakeholder Engagement

Prototyping is a powerful tool for engaging stakeholders, including sponsors, mentors, and end-users. It serves as a platform for:

1. **Value Illustration**: Clearly illustrates the value and potential impact of the design solution.
2. **Feedback and Guidance**: Receives constructive feedback and guidance, helping to refine and improve the design.
3. **Documentation and Estimation**: Aids in the documentation process and provides valuable insights into the resources and time needed for project completion.

13.4.5   Criteria for a Successful POC

A successful POC prototype should:

1. **Demonstrate Core Functionalities**: Show that the essential functions of the design solution work as intended.
2. **Performance Indication**: Provide an initial indication of the performance characteristics of the final product.
3. **Highlight Issues**: Identify any critical issues or limitations that need to be addressed before further development.
4. **Stakeholder Satisfaction**: Satisfy the primary concerns and requirements of stakeholders, even if it is not entirely free of bugs or fully optimized.

In summary, the viability and usability assessment during the POC phase is a pivotal step in the design process. The proof of concept presentation, in particular, is a key component of this phase. It allows design teams to validate their concepts, identify and solve problems, and refine their solutions. This process not only saves resources but also enhances the likelihood of project success by ensuring that the design solution is both feasible and valuable in real-world applications. Prototyping provides a clear path forward, helping design teams make informed decisions and secure the necessary support from stakeholders for the next stages of development.

13.6 Assignments

Assignment 13-1: Proof of Concept Presentation

- **Objective**

  The objective of your design team is to create a 15-minute presentation that clearly demonstrates the proof of concept for your design project. This presentation should provide compelling evidence that your design not only addresses the problem definition but also meets the customer requirements. It should showcase a functioning device or critical part of the design. All team members must participate in the presentation. You are required to upload your PowerPoint presentation to the course learning management site no later than 24 hours before your scheduled presentation time.

- **Guidelines for Preparing Your Presentation**
  1. **Understand the Requirements**
     - Review the problem definition and customer requirements.
     - Ensure your proof of concept addresses all key aspects and requirements.
     - Identify the specific evidence you will present to demonstrate functionality and compliance.
  2. **Plan the Presentation Structure**
     - Allocate time for each section of the presentation, ensuring you leave 3-5 minutes for Q&A.
     - Suggested time allocation:
       - Introduction: 1-2 minutes
       - Problem Definition and Customer Requirements: 2-3 minutes
       - Design Process and Methodology: 3-4 minutes
       - Prototype Demonstration: 4-5 minutes
       - Conclusion and Future Work: 1-2 minutes
       - Q&A: 3-5 minutes
  3. **Develop the Content**
     - **Introduction**
       - Briefly introduce your team members.
       - Provide an overview of the presentation.
     - **Problem Definition and Customer Requirements**
       - Clearly state the problem your design addresses.
       - Outline the critical customer requirements.
     - **Design Process and Methodology**
       - Describe the design process your team followed.
       - Highlight critical decisions, iterations, and any challenges

overcome.

- o **Prototype Demonstration**
  - ▪ Present your proof-of-concept prototype.
  - ▪ Explain the functionality and key features.
  - ▪ Demonstrate the working prototype or critical parts of it.
  - ▪ Use videos, live demonstrations, or simulations as needed.
- o **Conclusion and Future Work**
  - ▪ Summarize how your prototype meets the problem definition and customer requirements.
  - ▪ Discuss any future improvements or next steps.
- o **Q&A**
  - ▪ Prepare to answer questions from the audience and instructors.

4. **Design the PowerPoint Slides**
   - o Ensure slides are visually appealing, clear, and concise.
   - o Use bullet points, diagrams, and images to support your content.
   - o Avoid clutter and excessive text.
   - o Ensure consistent formatting and style throughout the presentation.

5. **Assign Roles and Rehearse**
   - o Assign specific sections of the presentation to each team member.
   - o Ensure each member knows their part and transitions smoothly between sections.
   - o Rehearse the entire presentation multiple times to stay within the time limit and ensure a cohesive delivery.
   - o Practice handling the Q&A session and preparing answers to potential questions.

6. **Technical Preparation**
   - o Test all equipment and software you will use during the presentation (e.g., projector, laptop, prototype).
   - o Ensure your PowerPoint file is compatible with the presentation equipment.
   - o Have backup copies of your PowerPoint file on a USB drive and cloud storage.

7. **Upload the PowerPoint**
   - o Upload your finalized PowerPoint presentation to the course learning management site no later than 24 hours before your scheduled presentation time.
   - o Verify the upload to ensure it was successful and accessible.

- **Checklist for Submission**
  - • Review problem definition and customer requirements.
  - • Plan presentation structure and allocate time for each section.

- Develop content for each section of the presentation.
- Design visually appealing and clear PowerPoint slides.
- Assign roles and rehearse the presentation.
- Test all equipment and software for compatibility.
- Upload the PowerPoint to the course learning management site 24 hours before the presentation.
- Verify the successful upload and accessibility of the PowerPoint file.

- **Evaluation Criteria**
  - **Content**: Completeness and accuracy of the information presented.
  - **Organization**: Logical flow and clarity of the presentation structure.
  - **Engagement**: Effectiveness of the presentation in engaging the audience.
  - **Team Participation**: Equal participation of all team members.
  - **Technical Quality**: Quality of the PowerPoint slides and demonstration.
  - **Q&A Handling**: Ability to answer questions clearly and accurately.

By following these guidelines, your team will be well-prepared to deliver a compelling and professional proof of concept presentation. Good luck!

14   Documentation of Capstone Design Projects

14.1 Introduction

Documentation in engineering design encompasses a broad scope and a diverse range of subject areas. It integrates writing, organizing, project management, critical thinking, analysis, and engineering problem-solving. Adequate documentation is governed by a set of rules, guidelines, and best practices that ensure clarity, consistency, and comprehensiveness. Proper documentation is crucial for capturing lessons learned, facilitating knowledge transfer, and enhancing the success of design projects. It is also essential for meeting the expectations of sponsors and preparing students for professional engineering practices.

14.1.1   Importance of Formal Documentation

A structured approach to documentation significantly increases the probability of a design project's success. Without formal guidelines, essential details and lessons from past designs may be lost, reducing the efficacy of future projects. Establishing clear rules for documentation in capstone design classes is vital for several reasons:

1. **Knowledge Retention**: Captures lessons learned and best practices for future reference.
2. **Learning by Doing**: Enhances students' skills in documentation, benefiting their professional careers.
3. **Sponsor Satisfaction**: Meets sponsor expectations through comprehensive design reports, presentations, and regular updates.
4. **Process Clarity**: Ensures all stakeholders can follow and understand the design process and decisions made.

14.1.2   Engineering Design Documentation

Documentation is a critical element of the engineering design process. It captures the complexity of engineering design, involving numerous analyses and decision-making points based on data and evidence. Adequate documentation ensures that all aspects of the design process are recorded, allowing others to follow the rationale behind design decisions and the final results.

14.1.2.1   Key Components of Documentation

1. **Capstone Design Workbook**
   o **Purpose**: Acts as a diary of the project, recording the evolution of ideas, problems encountered, drawings, schematics, readings, and assignments.
   o **Content**: Includes detailed information, readings, assignments, research notes, lecture notes, design concepts, solutions considered, decisions made, and guidelines for design steps.
2. **Meeting Minutes**
   o **Purpose**: Capture discussions and incremental work performed during meetings.
   o **Content**: Document design steps, critical decisions, and action items. Ensure clarity and detail to facilitate follow-up and accountability.
3. **Design Presentation Slides**
   o **Purpose**: Communicate critical information at various stages of the project.
   o **Content**: Include details on design reviews, proof of concept, and testing phases. Used for stakeholder presentations to provide updates and gather feedback.
4. **Preliminary Design Report (PDR)**
   o **Purpose**: Capture the design process details leading up to the proof of concept and prototype.
   o **Content**: Includes background research, design requirements, initial concepts, and development steps. Serves as a milestone document for sponsor review and approval.
5. **Brochures and Posters**
   o **Purpose**: Summarize the project for a broader audience, such as conferences or design showcases.
   o **Content**: Highlight critical aspects of the project, including objectives, methodologies, results, and potential impacts.
6. **Final Design Report**
   o **Purpose**: Comprehensive documentation of the entire design project.
   o **Content**: Captures all critical aspects of the design work, including final designs, detailed analysis, test results, and lessons learned. Serves as the

definitive record of the project for sponsors and future reference. Additionally, these reports serve as a part of an engineering student portfolio. During job interviews, capstone design projects often become the main focus, demonstrating the student's competency and ability to complete a major engineering project while working in a team.

14.1.3 Guidelines for Effective Documentation

1. **Consistency**: Use a uniform format and structure for all documents to ensure ease of reading and comprehension.
2. **Clarity**: Write clearly and concisely, avoiding jargon and technical language that all stakeholders may not understand.
3. **Detail**: Include sufficient detail to allow others to understand the design process, decisions made, and outcomes achieved.
4. **Accessibility**: Organize documents in a manner that allows easy access and retrieval of information.
5. **Regular Updates**: Maintain documentation regularly throughout the project to capture real-time progress and changes.

Adequate documentation is integral to the success of engineering design projects. It ensures that the design process is transparent, decisions are well-documented, and knowledge is preserved for future use. By adhering to established guidelines and maintaining thorough documentation, design teams can enhance project outcomes, meet sponsor expectations, and build valuable skills for their professional careers. Moreover, the capstone design reports serve as a valuable portfolio component, showcasing students' abilities and achievements to potential employers and affirming their competence in executing major engineering projects.

14.1.4 Design Binder Implemented as a Shared Electronic Drive

A design binder, traditionally a physical three-ring binder, can be efficiently implemented as a shared electronic drive. This method allows the project team to collect, organize, and share their project materials seamlessly. For a typical engineering capstone project conducted over two terms, an electronic design binder provides easy access to all relevant documents and facilitates collaboration among team members, professors, mentors, and sponsors.

14.1.4.1 Benefits of a Shared Electronic Design Binder:

1. **Accessibility**: Team members can access the binder from any location, promoting collaboration and real-time updates.
2. **Organization**: Digital folders and subfolders help keep materials well-organized

and easy to navigate.

3. **Security**: Cloud storage services offer secure backups and version control, preventing data loss.

14.1.4.2   Key Components and Organization:

1. **Index and Table of Contents**
   - **Purpose**: Provide a roadmap of the binder's content.
   - **Content**: Include headings for each section.
2. **Problem Definition**
   - **Purpose**: Detail the problem being addressed.
   - **Content**: Include supporting materials created by the team.
3. **Team Work**
   - **Purpose**: Document team interactions and progress.
   - **Content**: Include meeting minutes, progress reports, and email correspondence with sponsors.
4. **Engineering Analysis**
   - **Purpose**: Capture technical details and analyses.
   - **Content**: Include notes, sketches, calculations, drawings, charts, and schematics.
5. **Project Plan**
   - **Purpose**: Outline the project timeline and tasks.
   - **Content**: Include baseline and updated project plans, Gantt charts, and task schedules with milestones.
6. **Presentations**
   - **Purpose**: Archive all project presentations.
   - **Content**: Include PowerPoint slides from all stages of the project.
7. **Literature and Patent Searches**
   - **Purpose**: Record background research.
   - **Content**: Include articles, book sections, journal articles, and patents.
8. **Design Approaches**
   - **Purpose**: Document design methods used.
   - **Content**: Include descriptions of methods and their applications to the project.
9. **Design Specifications**
   - **Purpose**: Record detailed design specifications.
   - **Content**: Include the revision history and dates.
10. **BOM (Bill of Materials)**
    - **Purpose**: Track materials and suppliers.
    - **Content**: Include a cross-referenced BOM with supplier and contact lists.
11. **Systems Analysis**

- o **Purpose**: Capture system-level analyses.
- o **Content**: Include computational analyses, such as load-bearing capacity, strength, heat transfer, and materials behavior.

12. **References**
    - o **Purpose**: Record all external references.
    - o **Content**: Include competition rules, industry materials, standards, papers, and manuals.

13. **Modeling**
    - o **Purpose**: Archive all modeling efforts.
    - o **Content**: Include models and simulations using tools like CAD, Abaqus, Fluent, and Comsol.

14. **Trade-Off Analysis**
    - o **Purpose**: Document decision-making processes.
    - o **Content**: Include radar/spider charts, QFD models, and formal decision models.

15. **Financial Analysis**
    - o **Purpose**: Record cost analyses.
    - o **Content**: Include cost estimates, personnel effort, manufacturing costs, and facility usage.

16. **Critical Thinking and Analysis**
    - o **Purpose**: Capture broader analysis results.
    - o **Content**: Include analyses of economic, environmental, social, political, ethical, health and safety, manufacturability, and sustainability factors.

17. **Administrative**
    - o **Purpose**: Track administrative tasks.
    - o **Content**: Include purchase orders, bid sheets, quotations, registrations, and other paperwork.

18. **Resumes**
    - o **Purpose**: Record team member resumes.
    - o **Content**: Include the most recent resumes for all team members.

14.1.4.3   Practical Style Guide for Digital Organization:

- • **Use a well-structured folder hierarchy**: Organize your documents in folders by sections similar to the traditional binder format.
- • **Use consistent naming conventions**: Ensure all files and folders follow a consistent naming scheme for easy retrieval.
- • **Regularly update and backup**: To prevent data loss, keep the electronic design binder up-to-date and regularly backed up.

14.1.5   Electronic Files and Project Archive

Electronic files created during the design project are an essential part of the design work's documentation. The electronic files for the design project should be organized and available to all team members (including professors and teaching assistants) and possibly mentors and sponsors. A suggested organization of information is presented in the following list of folder names:

- Additional considerations
- Administrative
- Assessment
- Brochures
- CAD Files
- Concepts
- Cost Analysis
- Critical Design Review
- Design for X
- Design Specifications
- Final Design Report
- Manuals
- Meeting Minutes and Notes
- Paper
- Patent Search
- Photos
- Poster
- Preliminary Design Report
- Presentations
- Previous Project Information
- Problem Definition
- Project Management
- QFD Analysis
- References
- Resumes
- Testing
- Videos
- Weekly Progress Reports

Electronic files can be shared with cloud file systems such as Google Drive or Dropbox. Many universities have standardized the Google application suite (G Suite), which facilitates collaboration and document sharing.

Adequate documentation is integral to the success of engineering design projects. It ensures that the design process is transparent, decisions are well-documented, and knowledge is preserved for future use. By adhering to established guidelines and maintaining thorough documentation, design teams can enhance project outcomes, meet sponsor expectations, and build valuable skills for their professional careers. Moreover, the capstone design reports serve as a valuable portfolio component, showcasing students' abilities and achievements to potential employers and affirming their competence in executing major engineering projects.

14.2 Verbal Presentation with Slides

Verbal team presentations are a cornerstone of capstone design projects. Typically, student design teams are expected to make two or three presentations over an academic term. These presentations, which range from 10 to 30 minutes, must be meticulously prepared and delivered to communicate progress, accomplishments, and future directions effectively. To achieve a professional presentation, teams must practice extensively and adhere to specific guidelines.

14.2.1 Tips for Preparing an Excellent Presentation

1. **Plan Your Presentation Structure**

   o Ensure it includes an introduction, key points, and a conclusion.

   o Use a logical flow to guide the audience through your content.

2. **Enhance Visual Communication**

   o Use PowerPoint to organize and present complex information clearly.

   o Select a clean, professional template with consistent formatting.

3. **Use a Cohesive Color Scheme**

   o Ensure text and visuals are well-aligned.

   o Stick to a cohesive color scheme that enhances readability.

4. **Follow the 6x6 Rule**

   o Use no more than six bullet points per slide.

   o Each bullet should have no more than six words.

5. **Keep Content Concise**

   o Avoid overloading slides with information.

   o Make sure every slide serves a purpose and avoid redundancy.

6. **Control Your Timing**

   o   Allocate 1-2 minutes per slide.

   o   Practice to ensure you stay within the allotted time.

7. **Speak Clearly and Confidently**

   o   Make eye contact with the audience.

   o   Use clear and descriptive titles for each slide.

8. **Tailor to Your Audience**

   o   Ensure your content is relevant and engaging.

   o   State the purpose of your presentation clearly at the beginning.

9. **Tell a Compelling Story**

   o   Highlight the journey from problem definition to solution development.

   o   Focus on critical points and significant findings without getting lost in details.

10. **Structure Logically**

    o   Use transitions to connect different sections.

    o   Aim for a natural and engaging delivery with controlled pacing.

11. **Ensure Accessibility**

    o   Use high-contrast colors and large fonts.

    o   Test readability in different lighting conditions.

12. **Use Graphics Effectively**

    o   Include relevant, high-quality graphics to illustrate key points.

    o   Avoid overly bright or clashing colors and ensure good contrast.

13. **Make It Enjoyable**

    o   Use humor or interesting anecdotes where appropriate.

    o   Keep the audience engaged with interactive elements or questions.

14. **Refocus Attention**

    o   Periodically summarize key sections before moving on.

      o Recap the main takeaways of your presentation at the end.

15. **Proofread and Edit**

      o Carefully proofread all slides for clarity, accuracy, and conciseness.

      o Avoid common PowerPoint pitfalls like overloading slides with text.

16. **Utilize PowerPoint Features**

      o Use transitions, animations, and multimedia effectively.

      o Ensure they enhance, rather than detract from, your message.

17. **Prepare for Q&A**

      o Allocate time for a Q&A session at the end.

      o Prepare for potential questions and answer them confidently.

14.2.2   Guidelines for Capstone Design Presentations

Effective design presentations require adherence to specific guidelines that ensure all team members participate and the presentation is professional, efficient, and informative. Given the limited time available due to class size and scheduled meeting hours, the following guidelines can help teams prepare:

1. **Time Management**

      o Establish a maximum time limit for each presentation (e.g., 15, 20, 25, or 30 minutes), including setup, questions, and answers.

      o Aim to keep the presentation to 80% of the allotted time to allow for setup and Q&A.

2. **Inviting Sponsors**

      o Ensure sponsors are invited to the presentations.

      o Inform your professor if sponsors will attend.

3. **Team Participation**

      o All team members must participate in the presentation.

      o Assign specific slides to each member to practice and present.

4. **Note-taking**

      o Assign a team member to take notes on questions, comments, and suggestions from the audience.

5. **Introductions**

    o   Introduce all team members and their roles at the beginning of the presentation.

6. **Presentation Content**

    o   Include team organization and division of responsibilities.

    o   Define the problem and provide a design specifications table.

    o   Present design concepts and how they address the sponsor's problem.

    o   Conduct engineering analysis, testing, validation, and verification using relevant tools and software.

    o   Include Pugh and QFD analysis when appropriate.

    o   Conduct a trade-off analysis of alternatives.

    o   List problems and unresolved issues and explain plans to address them.

    o   Summarize and conclude the presentation, highlighting creativity and innovation.

7. **Preparation and Submission**

    o   Provide a copy of the presentation (e.g., PowerPoint file) to the professor and sponsors before the presentation.

8. **Assessment**

    o   The presentation will be evaluated on the team's ability to work together, the effectiveness of the presentation, and the probability of achieving a successful design solution.

9. **Audience Participation**

    o   Encourage the audience (students, professors, mentors, and sponsors) to provide meaningful critiques and comments.

10. **Positive Attitude**

    o   Maintain a positive attitude throughout the presentation, demonstrating enthusiasm and confidence in the project.

By following these guidelines and tips, capstone design teams can deliver effective, professional presentations that clearly communicate their project's progress and future direction.

14.3 Photos

Photographs have been utilized as sources of evidence and information in scientific endeavors for over a century. In recent years, the camera has become ubiquitous, integrated into virtually every cell phone, making it an ever-available tool to capture observations instantly. In the context of engineering, and particularly in capstone design projects, photographs are invaluable for documenting the progression of design work, capturing precise details that written descriptions alone may not fully convey.

14.3.1 Importance of Photographs in Engineering Design

1. **Precision and Clarity**
   - A photograph can convey complex design details more precisely than many words. It provides an exact likeness of the objects or processes captured, reducing ambiguity and enhancing understanding.
   - Visual documentation ensures that the exact features, dimensions, and conditions of the design are accurately recorded and easily interpreted by anyone reviewing the project.

2. **Persuasive Evidence**
   - Photographs serve as persuasive evidence, showing the actual state of a design at various stages. This visual record is compelling because it provides concrete proof of what was achieved and how it was done.
   - Unlike text descriptions that rely on the reader's imagination and interpretation, photographs present an objective reality that can be universally understood.

3. **Enhanced Understanding with Annotations**
   - Adding descriptive words and annotations to photographs further enhances comprehension. Labels, arrows, and notes can highlight critical components, explain functions, and illustrate procedures, making complex concepts accessible and transparent.
   - Annotated photographs can be used in presentations, reports, and documentation to provide detailed visual explanations that support and clarify written content.

4. **Documenting Design Evolution**
   - Capturing photographs at each stage of the design evolution, including assembly, parts, operation, maintenance, and fitting, provides a chronological record of the project's development.
   - This visual timeline helps track progress, identify changes, and reflect on the design decisions made throughout the project.

5. **Human Element**
   - Including photographs of design team members working on the

project adds value by documenting their participation and contributions. These images serve as historical records of who was involved, what roles they played, and where and when specific tasks were performed.

o People in photographs also provide a sense of scale and context, helping to understand the form, size, and function of the design elements in relation to human interaction.

14.3.2   Practical Applications of Photographs in Capstone Design Projects

1. **Capturing Design Stages**
   o Take photographs during critical stages of the design process to document initial concepts, prototypes, testing phases, and final products.
   o Use these images to illustrate the development and refinement of the design, showing the progression from idea to implementation.

2. **Recording Assembly and Maintenance**
   o Photograph the assembly process to document how components fit together, identify potential issues, and provide a visual guide for future assembly.
   o Capture images of maintenance procedures to create instructional materials that explain how to service and repair the design.

3. **Showing Operations and Functionality**
   o Use photographs to demonstrate how the design operates in real-world conditions. Highlight key features and mechanisms and show the design in action.
   o These images can be used in presentations and reports to provide a clear understanding of the design's functionality and performance.

4. **Including Annotations and Descriptions**
   o Annotate photographs with labels, arrows, and notes to explain specific features, parts, and functions. This helps clarify the images and provides detailed visual explanations.
   o Combine photographs with descriptive text to create comprehensive visual documentation that supports the written narrative of the project.

14.3.3   Photo Example

In a project focused on improving a magnetic seal demagnetizer, photographs can capture each stage of the design and testing process. Annotated images can highlight the components used, the setup of the testing apparatus, and the results observed. Including team members in these photographs adds context and demonstrates their active involvement in the project.

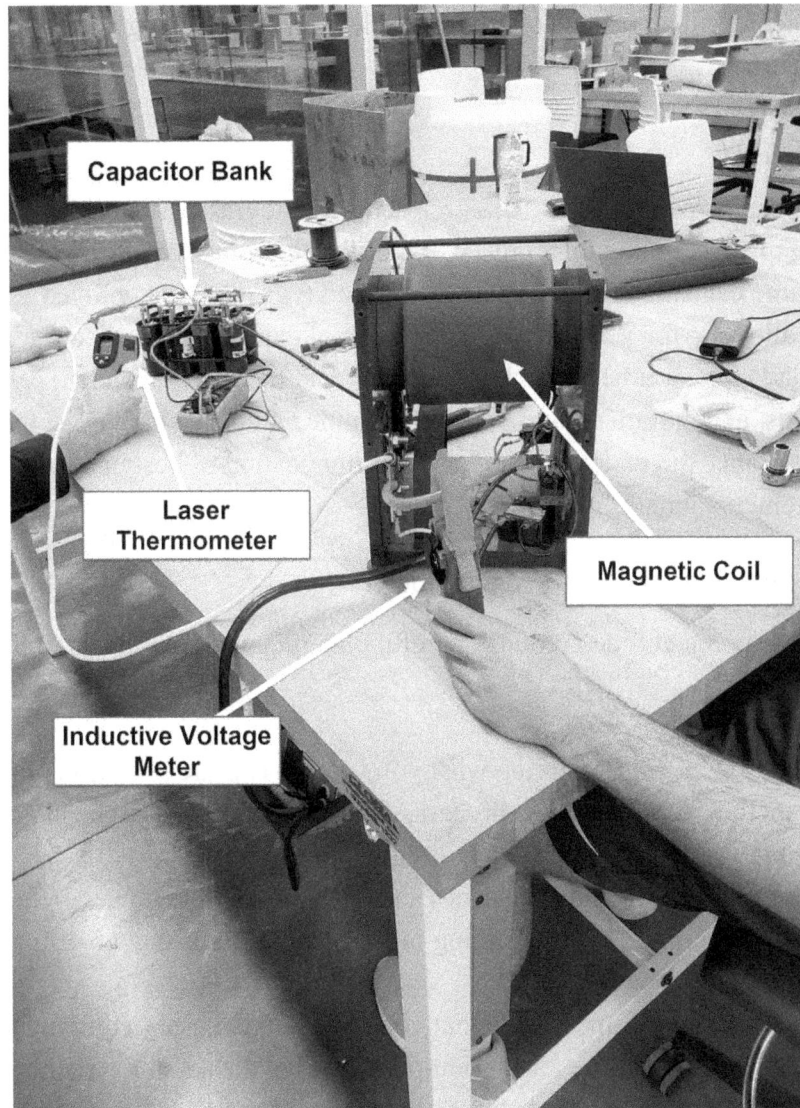

Figure 14-1. Example photograph showing magnetic seal demagnetizer.

Example Photograph Descriptions:

- **Initial Design Concept**: Photograph of the initial sketches and 3D models.
- **Prototype Assembly**: Images showing the step-by-step assembly of the prototype.
- **Testing Setup**: Photographs of the testing equipment and the prototype during testing, with annotations explaining the setup and measurements.
- **Team Involvement**: Pictures of team members working on different aspects of the project, such as assembling components, conducting tests, and analyzing results.

By systematically capturing and annotating photographs throughout the capstone design project, teams can create rich visual documentation that not only records their progress and achievements but also provides a clear and compelling narrative of their engineering journey.

14.4 Videography

Videography in engineering design extends photography's capabilities by adding the dimension of time, making it a powerful tool for capturing dynamic processes and events. The same cameras used for photography can typically be employed for videography, allowing for versatile documentation of engineering projects. Videos can capture processes in real time, providing valuable data for time-dependent calculations, process analysis, and motion studies. Properly planned and executed project videos can significantly enhance the understanding and communication of design concepts and their practical implementations.

14.4.1   Planning and Goals for Project Videos

Creating a project video requires careful planning and clear goals. Before filming, the team should:

1. **Define Objectives**: Determine what you want to capture and why. Objectives might include documenting a process, demonstrating a prototype in action, or recording user interactions with a design.
2. **Storyboard**: Outline the key scenes and shots required to achieve your objectives. This helps in organizing the filming process and ensures that no critical aspect is overlooked.
3. **Equipment and Setup**: Ensure that you have the necessary equipment, including cameras, tripods, lighting, and microphones. Plan the setup for each shot to capture clear and stable footage.
4. **Coordination**: If the video involves process operators or team members, coordinate with them to ensure smooth execution and minimal disruptions. Brief participants on their roles and the timing of their actions.

14.4.2   Timing Studies

Videography is particularly useful for time and motion analysis in process design and improvement. Software tools can be employed to accelerate the procedures for capturing and analyzing the timing of events, saving hours of manual effort and providing accurate documentation.

1. **Software Tools for Analysis**
   o   Use specialized software to analyze video footage and measure time intervals, motions, and actions within a process.

       o   These tools can automatically track and record the timing of each event, reducing the potential for human error and increasing precision.

2. **Process Documentation**

       o   Coordinate videography with process operators to ensure that all events, activities, or operational steps are captured comprehensively.

       o   Document the baseline state of the process before any design changes are made. This provides a reference for comparison and analysis.

3. **Benefits of Video Time and Motion Studies**

       o   Videos accurately document and time tasks while isolating non-value-added work content.

       o   Video-supported analysis creates a verifiable history of the current process state, which can be invaluable for process optimization and redesign.

14.4.3   Position, Velocity, and Acceleration Measurement

Videography is also practical for measuring dynamic parameters such as position, velocity, and acceleration. This is essential in various engineering applications where understanding motion is critical.

1. **Capturing Motion**

       o   Use high-frame-rate cameras to capture fast-moving objects or processes. This allows for a detailed analysis of motion that might be missed at standard frame rates.

       o   Ensure that the camera setup minimizes motion blur and captures clear, precise footage.

2. **Analyzing Motion Data**

       o   Employ video analysis software to track objects frame-by-frame, calculating position changes over time.

       o   Use the data to compute velocity (rate of change of position) and acceleration (rate of change of velocity). This can be visualized through graphs and charts for better understanding.

3. **Applications in Engineering Design**

       o   Measure the performance of mechanical components under operational conditions.

       o   Analyze the dynamics of robotic arms, conveyor systems, or any other automated machinery.

       o   Validate simulation models by comparing them with real-world motion data captured through video.

14.4.4   Practical Considerations for Effective Videography

1. **Lighting and Clarity**
   o   Ensure proper lighting to avoid shadows and enhance the visibility of details. Good lighting is crucial for capturing clear and usable footage.
   o   Use additional lights if necessary, and adjust the camera settings to optimize exposure.

2. **Sound Quality**
   o   If the video includes spoken commentary or requires capturing ambient sounds, use external microphones to ensure high-quality audio.
   o   Minimize background noise and ensure that speakers are close to the microphone for clear sound recording.

3. **Editing and Presentation**
   o   Edit the video to remove unnecessary parts, enhance clarity, and add annotations or captions where needed.
   o   Use editing software to splice together different clips, add transitions, and incorporate background music or voiceovers to improve the overall presentation.

4. **Storage and Sharing**
   o   Store video files in organized folders with clear labels and descriptions. Use a structured naming convention for easy retrieval.
   o   Share videos through cloud platforms like Google Drive or Dropbox to ensure accessibility for all team members, mentors, and sponsors.

14.4.5   Example Applications of Videography in Capstone Design Projects

1. **Prototyping and Testing**
   o   Document the assembly and testing phases of prototypes, capturing detailed footage of each step. This can be used to review and improve the design.
   o   Record tests to demonstrate the functionality and performance of the design, providing clear evidence for reports and presentations.

2. **Process Improvements**
   o   Conduct time and motion studies to identify inefficiencies in current processes. Use video analysis to support redesign efforts aimed at optimizing workflow and reducing waste.
   o   Measure and document improvements achieved through design changes, using before-and-after footage for comparison.

3. **Educational and Training Purposes**
   o   Create instructional videos that explain complex processes or demonstrate the operation of equipment. These can be used for

training new team members or for educational purposes in related courses.

- o Use videos to share knowledge and best practices within the team and with future capstone design students.

By incorporating videography into the documentation and analysis processes, engineering design teams can gain deeper insights into their projects, improve communication of their work, and create a robust visual record that complements other forms of documentation. This multimedia approach enhances the overall quality and effectiveness of the capstone design experience.

14.5 Preliminary Design Report

The Preliminary Design Report (PDR) is a formal engineering document that encapsulates the accomplishments of the design team during the first phase of the design project. Typically completed at the end of the first semester, the PDR documents the team's entire body of work, leading to the creation of a proof-of-concept design and a prototype. Given its complexity, the PDR requires a set of guidelines to ensure thorough and accurate documentation. The typical length of a PDR ranges from 50 to 150 pages.

14.5.1 Guidelines for Preliminary Design Report

14.5.1.1 General Format

- **File Format**: PDF format is recommended for the PDR, as it is easily shared and maintains consistency across various operating systems and printing systems.
- **Margins**: A one-inch margin on all sides of each page is recommended.
- **Page Numbers**: Include a page number on every page.
- **Outline**: The report should follow a structured outline as prescribed below.

14.5.1.2 Title Page

- **Content**: Project title, team number, team name, team logo, team members (and their roles), sponsor, faculty advisor(s), academic year, submission date, and a report identification number (if assigned).
- **Page Number**: No page number on the title page.

14.5.1.3 Abstract

- **Content**: A concise summary of the project, including objectives, accomplishments, and an overview of the design process.
- **Length**: Should not exceed 500 words and be limited to one page.
- **Page Number**: "ii."

14.5.1.4   Table of Contents

- **Content**: List of major and minor headings within the report along with page numbers.
- **Page Number**: "iii."

14.5.1.5   List of Acronyms

- **Content**: List all acronyms used in the report.
- **Page Number**: "iii."

14.5.1.6   List of Tables and Figures

- **Content**: Include a list of all tables and figures in the report, with page numbers for each. Each table and figure must be numbered.
- **Page Number**: "iii."

14.5.1.7   Introduction

- **Content**: Project description, problem definition, design requirements and expectations, a brief chronology of previous work, purpose, scope, and objectives of the design work.
- **Page Number**: "1."

14.5.1.8   Literature Searches

- **Content**: Each team member researched a complete list of literature and referenced it in the PDR. Explain what was learned from each source and its relevance to the design solution.

14.5.1.9   Patent Searches

- **Content**: List of patents researched and used as references. Explain the relevance of each patent to the design solution.

14.5.1.10   Project Planning

- **Content**: Description of project planning and management, including a Gantt Chart showing the timeline, tasks, milestones, and resources with team member task assignments. Specify tasks completed as a percentage, milestones achieved, and managerial tools implemented.

14.5.1.11   Design Specifications

- **Content**: Describe how customer requirements and engineering criteria were transcribed into engineering design specifications. List target values for specifications in a tabular format.

14.5.1.12 Conceptual Design

- **Content**: Describe the process for generating design solution concepts. Include sketches of all concepts generated. Analyze and evaluate each concept, perform a Pugh analysis, and include the decision matrix.

14.5.1.13 Competitive Analysis

- **Content**: Conduct a market analysis to identify and assess competition. Present a QFD analysis, a trade-off analysis, and a competitive analysis.

14.5.1.14 Project Specific Details & Analysis

- **Content**: Include any details, data collection, or engineering analysis specific to the project. For product design teams, include market analysis, demand forecasting, cost vs. price information, and surveys. For process design teams, include flow charts, floor plans, and time studies.

14.5.1.15 Product or Process Design Details

- **Content**: Describe how the chosen concept was developed into the prototype. Include drawings, dimensions, tolerances, and annotations. If needed, reference detailed drawings in the Appendices.

14.5.1.16 Engineering Analysis

- **Content**: Describe mathematical models and analyses (e.g., process, thermal, fluids, structural, vibrational, static, dynamic, materials, and finite element analysis). Ensure the design is sound, safe, and compliant with regulatory requirements, standards, and codes.

14.5.1.17 Proof of Concept

- **Content**: Discuss the proof of concept or prototype and how the design functioned as envisioned. Compare the prototype's features, functions, and performance against the design specifications.

14.5.1.18 Financial Analysis

- **Content**: Describe the cost analysis of implementing the design solution, including person-hours, costs for mass production, market demand, and return on investment. Discuss cost savings and efficiencies resulting from the project.

14.5.1.19 Conclusions

- **Content**: Summarize how the design meets the specifications and project objectives.

14.5.1.20   Further Work

- **Content**: Describe the remaining tasks and the plan for continuation and further development.

14.5.1.21   References

- **Content**: List all publications, papers, and reports cited in the report.
  Appendices
- **Content**: Include important supplementary information such as computer programs and detailed assembly drawings. Ensure all items in the appendices are referenced in the main text.

14.5.2   Professional Formatting Guidelines

14.5.2.1   Footnotes

- Number consecutively using superscript numbers.
- Position flush left at the bottom of the column/page in which the reference first appears.
- Text should be ten pt. with extra spacing between the text and footnote.

14.5.2.2   Equations

- Insert equations separately from the text and center them.
- Number equations consecutively, using Arabic numerals in parentheses, flush right.

14.5.2.3   Graphics

- Number consecutively and include a caption below the graphic.
- Annotations within the graphic should be no smaller than 9 pt.
- Position graphics close to their references in the text.
- Size graphics to fit 7.5 in. across the page or 9 x 7.5 in. for a full page.

14.5.2.4   Tables

- Number consecutively and include a centered caption above the table.
- Position tables close to their references in the text.
- Size tables as large as necessary for clarity, up to 7.5 in. across the page or 9 x 7.5 in.

14.5.2.5   References

- List references in numerical order according to their appearance in the text.
- Follow specific formats for different types of references (journal articles, books,

conference papers, theses, technical reports).

Adhering to these guidelines will ensure that the Preliminary Design Report is a comprehensive and professional document that effectively communicates the design team's work and accomplishments during the project's first phase.

Bibliography

A. Buzzetto-More, N., Julius Alade, A., 2006. Best Practices in e-Assessment. JITE:Research 5, 251–269. https://doi.org/10.28945/246

Adams, D.F., Odom, E.M., 1987. Testing of single or bundles of carbon/carbon composite materials. Composites 18, 381–385. https://doi.org/10.1016/0010-4361(87)90362-4

Al-Thani, S.B.J., Abdelmoneim, A., Cherif, A., Moukarzel, D., Daoud, K., 2016. Assessing general education learning outcomes at Qatar University. Jnl of Applied Research in HE 8, 159–176. https://doi.org/10.1108/JARHE-03-2015-0016

Applied Imagination - Wikipedia [WWW Document], n.d. URL https://en.wikipedia.org/wiki/Applied_Imagination (accessed 7.12.20).

Atadero, R.A., Rambo-Hernandez, K.E., Balgopal, M.M., 2015. Using Social Cognitive Career Theory to Assess Student Outcomes of Group Design Projects in Statics: SCCT and Student Outcomes of Statics Projects. J. Eng. Educ. 104, 55–73. https://doi.org/10.1002/jee.20063

Benavides, E., 2011. Advanced Engineering Design. Woodhead Publishing.

Buchsbaum, A., Rey, C., n.d. Jean-Yves Beziau Federal University of Rio de Janeiro Rio de Janeiro, RJ Brazil 623.

Caple, M., Wild, J., Maslen, E., Nagel, J., 2015. Design of a controller for a hybrid bearing system, in: 2015 Systems and Information Engineering Design Symposium. Presented at the 2015 Systems and Information Engineering Design Symposium, IEEE, Charlottesville, VA, USA, pp. 236–239. https://doi.org/10.1109/SIEDS.2015.7116980

Carnevalli, J.A., Miguel, P.A.C., Calarge, F.A., 2010. Axiomatic design application for minimising the difficulties of QFD usage. International Journal of Production Economics 125, 1–12. https://doi.org/10.1016/j.ijpe.2010.01.002

CATIA [WWW Document], n.d. URL https://www.3ds.com/products-services/catia/?wockw=card_content_cta_1_url%3A%22https%3A%2F%2Fblogs.3ds.com%2Fcatia%2F%22

Cordon, D., Clarke, E., Westra, L., Allen, N., Cunnington, M., Drew, B., Gerbus, D., Klein, M., Walker, M., Odom, E.M., Rink, K.K., Beyerlein, S.W., 2002. Shop orientation to enhance design for manufacturing in capstone projects, in: 32nd Annual Frontiers in Education. Presented at the Conference on Frontiers in Education, IEEE, Boston, MA, USA, pp. F4D-6-F4D-11. https://doi.org/10.1109/FIE.2002.1158229

Council on Higher Education (South Africa), 2011. Work-integrated learning: good practice guide. Council on Higher Education, Pretoria, South Africa.

Creativity Unbound [WWW Document], n.d. . FourSight. URL https://foursightonline.com/product/creativity-unbound/ (accessed 8.16.20).

Criteria for Accrediting Engineering Programs, 2019 – 2020 | ABET [WWW Document], n.d. URL https://www.abet.org/accreditation/accreditation-criteria/criteria-for-accrediting-engineering-programs-2019-2020/ (accessed 6.10.20).

Davis, D.C., Gentili, K.L., Trevisan, M.S., Calkins, D.E., 2002. Engineering Design Assessment Processes and Scoring Scales for Program Improvement and

Accountability. Journal of Engineering Education 91, 211–221.
https://doi.org/10.1002/j.2168-9830.2002.tb00694.x

Deveci, T., Nunn, R., 2018. COMM151: A PROJECT-BASED COURSE TO
ENHANCE ENGINEERING STUDENTS' COMMUNICATION SKILLS.
ESPEAP 6, 027. https://doi.org/10.22190/JTESAP1801027D

Diefes-Dux, H.A., Moore, T., Zawojewski, J., Imbrie, P.K., Follman, D., 2004. A
framework for posing open-ended engineering problems: model-eliciting activities,
in: 34th Annual Frontiers in Education, 2004. FIE 2004. Presented at the 34th
Annual Frontiers in Education, 2004. FIE 2004., IEEE, Savannah, GA, USA, pp.
455–460. https://doi.org/10.1109/FIE.2004.1408556

Dollar, A.M., Kerdok, A.E., Diamond, S.G., Novotny, P.M., Howe, R.D., 2005. Starting
on the Right Track: Introducing Students to Mechanical Engineering With a Project-
Based Machine Design Course, in: Innovations in Engineering Education:
Mechanical Engineering Education, Mechanical Engineering/Mechanical
Engineering Technology Department Heads. Presented at the ASME 2005
International Mechanical Engineering Congress and Exposition, ASMEDC, Orlando,
Florida, USA, pp. 363–371. https://doi.org/10.1115/IMECE2005-81929

Duarte, B.B., de Castro Leal, A.L., de Almeida Falbo, R., Guizzardi, G., Guizzardi,
R.S.S., Souza, V.E.S., 2018. Ontological foundations for software requirements with
a focus on requirements at runtime. AO 13, 73–105. https://doi.org/10.3233/AO-
180197

Dynn, C.L., Agogino, A.M., Eris, O., Frey, D.D., Leifer, L.J., 2006. Engineering design
thinking, teaching, and learning. IEEE Eng. Manag. Rev. 34, 65–65.
https://doi.org/10.1109/EMR.2006.1679078

Eberle, B., 2008. Scamper: Creative Games and Activities for Imagination Development.
Prufrock Press.

Egelhoff, C., Odom, E., 2001. Advanced learning made as easy as ABC: an example
using design for fatigue of machine elements subjected to simple and combined
loads, in: 31st Annual Frontiers in Education Conference. Impact on Engineering
and Science Education. Conference Proceedings (Cat. No.01CH37193). Presented at
the 31st Annual Frontiers in Education Conference. Impact on Engineering and
Science Education., IEEE, Reno, NV, USA, pp. T4B-7-T4B-12.
https://doi.org/10.1109/FIE.2001.963934

Egelhoff, C.F., Odom, E.M., 1999. Machine design: where the action should be, in:
FIE'99 Frontiers in Education. 29th Annual Frontiers in Education Conference.
Designing the Future of Science and Engineering Education. Conference
Proceedings (IEEE Cat. No.99CH37011. Presented at the IEEE Computer Society
Conference on Frontiers in Education, Stripes Publishing L.L.C, San Juan, Puerto
Rico, p. 12C5/13-12C5/18. https://doi.org/10.1109/FIE.1999.841644

Egelhoff, C.J., Odom, E.M., Wiest, B.J., 2010. Application of modern engineering tools
in the analysis of the stepped shaft: Teaching a structured problem-solving approach
using energy techniques, in: 2010 IEEE Frontiers in Education Conference (FIE).
Presented at the 2010 IEEE Frontiers in Education Conference (FIE), IEEE,
Arlington, VA, USA, pp. T1C-1-T1C-6. https://doi.org/10.1109/FIE.2010.5673504

Facebook, Twitter, options, S. more sharing, Facebook, Twitter, LinkedIn, Email,
URLCopied!, C.L., Print, 1999. Mars Probe Lost Due to Simple Math Error [WWW

Document]. Los Angeles Times. URL https://www.latimes.com/archives/la-xpm-1999-oct-01-mn-17288-story.html (accessed 8.16.20).

Farid, A.M., Suh, N.P. (Eds.), 2016. Axiomatic Design in Large Systems. Springer International Publishing, Cham. https://doi.org/10.1007/978-3-319-32388-6

Griffin, P.M., Griffin, S.O., Llewellyn, D.C., 2004. The Impact of Group Size and Project Duration on Capstone Design. Journal of Engineering Education 93, 185–193. https://doi.org/10.1002/j.2168-9830.2004.tb00805.x

Howe, S., Rosenbauer, L., Poulos, S., n.d. 2015 Capstone Design Survey – Initial Results 4.

Ibrahim, A., Kurfess, T.R., Agogino, A., Alani, F., Burgess, S., Chen, W.-F., Lantada, A.D., Ertugrul, N., Felder, R., Genalo, L., Gillet, D., Hernandez, W., n.d. FOUNDER AND FIRST EDITOR-IN-CHIEF MICHAEL S. WALD (1932–2008) 1.

Iowa State University, Carter, R., Strader, T., Drake University, Rozycki, J., Drake University, Root, T., Drake University, 2015. Cost Structures of Information Technology Products and Digital Products and Services Firms: Implications for Financial Analysis. JMWAIS 2015, 5–19. https://doi.org/10.17705/3jmwa.00002

Leaman, E.J., Cochran, J.R., Nagel, J.K., 2014. Design of a two-phase solar and fluid-based renewable energy system for residential use, in: 2014 Systems and Information Engineering Design Symposium (SIEDS). Presented at the 2014 Systems and Information Engineering Design Symposium (SIEDS), IEEE, Charlottesville, VA, USA, pp. 193–197. https://doi.org/10.1109/SIEDS.2014.6829917

Morell, L., 2015. Disrupting engineering education to better address societal needs, in: 2015 International Conference on Interactive Collaborative Learning (ICL). Presented at the 2015 International Conference on Interactive Collaborative Learning (ICL), IEEE, Firenze, Italy, pp. 1093–1097. https://doi.org/10.1109/ICL.2015.7318184

Nagel, J., 2013. Guard Cell and Tropomyosin Inspired Chemical Sensor. Micromachines 4, 378–401. https://doi.org/10.3390/mi4040378

Nagel, J.K.S., Nagel, R.L., Eggermont, M., 2013. Teaching Biomimicry With an Engineering-to-Biology Thesaurus, in: Volume 1: 15th International Conference on Advanced Vehicle Technologies; 10th International Conference on Design Education; 7th International Conference on Micro- and Nanosystems. Presented at the ASME 2013 International Design Engineering Technical Conferences and Computers and Information in Engineering Conference, American Society of Mechanical Engineers, Portland, Oregon, USA, p. V001T04A017. https://doi.org/10.1115/DETC2013-12068

Nagel, J.K.S., Nagel, R.L., Stone, R.B., 2011. Abstracting biology for engineering design. IJDE 4, 23. https://doi.org/10.1504/IJDE.2011.041407

Nagel, Jacquelyn K.S., Nagel, R.L., Stone, R.B., McAdams, D.A., 2010. Function-based, biologically inspired concept generation. AIEDAM 24, 521–535. https://doi.org/10.1017/S0890060410000375

Nagel, J.K.S., Stone, R.B., 2012. A computational approach to biologically inspired design. AIEDAM 26, 161–176. https://doi.org/10.1017/S0890060412000054

Nagel, Jacquelyn K. S., Stone, R.B., McAdams, D.A., 2010. An Engineering-to-Biology Thesaurus for Engineering Design, in: Volume 5: 22nd International Conference on Design Theory and Methodology; Special Conference on Mechanical Vibration and Noise. Presented at the ASME 2010 International Design Engineering Technical Conferences and Computers and Information in Engineering Conference, ASMEDC, Montreal, Quebec, Canada, pp. 117–128. https://doi.org/10.1115/DETC2010-28233

Odom, E.M., Adams, D.F., 1990. Failure modes of unidirectional carbon/epoxy composite compression specimens. Composites 21, 289–296. https://doi.org/10.1016/0010-4361(90)90343-U

Odom, E.M., Beyerlein, S.W., Tew, B.W., Smelser, R.E., Blackketter, D.M., 1999. Idaho Engineering Works: a model for leadership development in design education, in: FIE'99 Frontiers in Education. 29th Annual Frontiers in Education Conference. Designing the Future of Science and Engineering Education. Conference Proceedings (IEEE Cat. No.99CH37011. Presented at the IEEE Computer Society Conference on Frontiers in Education, Stripes Publishing L.L.C, San Juan, Puerto Rico, p. 11B2/21-11B2/24. https://doi.org/10.1109/FIE.1999.839217

Odom, E.M., Egelhoff, C.J., 2011. Teaching deflection of stepped shafts: Castigliano's theorem, dummy loads, heaviside step functions and numerical integration, in: 2011 Frontiers in Education Conference (FIE). Presented at the 2011 Frontiers in Education Conference (FIE), IEEE, Rapid City, SD, USA, pp. F3H-1-F3H-6. https://doi.org/10.1109/FIE.2011.6143039

Online, S.B., n.d. Advanced Engineering Design [WWW Document]. URL https://learning.oreilly.com/library/view/advanced-engineering-design/9780857090935/ (accessed 6.6.20).

ORACLE PRIMAVERA P6 SOFTWARE [WWW Document], n.d. URL https://globalpm.com/products/oracle-primavera-p6-software/

Osborne, A.F., 1953. Applied imagination; principles and procedures of creative thinking. (Book, 1953) [WorldCat.org] [WWW Document]. URL https://www.worldcat.org/title/applied-imagination-principles-and-procedures-of-creative-thinking/oclc/641122686 (accessed 7.12.20).

Pahl, A.K., Newnes, L., McMahon, C., 2007. A generic model for creativity and innovation: overview for early phases of engineering design. JDR 6, 5. https://doi.org/10.1504/JDR.2007.015561

Pembridge, J.J., Paretti, M.C., 2019. Characterizing capstone design teaching: A functional taxonomy. J. Eng. Educ. 108, 197–219. https://doi.org/10.1002/jee.20259

Ribeiro, A.L., Bittencourt, R.A., 2018. A PBL-Based, Integrated Learning Experience of Object-Oriented Programming, Data Structures and Software Design, in: 2018 IEEE Frontiers in Education Conference (FIE). Presented at the 2018 IEEE Frontiers in Education Conference (FIE), IEEE, San Jose, CA, USA, pp. 1–9. https://doi.org/10.1109/FIE.2018.8659261

Ritter, S.M., Mostert, N., 2017. Enhancement of Creative Thinking Skills Using a Cognitive-Based Creativity Training. J Cogn Enhanc 1, 243–253. https://doi.org/10.1007/s41465-016-0002-3

Romaniuk, R.S., 2012. Communications, Multimedia, Ontology, Photonics and Internet Engineering 2012. International Journal of Electronics and Telecommunications 58, 463–478. https://doi.org/10.2478/v10177-012-0061-z

S. Nagel, J.K., W., F., 2012. Hybrid Manufacturing System Design and Development, in: Abdul Aziz, F. (Ed.), Manufacturing System. InTech. https://doi.org/10.5772/35597

Salustri, F.A., Eng, N.L., Weerasinghe, J.S., 2008. Visualizing Information in the Early Stages of Engineering Design. Computer-Aided Design and Applications 5, 697–714. https://doi.org/10.3722/cadaps.2008.697-714

Stroble, Jacquelyn K., Stone, R.B., Watkins, S.E., 2009. Assessing How Digital Design Tools Affect Learning of Engineering Design Concepts, in: Volume 8: 14th Design for Manufacturing and the Life Cycle Conference; 6th Symposium on International Design and Design Education; 21st International Conference on Design Theory and Methodology, Parts A and B. Presented at the ASME 2009 International Design Engineering Technical Conferences and Computers and Information in Engineering Conference, ASMEDC, San Diego, California, USA, pp. 467–476. https://doi.org/10.1115/DETC2009-86708

Stroble, J.K., Stone, R.B., Watkins, S.E., 2009. An overview of biomimetic sensor technology. Sensor Review 29, 112–119. https://doi.org/10.1108/02602280910936219

Swartwout, M., Kitts, C., Twiggs, R., Kenny, T., Ray Smith, B., Lu, R., Stattenfield, K., Pranajaya, F., 2008. Mission results for Sapphire, a student-built satellite. Acta Astronautica 62, 521–538. https://doi.org/10.1016/j.actaastro.2008.01.009

The Gantt chart, a working tool of management : Clark, Wallace, 1880-1948 : Free Download, Borrow, and Streaming [WWW Document], n.d. . Internet Archive. URL https://archive.org/details/cu31924004570853 (accessed 7.6.20).

The_effectiveness_of_cooperative_problem.pdf, n.d.

Todd, R.H., Magleby, S.P., Sorensen, C.D., Swan, B.R., Anthony, D.K., 1995. A Survey of Capstone Engineering Courses in North America. Journal of Engineering Education 84, 165–174. https://doi.org/10.1002/j.2168-9830.1995.tb00163.x

Truong, H.T.X., Odom, E.M., Egelhoff, C.J., Burns, K.L., 2011. Using Modern Engineering Tools to Efficiently Solve Challenging Engineering Design Problems: Analysis of the Stepped Shaft 10.

Tuckman, B.W., Jensen, M.A.C., 1977. Stages of Small-Group Development Revisited. Group & Organization Studies 2, 419–427. https://doi.org/10.1177/105960117700200404

Wenhao Huang, D., Diefes-Dux, H., Imbrie, P.K., Daku, B., Kallimani, J.G., 2004. Learning motivation evaluation for a computer-based instructional tutorial using ARCS model of motivational design, in: 34th Annual Frontiers in Education, 2004. FIE 2004. Presented at the 34th Annual Frontiers in Education, 2004. FIE 2004., IEEE, Savannah, GA, USA, pp. 65–71. https://doi.org/10.1109/FIE.2004.1408466

Why ABET Accreditation Matters | ABET [WWW Document], n.d. URL https://www.abet.org/accreditation/what-is-accreditation/why-abet-accreditation-matters/ (accessed 6.10.20).

Williams, C.B., Gero, J., Lee, Y., Paretti, M., 2010. Exploring Spatial Reasoning Ability and Design Cognition in Undergraduate Engineering Students, in: Volume 6: 15th Design for Manufacturing and the Lifecycle Conference; 7th Symposium on International Design and Design Education. Presented at the ASME 2010 International Design Engineering Technical Conferences and Computers and

Information in Engineering Conference, ASMEDC, Montreal, Quebec, Canada, pp. 669–676. https://doi.org/10.1115/DETC2010-28925

Xiao, A., Park, S.S., Freiheit, T., 2011. A COMPARISON OF CONCEPT SELECTION IN CONCEPT SCORING AND AXIOMATIC DESIGN METHODS. PCEEA. https://doi.org/10.24908/pceea.v0i0.3769

zotero / Insert Items from Amazon [WWW Document], n.d. URL http://zotero.pbworks.com/w/page/5511972/Insert%20Items%20from%20Amazon (accessed 8.16.20).

www.ingramcontent.com/pod-product-compliance
Lightning Source LLC
Chambersburg PA
CBHW080128220326
41598CB00032B/4995